There is a need to reconstruct project management, both as a scientific cross-disciplinary research domain and as a practical area of application. Peter Morris's book helps our thinking enormously regarding both these needs. The structure and content of Part 1 skilfully introduces the theoretical and practical components that constitute the field. In outlining its history, Peter builds a solid foundation for exploring its elements: systems, integration, efficiency, effectiveness, performance, requirements, 'solutioneering', contracting, procurement, risk and opportunities, value and benefits, people. New perspectives are meanwhile introduced: agility, competences and capabilities, governance and the sponsor, the front-end, technology, sustainability, innovation, partnering, context, philosophy. The final part is ground-breaking in its future projection of project management (mop/p^{3}m). Mankind faces many challenges, be they ecological, demographic, violence or other. Peter's book explores possible interactions between project management and these futures. No other book is the same.

Karlos Artto
Professor of Project Business
Aalto University, Helsinki, Finland

* * *

Bravo! *Reconstructing Project Management* is a tour de force on the philosophy, methods and practices of project and program management; a feast of PM lore, knowledge and insight.

Peter Morris' long and incredibly productive career has uniquely straddled project management research, teaching, practice, and consulting on some of the most important projects and biggest companies of the last few decades. *Reconstructing Project Management* provides the most complete and well-integrated coverage of the evolution of project management written to date. But, more importantly, Morris has extracted a wealth of insights from his broad and deep knowledge and experience regarding the shortcomings of conventional project management in addressing the daunting challenges that currently face managers of projects worldwide, and the even greater ones that the next generation of project managers must face.

This book is a must-read for teachers, students and reflective practitioners of the art and craft of project management.

Raymond E. Levitt
Kumagai Professor of Engineering
Director, Stanford Global Projects Center
University of Stanford, California, USA

* * *

Peter Morris' writings have influenced and informed the world of project management for over 40 years, not least his powerful idea of 'The Management of Projects': the thesis that management should be responsible for the development and delivery of the whole project and that managing the front-end of projects is key to projects' success.

Reconstructing Project Management continues his thought leadership. It starts with an absorbing history of project management. It then addresses the discipline's constituent parts, whether related to processes, tools, people or context. Finally, it re-assembles these parts in ways that can add value given tomorrow's business and social needs.

This book has something for everyone – facts, ideas, concepts and theories that will be of interest to students, practitioners and managers alike. Through whatever lens you are looking at project management, whether past, present or future, you will almost certainly find the answer in this book.

Mike Brown
Director of Project and Programme Management,
Rolls-Royce plc, Derby, UK

* * *

In *Reconstructing Project Management* Peter Morris demonstrates a profound under-standing of the nature of the discipline, its complexity and its component parts. The analysis is both broad and deep. After chronicling the growth of modern project management, he deconstructs it into its component parts. He then recombines those elements to address today's needs and tomorrow's challenges.

As a practitioner, I often witness project managers becoming so consumed in delivery that we forget the need to generate value. Peter reminds us of this need, within a contemporary context. His book challenges us to consider the rapidly changing envi-ronment we face in the 21st century. Fewer resources, less water, more carbon, a rapidly changing climate – the management of projects can only get tougher, and more important.

For anyone involved in thinking about projects, whether as deliverers, teachers or researchers, this book will fascinate and challenge in equal measure.

Robbie Burns
Regional Director for Network Rail's Western and Wales Region, Infrastructure Projects Directorate

* * *

This book begins by showing how our conception of project management evolved to what we today recognise as the project management body of knowledge. Professor Morris then deconstructs that body of knowledge and challenges the reader to think. Is project management merely a tactical approachto delivery or is it an organising principle with a distinctive philosophy? Should it have a role in corporate management or is it only about meeting project targets given to the p.m.team? Peter Morris dem-onstrates that seeing the discipline as the *'management of projects'* provides a com-pletely new approach.

Reconstructing Project Management is designed for the purist, the academic, the project practitioner and the project organisation in equal measure. It is at times a challenging read – a some of what it says may clash with traditional thinking – but as an aid to developing the profession and as food for the mind, it is a must-read.

Paul Hodgkins
PM@Siemens, Programme Executive – UK and North West Europe

* * *

This book is a must for all those who would like to 'awaken their dreams' or 'realise their vision' or 'concretise their abstract ideas'. It covers the entire gamut of project management in a lucid manner.

Part 1, 'Constructing Project Management', covers the discipline's history from ancient times to today showing that projects and their management have always been at mankind's centre. In Part 2, 'Deconstructing Project Management', Peter covers exceedingly well the complexity of the subject, simplifying and integrating the various conceptual elements relevant to the broad discipline of managing projects. Peter then does a marvellous job in Part 3, 'Reconstructing Project Management', unravelling the challenges in the 21st century as well as the opportunities opening up for us.

The art and science of project management is enunciated here in a way that can be grasped by all those wishing to manage change in our society.

Adesh Jain, New Delhi, India
President, Project Management Associates (PMA-India) and President, International Project Management Association, 2005

* * *

Reconstructing Project Management starts with a fascinating and highly readable history of the study and practice of managing projects. Starting with early efforts by the Pentagon to codify and structure its approach to projects in the 1960s, Professor Morris puts many of the landmark studies of project delivery approaches in perspective. He then proceeds to rebuild the readers' understanding of project management in a new way.

If you are a serious student of project management, you will find Morris's 'reconstruction' provocative and thought-provoking, whether you agree with every aspect or not. His critique of the literature and our efforts to date is devastating and his plea for more disciplined critical thinking is spot-on. Reading this book will be enjoyable to anyone interested in the broad field of managing work through projects and is required reading for those interested in contributing to the discipline.

Ed Merrow
Founder and CEO, Independent Project Analysis, Inc.
Ashburn, Virginia, USA

Reconstructing Project Management is a tour de force! Peter Morris describes the past, present and future of project management! The big picture! The world of project management, not only as it is but as it ought to be! This is one of the best books on project management that I have ever read. I believe *Reconstructing Project Management* will be recognized as one of the seminal books of the 21st Century on modern project management. Every thinking professional in the field should read it; every serious library must contain a copy. This book confirms Peter's place as THE world's leading critical thinker on the increasingly important topic of managing projects.

David L. Pells
Managing Editor, *PM World Journal*, Houston, Texas, USA

Scholarship is the word that comes to mind reading this truly comprehensive and impressively thorough volume. This is a book about and for the project management profession. It is rigorously researched and extremely readable. It challenges you to think, the need for which in project management has probably never been greater. Despite its scholarly nature, it is completely practical and covers an intriguing past, a diverse present, and a pragmatic future.

The role of universities in providing the profession with such thinking should be fully acknowledged. Knowledge of this kind is extraordinarily valuable.

Two lifetimes are reflected in the book. One is the long gestation of the discipline which, with its new 'construction', is now helping mankind address major challenges. The other is the professional lifetime of Peter Morris, who has dedicated much effort in practice and in education to establishing the principles of the effective management of projects.

Tom Taylor
President, Association for Project Management, UK

Peter Morris has been a critical commentator on, and an important contributor to, project management for over 40 years. In this important, summative book he draws upon his in-depth knowledge of the field to describe and reflect upon its emergence, both as a practical discipline and as a domain of academic study. His analysis is rich, compelling and robust. It highlights project management's inherent strengths and its current weaknesses, proposing ways in which it could be restructured to better fill its roles in society, today and in the future. The book makes observations and provides recommendations that will be provocative to many practitioners and academics, yet its proposals are profound and needed, and should be actionable by everyone who is concerned, whether directly or indirectly, with the effective management of projects.

Dr. Brian Hobbs, Project Management Chair at the University of Quebec at Montreal, and recipient of PMI's Research Achievement Award for 2012

Reconstructing Project Management

Peter W.G. Morris

Professor of Construction and Project Management
University College London

A John Wiley & Sons, Ltd., Publication

Library of Congress Cataloging-in-Publication Data
Morris, Peter W. G.
  Reconstructing project management / Peter W.G. Morris, professor of construction and
project management, The Barlett, University College London.
     pages cm
  Includes bibliographical references and index.
  ISBN 978-0-470-65907-6 (hardback : alk. paper)   1. Project management.   I. Title.
  T56.8.M725 2014
  658.4'04–dc23
                                              2012037674

A catalogue record for this book is available from the British Library.

Set in 10/12 pt Sabon by Toppan Best-set Premedia Limited
Printed and bound in Malaysia by Vivar Printing Sdn Bhd

Cover design by Steve Thompson

1   2013

Contents

Figures

Tables

About the Author

Peter Morris is Professor of Construction and Project Management at University College London (UCL). He is the author of *The Management of Projects* (Thomas Telford, 1994); with George Hough, of *The Anatomy of Major Projects* (John Wiley & Sons, 1987); and, with Ashley Jamieson, of *Translating Corporate Strategy into Project Strategy* (PMI, 2004). He is co-editor with Jeffrey Pinto of *The Wiley Guide to Managing Projects* (Wiley, 2005); and, with Jeffrey Pinto and Jonas Söderlund, of *The Oxford Handbook of Project Management* (OUP, 2011).

He was Chairman of the Association for Project Management (APM) from 1993 to 1996 and Deputy Chairman of the International Project Management Association (IPMA) from 1995 to 1997. He received the Project Management Institute's 2005 Research Achievement Award, IPMA's 2009 Research Award and APM's 2008 Sir Monty Finniston Lifetime Achievement Award.

Preface

A variety of titles suggested themselves for this book: *Project Management Past, Present and Future* was an obvious one. *The Book behind the BOK* was another, though a bit obscure. Something to do with construction seemed more appropriate for a professor of construction and project management.

I have been trying to formalise what one needs to know, and can say about, the management of projects since the late 1960s. Here is what will perhaps be my last such effort: *Reconstructing Project Management*. What is it; how did it get to where it is; where is it going?

My perspective has shifted on a number of occasions as I have climbed this mountain over the years.

- In 1967, when newly beginning work as a site engineer at Dungeness nuclear power station, I was amazed at the sudden, very late arrival of design changes. Why was nobody scheduling the release of information? From this came recognition of the horizontal (cross-function) project dimension and my PhD on how the amount and type of integration required varies for projects of differing size, speed and complexity (the matrix and contingency theory – Chapter 10).
- In 1976, in Brazil advising on the management of a $4 bn. steel mill, my colleague Joe Graubard, recently of US Steel project management, explained it was all about delivering 'on time, in budget, to scope' and showed me how to set out to do this and to control 'earned value' (Chapter 9).
- In 1980, advising the owners of the TransAlaskan Pipeline, I discovered that the overrun record of projects was truly appalling, and that the reasons for this had little to do with planning and monitoring but were rather contextual things like stakeholders, geophysical conditions, weather and other exogenous factors, as well as poor technology management, commercial issues, and people (Appendix 1). This lead me to suggest that the discipline should really be about 'the management of projects' where the project is the unit of analysis and the challenge is to develop and deliver it successfully. That the real place to focus is the front-end, where the project targets are set and value is built. And that we need to be quite sophisticated in choosing measures of success (Chapter 4).
- In the early 1990s, I was asked to produce the first draft of the Association for Project Management's 'Body of Knowledge'. Using the 'management

of projects' perspective, we developed a structure for the discipline which, I believe, more properly reflects the knowledge that one needs to have in order to manage projects effectively than was to be found in other BOKs at the time.

- Through my research and consulting with major companies in the 1990s, I began to realise just how difficult it is to build enterprise-wide project management competencies. How difficult it is to get people to seek out knowledge and learn. How difficult it is to acquire, build and deploy competent people who will be supported by an organisation having a mature project management infrastructure, in a manner that is appropriate both to the characteristics of the project or program, and to the context in which it operates (Chapters 10 and 19) – quite a mouthful; and in reality, it is!
- In the late 1990s/early 2000s, I was persistently reminded of two variants of project management and the challenges they pose to our understanding of the discipline: namely, program management and agile project management. Program management had a vociferous following that promoted it as a form of strategic change management, yet it has several other interpretations. Agile sounds attractive but defies the traditional definitions of the discipline. Our knowledge of both seems as yet not wholly worked out (Chapters 5 and 21).
- In the first decade of the 21st century, as part of project assurance activities, I did a number of Root Cause analyses for an oil and gas major on projects that had been significantly late and over budget. In all cases, we traced the root cause to People: human beings making the wrong decisions, either because they didn't have the requisite experience, missed things, or were overly influenced by inappropriate drivers. 'Projects are built by people, for people, through people' yet too often we give too little attention to getting our people right. As we'll see (Chapter 15): 'one should never compromise on people – but one always does'.
- Working with a number of companies at about the same time made me acutely aware of the huge impact that the sponsor can have on the conduct of the project, yet how little professional knowledge of the subject they may have.
- Finally, research at UCL around 2010 highlighted how serious many of the trends in our society are: the world in 2050 (one professional generation away) offers much to worry about. But what is the academic community in project management doing about it? Almost nothing: nearly all project management writing and research seems to be about means rather than ends. Very little connects with performance (Chapter 21). This seems to be a major shortcoming, challenge and opportunity.

The book explores the events on this journey from simple, classical project management to 'the management of projects' (mop) and to p^3m – portfolio, program and project management. It does so from two perspectives, reaching out to two or three sets of readers: practitioners who practise the discipline of project management (p^3m or whatever we call it); academics who research and teach in the domain; and of course, students. Maybe neither the practitioners nor the academics will feel the material is always quite in their comfort zone. That's too bad, for I believe that the language and thinking about the practice and theory of managing projects needs to be better integrated, and that it's worth struggling to help achieve this.

With Thanks, and More

This book is the product of two partnerships which have dominated my life: my work and my family. Without the former the latter would not be what it is, and vice versa. But without Carolyn, my wife, neither would have been what they are. I dedicate the book to her, therefore, with all my love.

Many people have helped me develop the ideas put forward in the book. From UMIST (the University of Manchester Institute of Science and Technology), Professors Dennis Harper, Roger Burgess and Stephen Wearne; from Sir Robert McAlpine, David Rolt; from Booz Allen & Hamilton, Joe Graubard, John Smith and George Steel; from Arthur D. Little, Dr. Albert J. Kelley and Ivars Avots; from the Major Projects Association and Templeton College, Oxford, Allan Sykes, Derek Fraser, Uwe Kitzinger and Dr. George Hough; from Bovis, Sir Frank Lampl; from INDECO, George Steel (again) and James Young; and from my time at UCL, Dr. Andrew Edkins, Dr. Joana Geraldi, Dr. Stephen Pryke and Dr. Hedley Smyth, and Professor Jeff Pinto of Penn State University and Professor Jonas Söderlund of BI Norway, Thanks especially to Dr. Sulafa Badi, who helped me enormously in marshalling the figures and tracking down several of the references. Thanks, too, to those who read early drafts of the book and who made many invaluable comments, not least Dr. Joana Geraldi, Dr. Peter Harpum, Dr. Efrosyni Konstantinou, Paul Mansel, Dr. Tyrone Pitsis, Miles Shepherd and Dr. Yannis Zoiopoulos. They invariably caused me to reflect and often to amend.

Mrs. Beeton opens the preface to her famous and eponymous guide to household management with the memorable sentence "I must frankly own, that if I had known, beforehand, that this book would have cost me the labour which it has, I should never have been courageous enough to start it". In the case of this book, I have to admit that I had a pretty good idea of the labour that was to be required, but I neither recoiled from the prospect nor regretted its expenditure in execution. I only hope that the pleasure, if such there is, and reward that you may derive in reading it is proportional to that which I found in writing it.

Peter W.G. Morris
University College London
2 June 2012

Introduction

There once was a little girl who was found staring at a can of orange juice. When asked what she was doing, she said, it says 'Concentrate'. Seems a funny way to begin a book on project management? Well, this little literary cartoon has several lessons directly relevant to this book. Don't take words necessarily at their face value. Don't act on them without thinking critically about their meaning. Beware of following normative or prescriptive advice too literally. Don't underestimate the young, neither their determination nor, like the rest of us, their ability to misread. Sometimes, a sense of humour may not be amiss! And lastly, concentrate!

This book is an essay on the knowledge needed to manage projects (and programs) effectively*. We'll see the little girl's lessons apply richly in tackling the subject. Management is an eminently practical business. Managers are essentially concerned with achieving results effectively. By and large, managers respect 'academic' rules and practices but may not feel too constrained by them unless there are clear reasons for doing so. Project management has chosen to put considerable weight on normative rules, adopting a persona of professionalism; a persona where knowledge is codified to the point where it has its own unique 'body of knowledge', or bodies, which can be taught and examined.

But management knowledge needs to be tailored in its application. Doing this requires judgement. It would be naïve and could be dangerous to imagine that project and program management could be applied straight 'out of the box'. This book aims to help form the judgement needed for such tailoring by addressing the principles of managing projects and programs: looking at how they may be applied in practice in different ways,

---

*I explain why I spell program only with one 'm' on page 93 – briefly, it is the spelling recommended by the *Oxford English Dictionary*.

Reconstructing Project Management, First Edition. Peter W.G. Morris.
© 2013 John Wiley & Sons, Ltd. Published 2013 by John Wiley & Sons, Ltd.

according to need and context. It thus approaches the subject in the spirit of education rather than as a 'how-to' handbook or a form of training; it is concerned with thinking about the management of projects rather than simply with applying its practices mechanistically. It is about both the theory and the practice of managing projects and programs.

The book is therefore written for those who wish to develop both their knowledge of, and their ability to think about, the management of projects, and programs. Specifically, it is written for:

- tomorrow's leaders in project and program management (as an area of management practice and as a domain of management study);
- project management professionals and practitioners who ought to be, as professionals, or might be, as practitioners, interested in a slightly more ambitiously theoretically-grounded treatment of the area they are working in;
- post-graduate students who are looking for a significant resource book, addressing, hopefully, just about all the principle areas of knowledge relevant to the subject;
- my colleagues in the academic project management community, as a contribution to the ongoing discussions and debates about the knowledge area; and
- anybody else who is interested in projects or programs and their management and who wants a good read!

Structure and Thesis of the Book

The book is in three parts. Part 1 traces how our knowledge of the field developed, how the subject has come to be *constructed* in the way we think of it today. It chronicles the growth of a discipline for managing projects; a discipline that, as many envision it today, needs to be enlarged. In parallel it chronicles the development of what many scholars see more as a field of enquiry – a knowledge domain – than as a discipline.

Part 2 takes this construct apart – *deconstructs* it – describing and presenting back to the reader the range of practices that are available for managing projects.

Part 3 then looks at how these elements of project management may be recombined – *reconstructed* – when deployed under different circumstances to meet today's needs and tomorrow's challenges. As a discipline, how and to what ends should it be deployed? Looking forward, it is argued, it needs refocusing around value and context. It could, and should, engage more with the major issues facing society today – issues such as climate change, profound demographic changes, hugely expanded data and information pools, and the need for massive infrastructure replenishment, all of which will require a project or program approach if they are to be tackled effectively.

More specifically:

Part 1: the development of the domain

Some of the early chapters of Part 1 have been covered in my earlier book, *The Management of Projects* (1994) and in chapter 1 of *The Oxford Handbook of Project Management* (2011). I have kept largely to the events used in these previous accounts, albeit with significant additions and some omissions, because I believe that they do truly illustrate the story of how project management developed from something instinctive and *ad hoc* to something formal and well structured. No part of Part 1 (or any other part) is merely a copy of these previous publications. It is freshly written and draws on much writing and research that has been done since *The Management of Projects* was originally published in 1994 and is almost four times the length of the *Handbook* chapter, having new information, new case studies, and more to say on several matters – the 19th century; the development of project management in the US aircraft and missile programs between 1920 and 1960; and the origins of PMBOK, Agile, and the Scandinavian School, amongst other topics.

It is important to acknowledge that the events reviewed here are not the only events that could be chosen to illustrate the emergence of the discipline. There is potentially an infinitude of possible candidates and there are clearly some obvious omissions. I dwell on the rise of coordination in the nascent aircraft industry of the 1920s but suspect similar evidence could be found in, for example, shipbuilding, and of course the military. I have chosen aircraft manufacturing because there is evidence readily to hand (Benjamin Pinney's PhD). I simply don't have the time now to do the primary research that would lead to data for other examples. Similarly, a comprehensive history of the development of the discipline would search out examples from other countries and regions, but this is beyond the scope of this book. These omissions, however, represent opportunities for other researchers.

This said, even as it stands, I believe Part 1 is very important in setting the context in which Parts 2 and 3 need to be seen: the 'state of play' today and thoughts for the future. Reading a history of the development of the subject – of the issues faced and responses taken – as given in Part 1 provides a means of seeing the scope of knowledge that has been deployed, as time and events have gone by, in managing projects and programs effectively, and to reflect on the coherence and robustness of this knowledge and the challenges in deploying it.

This re-acquainting with the true scope of the subject is especially important in project and program management as a result of the many guidance manuals and 'standards' that have been published by the professional associations, government agencies, and national and international standards bodies over the last 30–40 or more years and the impact these have had on the discipline. For this codification can become troublesome if, as is currently the case, it is in some way not an adequate representation of the knowledge that is really needed. (And matters are made worse if educational programs are then accredited against the very standards which their researchers and teachers should be critiquing.)

This may seem a little over the top, but there are places where the official view is at odds with what this book will present. Take two examples. First, the most popular conception of project management is that it is essentially about delivery execution, and about outputs rather than outcomes[1]. This conception is inadequate, as Part 1 demonstrates. History, and research, shows that the effective management of projects involves the management of the project's development and definitional 'front-end' stages as much as its downstream execution. Second, much of project and program management is portrayed as centring on monitoring and controlling. Yet a perfectly valid alternative, and a logically preferable one, would be to centre it on improving value, for example, to the sponsor, which is an approach argued for in Part 3*. These are not the conceptions promoted by some of the more influential standards-setting bodies. They come from university-led research[2]. They are cogent and they make a substantial difference to the way the discipline can be viewed. Their omission dramatically diminishes what I believe is a proper and richer perception of the subject. It profoundly and adversely misshapes its essence and reduces its potential effectiveness.

This said, Part 1 tells an outstanding story of intellectual and professional development. It paints a picture of a robust, valuable, healthy, popular, coherent subject which has become a major force in society and business, even if it still has some issues, such as the ones discussed above, unresolved.

Part 2: analysing the elements

The middle portion of the book, Part 2 – the deconstruction of project management – is the largest section. It is essentially a comprehensive reference work on the functions of project and program management: on the competencies required and on the tools and techniques, and processes and practices available.

Some thinkers are sceptical about the possibility of identifying generic 'good practices' (or more precisely, of identifying context-free prescriptive practice) for the reasons discussed at the beginning of this introduction. I maintain that, to an extent, however, one can; that indeed our research and teaching *should* have a 'should' about it – because we do want to learn about practices which are likely to lead to improved performance rather than just describe in a non-teleological, value-free way who did what to whom and how (as with 'practice-based' research[3]) – but that when doing so one needs to make clear the epistemological base that one is using and the methodology that one is adopting. Merely asserting something as 'the way to do project or program management' is often not convincing. Indeed, its very superficial plausibility may obscure issues and be positively dangerous. (Epistemology and methodology are addressed in Part 3.)

---

*The role of the sponsor, by the way, is another example of a key actor – a key piece of knowledge – being poorly, if at all, treated in these standards. See pages 145–7.

Part 3: putting it together again

The final part of the book looks at how the elements of project management may be *reconstructed* to meet society's future needs. It does so from four perspectives: its conceptual and philosophical roots; shaping context; the ethos of the discipline (enhancing sponsor value rather than just planning and monitoring); and responding to organizational, technological and societal developments and to the changing challenges facing our world. Part 3 suggests that the discipline needn't be limited just to responding reactively. In doing so, it notes that much of project and program management scholarship has tended to ignore application and impact – to be more concerned with means than with ends, with theory than with practice. Part 3 argues that the academic value of project, program and portfolio management would be greater if we could relate theory more directly to practical benefits, that it should concern itself more with how project practices can make a difference to society's issues.

Take-Aways

So much the structure. What do I expect you, the reader, to get out of reading the book? I would hope:

1. An understanding of how the modern discipline (domain) developed.
2. A better understanding of what it comprises – what its scope and subject matter really are.
3. A recognition that the benefits of seeing the domain as the broader field of the 'management of projects' covering the development and shaping of the project in its front-end stages, and of addressing topics such as strategy and governance, context, technology, commercial and supply chain issues, control, and above all, people, greatly outweigh the benefits of seeing it pre-eminently merely as a control-driven, delivery-execution management function.
4. An outline at a more detailed level of what we know about the management of projects, and programs, and where we are less sure – including, therefore, some suggestions for research.
5. A better appreciation of how project and program management can add value in an organization.
6. A more detailed understanding of how context shapes application, and how management can influence context.
7. A cry for rigour in methodology when putting forward new ideas.
8. A recognition that the academic community could focus more on the application of project and program management, looking at what impact it could have in the second half-century of its life.

It's important to recognise that the book is not a summary of all the research available in the field. To attempt this would be invidious and the result would be overwhelmingly unusable. The literature is instead called on to support these eight aims.

In pursuing these aims, but particularly with respect to the last two, I ask you, the reader, to engage and read critically: with an alertness, commitment and awareness that the little girl would be proud of; with a concern for purpose and evidence that is appropriate to a discipline that, in its contribution to society, past, present and to come, is truly important.

References and Endnotes

[1] As in the Project Management Institute (2013), *A guide to the project management body of knowledge*, fifth edition, Project Management Institute: Newton Square: PA and Office of Government Commerce (2007) *Managing successful programmes*, The Stationery Office: Norwich.

[2] For example, Flyvbjerg, B., Bruzelius, N. and Rothengatter, W. (2003), *Megaprojects and risk: An anatomy of ambition*, Cambridge University Press: Cambridge; Miller, R. and Lessard, D. R. (2000), *The strategic management of large engineering projects*, MIT Press: Cambridge; Morris, P. W. G. and Hough, G. H. (1987), *The anatomy of major projects*, John Wiley and Sons: Chichester.

[3] This is a mild dig at the 'projects-as-practice' approach: Hällgren, M. and Söderholm, A. (2011), Projects-as-practice: New approach, new insights, in: Morris, P. W. G., Pinto, J. K. and Söderlund, J. (eds.) *The Oxford handbook of project management*, Oxford University Press: Oxford, pp. 500–518.

Part 1

Constructing Project Management

Its history constructed

Reconstructing Project Management, First Edition. Peter W.G. Morris.
© 2013 John Wiley & Sons, Ltd. Published 2013 by John Wiley & Sons, Ltd.

Introduction to Part 1

Part 1 is a description of how the elements of what we call project management evolved over many years, but particularly since the early 1950s, and were slowly constructed into the thing that most project managers would recognise by the term today.

It is not an account of the management of projects through history; such a thing would be huge and probably meaningless. It does not claim – indeed it positively challenges the notion – that project and program management is now all defined and textbook clear. It shows rather that there are points of divergence and contradiction in the way we describe it and present our knowledge of it.

Some argue that such pluralism of knowledge is no bad thing since it shows vigour and reflects widespread adoption under differing conditions[1]. Maybe. Such a thought is at least comforting. But it doesn't diminish the concern where one believes misperceptions or mistakes are being propagated.

It is not the intent of this first section of the book to enter into any real or detailed critical discussion of the theory of the subject. This will be more the aim of Parts 2 and 3. It is instead intended as a description of the major actions that have contributed to the development of what passes for the discipline: an account of the major insights which slowly have built up our knowledge of the domain.

Historical Method

In presenting this chronology, I have endeavoured to be scholarly, respecting original texts (though admittedly much of the source material is secondary) and reflecting the thinking of the actors of the time and the contexts in which they were operating.

Reconstructing Project Management, First Edition. Peter W.G. Morris.
© 2013 John Wiley & Sons, Ltd. Published 2013 by John Wiley & Sons, Ltd.

All historians face the twin challenges of how to choose – how to frame – the object to be investigated, and then how to evaluate the data that are available and relevant. Scholarship requires absolute respect for the data, rigour and lucidity of analysis, and clarity of exposition. But judging relevance is not a value-independent exercise: it reflects a perspective. History today is rarely seen as an objective, disinterested enquiry but rather as socially constructed. My personal concern is how best to manage projects but critically *my unit of analysis is the project, not management processes and practices*. So I look for examples of how projects were, or were not, successfully managed. My history is thus different in scope and purpose from much of the more traditional project management preoccupation with planning and control.

The trouble is, the field is vast. Selecting events to illustrate the evolution of the discipline and, to a degree, in describing them, will inevitably reflect my own views, despite the desire for objectivity. But contemporary history acknowledges this: we are long past the time when we claimed that history was based on hard facts from which 'objective' truth was inductively drawn. Historians create historical facts, as the eminent historian E.H. Carr put it, according to their interests – feminism, gender, poverty, Marxism, colonialism, etc.[2] Study the historian to understand the history.

Bespeaking Relevant Knowledge

The examples I have chosen reflect major learning cases: one extraterrestrial (the Apollo Moon program); some international (Concorde); some national but private sector (the Andrew North Sea oil project); and others public (the US Department of Defense programs or the UK 'New Accommodation Program' (NAP) – the relocation of the UK's intelligence services). Were I say German, Japanese, Brazilian or Ghanaian, to pick a few nationalities at random, my examples would doubtless be different. Apollo would figure, though I am not so sure about the others. But I am not. I am an English academic with a strong practitioner bias who has spent a lot of time working in the Americas, Europe and the Middle East, and who believes passionately that there are things one can say about good practice in managing projects and programs.

And I also recognise the importance of context. Management, as we noted in the Introduction and as we shall see reiterated often, as a subject is inherently contextual[3]. One of the very strong aims of Part 1 is to illustrate this, showing how different contexts create the need for different management responses.

Aristotle said the mark of an educated man is to recognize in every field as much certainty as the nature of the matter allows. Context and personal perspectives shield us from ever attaining pure truth, be this historical or operational. Pure, whole truth is, in the social sciences, epistemologically impossible given the types of knowledge potentially in play and the effect of context, topics we shall discuss in Part 3.

Practising project and program managers must therefore shape their own version of 'what we need to know to manage projects effectively'. Part 1 is presented in the belief that reading a chronological account of how the project and program body of knowledge came into being will provide a foundation to help do this.

So, read and reflect; evaluate and adjust; modify and apply! Conjure your own account of what has made project management what it is. Most importantly, ask yourself, what in fact it – this knowledge – is.

References and Endnotes

[1] Söderlund, J. (2011), Theoretical foundations of project management, in: Morris, P. W. G., Pinto, J. K. and Söderlund, J. (eds.) *The Oxford handbook of project management*, Oxford University Press: Oxford, pp. 37–42.

[2] Carr, E. H. (1961), *What is history?* Cambridge University Press: Cambridge.

[3] Griseri, P. (2002), *Management knowledge: A critical view*, Palgrave Macmillan: Basingstoke.

Project Management before it was Invented

Projects have been around since man first walked on Earth, ranging from the informal, such as cooking or hunting, to the large formal construction or military ones. Often – maybe generally – they were accomplished very well. But there was no formal discipline of 'project management'. Indeed, there was no formal discipline of management at all until the 20th century – there was no 'village Project Manager' in the way there was a village butcher or baker, parish priest or possibly doctor. Not until the early 1950s did anyone even suggest that there might be a formal discipline called 'project management'.

This chapter, after acknowledging the enormous organisational abilities of man even at the dawn of civilisation, identifies some of the early formal techniques for managing projects – Plutarch, Vauban and Perronet on contracting; Wren and Hooke on construction organisation; the tools of Scientific Management (Gantt, Adamiecki, DuPont); and the rise of formal integration (the Bureau of Land Reclamation, aircraft manufacturing in the 1920s). The Manhattan Project – the program to build the USA's atomic bomb – often quoted as the originator of the discipline – is, however, no more an early example of project management than the pyramids four and a half thousand years earlier.

Pre-History: Projects and Society

Projects are undertakings to realise an idea. 'Project' (noun) means something thrown forth or out; an idea or conception[1]. People have, of course, been doing this since time began. The cave paintings of the Upper Palaeolithic era (35,000 BC) both reflect projects (hunting) and are themselves the result of complicated shamanistic belief systems and practices[2].

Projects are organisational entities. They differ from non-project organisations in that they all follow the same generic development sequence.

Reconstructing Project Management, First Edition. Peter W.G. Morris.
© 2013 John Wiley & Sons, Ltd. Published 2013 by John Wiley & Sons, Ltd.

Something like: (1) idea; (2) outline concept and strategy; (3) detailed planning; (4) execution; and (5) completion/close-out. *All projects, no matter how complex or trivial, large or small, follow this development sequence.* Non-projects (for example, running a production line in a bottling plant or a business) do not. Cooking my dinner is a project: it follows the same sequence: idea (menu), preparation, execution (eat), and wash-up*. Hunting is a project in that it follows this sequence, though its conception and execution may, in large part, be instinctive. Animals hunt. *Homo erectus*, 1.8 million years ago, hunted. Managing projects is partly instinctive.

As groups grew and society formed, projects became more complex. Many were unremarkable, as many are today. Others were spectacularly large and complex, designed to reflect temporal power, acknowledge deity, and witness life through death. The Giza pyramids of 2600–2700 BC probably involved the labour of some 70,000 people[3] – essentially the whole community – quarrying, hauling, dressing and laying the 25 million tons of giant limestone pieces that make up the three major pyramids; in effect, nation building. Clearly this huge effort required managing. Imhotep, Giza's architect, the builder before Giza of the pyramid at Saqqara, the first two-storey pyramid ever, was the organising genius. (He later died of cancer and was deified.)

Stonehenge, approximately 3000–1600 BC, was no less spectacular a feat of organisation, more so perhaps in both its size, relative to the population, duration, and given the less benign weather (frequently foul and quite un-Egyptian in its variability) and the long haulage over rough ground – 200 miles for each of the 82 four-ton bluestones that were probably dragged from south Wales to Wiltshire used for Stonehenge II/3.1 (2400 BC), and later, in Stonehenge III/3.2 (2400–2200 BC), and some 20 miles for the 77 huge twenty-ton sarsens brought from their quarry on the Marlborough Downs[4]. (Not only were the stones placed incredibly accurately in plan with regard to the Solstice, in elevation they were level despite being placed on an inclining site.) This enormous project – actually many projects, reflecting at least two belief systems[5] – must have involved the entire population of Wessex, some 50,000 people, working and supporting a thousand labourers, for many generations[6]. It's been estimated that hauling the stones and building Stonehenge required about 30 million man-hours of labour[7].

Projects (and programs of projects[†]) clearly existed then from the times of pre-history – since man has been on Earth – and were managed as such, often very effectively. But no-one thought of their management as a formal

---

*Cooking, as it happens, is a complex and by no means trivial example, either in today's terms or in the past's. No animal other than man cooks. Fire, the prerequisite development of cooking, has been controlled for a quarter to half a million years. Cooking has shaped the human form (Wrangham, R. [2010], *Catching fire: How cooking made us human*, Profile Books: London). Cooking, like gardening, is actually a close analogue of project management, as we shall see in Chapter 18.
†Programs are defined as projects having shared objectives or goals and having interdependent activities, often with shared resources.

activity called project management. Indeed, no-one thought abstractly of management as a subject at all. For although the management terms 'supervisor' and 'vizier' were in use, for example in Egypt, and we find discussion of organisation and leadership by, *inter alia*, Socrates, Aristotle and Xenophon, and of course Sun Tzu also on strategy[8], the topic as such would have to wait until the early 20th century to appear as a subject of credibility and substance.

Nor was management alone in not yet being a subject of intellectual enquiry: the engineering of projects was also quite rule of thumb. Until the Enlightenment and really well into the 17th century, engineering was preeminently based on intuition and experience rather than science. Despite the contributions of a few notable theoreticians such as Archimedes, Hero, Vitruvius and Frontinus, "the master builder remained a craftsman who, even in the design of important structures, was mainly guided by intuition"[9]. This is as true for the large irrigation projects and programs, for example of Iraq or Spain, as of the magnificent Roman aqueducts, roads and bridges, the vaulted buildings in Roman and Arabic architecture, or of the use of buttresses in Romanesque and Gothic architecture. (With often calamitous results: collapsing towers, domes and arches were frequent mishaps.)

The organisation of project work followed a similar practical bent. We see much that resonates with contemporary 'good management practice' – for example there is evidence as far back as Greek times of projects being divided into work packages, as in the Long Walls of Athens, divided amongst 10 contractors, or the building of the Coliseum in Rome, divided amongst four. Contracting strategies seem to have changed but little since then, at least until recently. Plutarch could be describing some of today's procurement practices when he writes: "when the local authorities intend to contract the construction of a temple or the erection of a statue, they interview the artists who apply for the job and submit their estimates and drawings; whereupon they select the one who, at the lowest price, promises the best and quickest execution"[10].

Managing engineering projects: the master mason

So, for many hundreds of years specialisation was by craft (hence the guilds) rather than by function (e.g. engineer). It was not until the 18th century that scientific theory began to be developed, enabling explanation and prediction, first, of the technical basis of project work, and later, more slowly and infinitely more tentatively, of some of their management needs and consequences.

Sir Christopher Wren exemplified the changing role of knowledge, both in engineering and architecture and in management. A mathematician turned astronomer and architect/engineer, Wren was a founding Fellow of the British Royal Society ('for Improving Natural Knowledge', i.e. for Science) which was granted its Royal Charter in 1663. An outstanding example of Wren's project management skills was his role in responding to the catastrophic Great Fire of London of 1666, which reduced 13,000 houses, 90 churches and many other buildings, some quite magnificent, to ash and ruin. Wren and Robert Hooke, the Oxford physicist, another Royal Society member, were commissioned to survey the blasted site, plan the new city, design the churches, houses and other buildings, and oversee the rebuilding works, Wren being appointed chief architect to the Crown in 1669. So great were the organisational challenges that Wren and Hooke abandoned the traditional approach to building – of craftsmen constructing the architect's design[11]. "The huge amount of materials and personnel necessary called for careful management of the work and control of costs. The beginnings of modern construction management can be seen in the way Wren's office was organized. The complimentary roles of architect, engineer, surveyor and contractor emerged"[12].

Military projects

Not that project management skills were limited to construction. Moving thousands of soldiers over large expanses of territory inevitably poses considerable challenges of project planning and control. William of Normandy's invasion of England in 1066 involved the shipment of 8,000 to 10,000 men, as well as their horses, across the English Channel in 400 boats over several days:, an immense undertaking[13]. We have little idea of how detailed William's inventorying and scheduling were but there was certainly considerable overall planning for this huge project. Loose planning under an overall schema was typical of armies on the move well up to the American Civil War and beyond. Often it was done badly.

Napoleon, while recognising the need for substantial logistical support in his invasion of Russia in 1812, established immense supply depots at Könisberg and other towns and provided additional transport battalions, yet failed to appreciate how wholly his *grande armée* would need to forage off the land, which the Russians destroyed as they retreated in front of his famished troops.

But if Bonaparte was rather careless of life (380,000 of his men died in the invasion), John Churchill, Duke of Marlborough, the greater soldier, who never lost an engagement let alone a battle, wasn't*. His famous march down the Rhine and Danube in 1704, ending in the Battle of Blenheim, is a fine example of project planning and leadership. "Vital stores and lighter canon were conveyed by river barges as far as possible. . . . Initially Marlborough's 21,000 men were accompanied by 1,700 supply carts carrying 1,200 lbs. of stores apiece, drawn by 5,000 draught-horses. The artillery needed as many more. At Heidelberg a new pair of shoes awaited every

---

*Marlborough was Captain General (i.e. Commander-in-Chief) of the Allied Forces (the Dutch, English, Hanoverians etc. and the Holy Roman Empire [the Austrians/Hapsburgs]), fighting Louis XIV of France and the Elector of Bavaria for the succession of the Spanish throne after the Spanish King, Charles II [a Bourbon], died childless in 1700 – the War of Spanish Succession (1701–1711). See Churchill, W. S. (1933), *Marlborough: His life and times*, London: Harrap.

man. . . . These measures were supplemented by a carefully-planned march time-table designed to confuse the enemy scouting parties and save the men from undue wear and tear"[14].

These examples of military, and civil, projects, managed at times superbly, at others not, are offered not gratuitously but simply to make the point that people have been managing projects, of all kinds, many very challenging, well before the advent of the tools, the language, the concepts that we today associate with the contemporary discipline of project management. And this was to remain the case, as we shall see, at least until the early 21st century, but really up until the 1950s.

The rise of the professional engineer: integration is threatened

It was Marlborough's opponent in the War of Spanish Succession, the French, under Sébastien Le Prestre de Vauban, who now advanced engineering onto a new, more scientific and professional basis. Vauban was a soldier first, a master builder of redoubtable fortresses, but later also of civil public works. Vauban promoted the role of 'Ingénieur' as a professional title for a scientifically trained technician in, effectively, civil engineering, with the "Corps des ingénieurs du Génie Civil" being founded in 1675 and the "Corps des ingénieurs des ponts et chausées" following in 1720. Vauban's approach to projects was a masterpiece of clarity, comprising a covering letter, a 'Mémoire' and drawings, the Mémoire giving information on the background to the project, a detailed description of the works, cost estimates, and a description of any special features. Standard rules were laid down for tendering and contracting including the preparation of an 'execution plan', definition of the duties and responsibilities of the contractor, stipulation of required records, etc.[15]. Jean-Rodolphe Perronet, appointed Chief Engineer of the Corps in 1763, equally put great emphasis on the organisation of the project: specifications were supremely important and a construction program(me) and site organisation plan had to be submitted.

The rise of the professional civil engineer in Great Britain in the late 18th/ early 19th centuries in the new era of the canals and railways however now introduced a new obstacle to the effective management of projects by breaking the integration between design and construction that had, up until this time, been provided by the master masons, and which Wren and Hooke had begun to develop further* – integration that we shall see is so much at the heart of effective project management.

The advance of steam power and new materials such as cast iron in Britain in the 18th century revolutionised methods of working, shifting society from its dominant rural base to an industrial one. Engineers, beginning with Brindley's design of the Bridgewater Canal in 1759, were engaged

---

*The Institution of Civil Engineers was founded in 1818, University College London, appointing the first Professor of Civil Engineering in 1841, the same year it created a Chair in Architecture. It appointed one of the first Chairs in the United Kingdom in Project Management in 2002.

to produce designs and to present (or oppose other competing) plans in Parliament for planning approval. So busy did engineers like John Smeaton, Thomas Telford and John Rennie become in doing the front-end planning that they effectively disengaged from the day-to-day building of the project, leaving this to the contractor. This trend was magnified with the building of the railways, not just in Great Britain but all over the world[16]. Contractors like Thomas Brassey assembled and ran huge organisations with, in his case, a labour force of up to 100,000 men scattered around the globe*. Brassey's organisation was pyramidal in nature, with his agents managing subcontractors[17]. The quality of these agents – project managers in deed if not in name – was absolutely central to the effective functioning of the organisation.

The divorce of design from construction, in the United Kingdom at least, was made worse when the Royal Institute of British Architects (RIBA) in 1887 prohibited its members from being in a profit-making position with respect to the organisation of building work[18]. This breaking of the integration between design and construction – of the front-end of the project from production – was then not addressed in construction for almost a hundred years, till the advent of modern construction management in the 1960s[19] (see below p. 78).

Interestingly, just when the Victorians were de-integrating their construction project management processes, technology, in the form of rising specialisation amongst engineers and others – civil, mechanical, structural, electrical, sewerage, as well as town planners and building and landscape architects – was creating greater specialisation and differentiation, and in doing so bringing with it the obvious need for enhanced integration.

Health and safety

In fact, the architects' professional position of non-involvement, commercially, in production was not a requirement placed on any other profession, such as the engineers, and many engineers did involve themselves directly in both design and construction. Isambard Kingdom Brunel, for example, was intimately involved in overseeing the construction of the Great Western Railway (GWR) and other projects. Charging around in a specially appointed carriage (the "flying hearse"), he worked "rarely much under 20 hours a day"[20]. Great engineer though he was, not everything was a success (the GWR locomotives and *The Great Eastern* were disasters) and his views, for example on Health and Safety, were, unsurprisingly, well, Victorian: "any change that made the master liable for accidents that befell those occupying

---

*By 1847, Brassey had built about one-third of the railways in Britain, and by the time of his death in 1870 he had built one in every 20 miles of railway in the world. This included three-quarters of the lines in France, major lines in many other European countries and in Canada, Australia, South America and India. He also built many of the structures associated with those railways, including docks, bridges, viaducts, stations, tunnels and drainage works.

the lower rungs would inevitably alter the present position of things, by which every department, from highest to lowest, is sublet to men, who are free agents, and seek to execute the work in the cheapest way"[21].

This apparent carelessness of life was not so unusual of the time. Forced labour was still being used on rail projects built by British, American and German companies in Latin America and Africa throughout much of the 19th century. Around 60,000 *corvé* labourers were supplied by the Egyptian government to Ferdinand de Lesseps to dig the Suez Canal during the years 1859–1869, thousands of whom died[22]. The disruption caused to the domestic economy by taking this many fellaheen off the land was huge: socially and economically, the project was the inversion of building the pyramids two and a half thousand years earlier. De Lesseps then went on to dig the Panama Canal where, during the initial French construction period of 1881–1889, over 20,000 labourers died, most from disease; a further 6,000 to 10,000 died under the period of American management, 1904–1914.

Lest we get too smug, as late as the early 20th century, forced labour was being used by the British in Kenya[23], to say nothing of the Nazis (Peenemünde, say, the V1/V2 rocket base, where 20,000 died between 1943 and 1945 hacking underground facilities out of solid rock in appalling conditions under murderously brutal supervision); or Stalin's White Canal, where 25,000 died between 1930 and 1933 under carelessly inhuman direction*. Projects have been built and managed not just inefficiently but under very different value systems.

Supra projects

Programs (of interconnected projects having shared aims and objectives) existed too, though there is little evidence of the term being used. London's public health reforms in the mid-19th century, including the much needed re-building of its sewers, are an important example of a major public sector program. Sir Edwin Chadwick, the English civil servant who led the call for reforms to public health, deployed a range of instruments which illustrates the diversity that can and may need to be called on to address large-scale social change – instruments ranging from legislation to 'hard' engineering projects (re-building the sewers). His massive survey, *Inquiry into the Sanitary Conditions of the Labouring Population of Great Britain* of 1842, compared the urban environments in London, Glasgow, Manchester, Birmingham, Leeds, and other large centres. To implement the recommendations from this report, Chadwick led the campaign for England's first Public Health Act. He pushed for government health inspectors, better ventilated and less crowded housing, wider streets, workplace health and safety legislation, increased use of indoor plumbing, and limits on the employment of

---

*Awful though these figures are, they pale in comparison with the 2 million labourers that are said to have died – nearly 40% of the total workforce – building the 1,200 mile-long Grand Canal of China around 600 A.D.

children in factories. Neither the term project management nor program(me) management appear in his work.

Similarly, Thomas Hughes, in his much later, extensive study of the electrification of Germany, the UK and the United States, makes no reference to program or project management (though 'systems' figures extensively in his analysis[24]). As in several industries at this time – rail, mining, petrochemicals; medicine, shipping, banking – the 'system' had to be designed at a 'supra' system level as well as at a city or subregional 'utility' level, and at a project level. Standards had to be set; growing capacities provided; and reliability and safety assured. The role of the engineer was central, but neither that of the program nor the project manager had yet formally crystallised out. 'Management' was, however, beginning now, in the late 19th and early 20th centuries, to be discussed as a valid, general field of enquiry. And as this happens, we begin to see the emergence, if only sparsely, of formally recognised project roles – project engineers, project coordinators and the like.

Early Attempts at Formal Project Integration

The incubator for the growth of interest in management as a subject was the industrialisation of work – the Industrial Revolution. The rail, steel and construction industries especially provided, particularly in the USA, the conditions and incentives for management to rise, both as a practice and as a domain of intellectual enquiry. Though writers on organisation (e.g. Max Weber) and general management (e.g. Henri Fayol) were at work across the turn of the centuries, 'scientific management' for many years at this time dominated the field of management's intellectual development – and it certainly had a strong influence on the emergence of project management in the areas of project control and coordination.

'Scientific management', as pioneered by Frederick Taylor around 1881–1914, applied measurement and analysis to the organisation of repetitive operations, such as bricklaying. Coupled with the setting of piece-rate targets, such analysis was able to improve productivity and output reliability significantly. Frank Gilbreth extended these principles, applying motion study to Taylor's time studies, particularly in construction. Henry Gantt developed his famous planning and tracking bar chart (Figure 2.1) for the US Army at its Frankford Arsenal in 1917, though he had been using earlier versions for analysing irregular activities such as plant maintenance since the first years of the century. Similar graphical tools had been used from 1911 to 1912, for example for road construction by the contractor Stone & Webster, and no doubt others[25]. And the Polish academic, Karol Adamiecki, had developed his Harmonygraph (effectively a vertical bar chart) in 1903 (Figure 2.2). It was used almost exclusively in Poland and Russia, however, and was only published in the West in 1930[26]. In fact, the idea of laying activities out in horizontal 'time lines' had been around since at least the time of Joseph Priestley, who published his Chart of Biography in 1765, and the idea was incorporated in other tools for example in building The

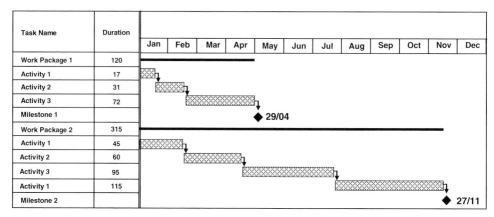

Figure 2.1   The Gantt chart.
Source: Author's own.

Time	From	-	-	-	A-1	B-1	B-1	D-1	A-2	B-2	C, D-2	A-3 E-1DUM	C, D-2
	To	A-2	B-2, C	D-2	A-3	E-1	D-3 DUM	D-3 DUM	E-2	E-2	E-2	-	-
	Activity	A-1(4)	B-1(4)	D-1(2)	A-2(4)	B-2(3)	C(3)	D-2(3)	A-3(1)	E-1(4)	DUM(0)	E-2(e)	D-3(8)
1		▨	▨	▨									
2		▨	▨	▨									
3		▨	▨					▨	Sliding tab for activity D-2				
4		▨	▨					▨					
5					▨	▨	▨						
6					▨	▨	▨						
7					▨	▨	▨						▨
8					▨					▨			▨
9									▨	▨			▨
10										▨			▨
11										▨		▨	▨
12												▨	▨
13												▨	▨
14												▨	▨
15												▨	▨

Figure 2.2   Adamiecki's Harmonygraph.
Source: Morris (1994), p. 7. (© Moder, J. J., Philips, C. R. and Davis, E. W. [1983], *Project Management with CPM, PERT and precedence diagramming*, 3rd edition, Van Nostrand Reinhold: New York.)

Crystal Palace (Figure 2.3), Joseph Paxton's huge prefabricated glass building, over 560 meters long, 123 meters wide and 33 meters high, built for Great Britain's Great Exhibition in Hyde Park in just eight and a half months in 1850–1851.

By the 1920s, cost control was likewise receiving increased attention. Pierre DuPont, one of the three brothers who had bought the DuPont

Figure 2.3   Crystal Palace.
Source: Public domain.

Powder Company in 1902 to create E.I. DuPont de Nemours & Company – later the company that developed the Critical Path Method in 1957 – took forward Taylor's ideas, establishing a formal system of project accounting from 1911. Crucially to this history, "the system used the term 'project' to cover the entirety of the administrative and construction work required to make any addition to DuPont's fixed assets". By 1920, project-based accounting was being used for laboratory research in DuPont and several other large, research-based firms such as Bell Labs and Arthur D. Little[27].

In fact, a more ambitious planning and control 'system' in which the term 'project' was central had been established earlier than this, in 1902, in the US Bureau of Reclamation. Given the geographical diversity of its operations "the Bureau of Reclamation built their control system around a choke point called "the project office"[28]. In establishing administrative control, the Bureau assigned a "project engineer", the officer in charge in the field on one complete project"[29]. This surely is one of the earliest examples of the term 'project' being used in its modern management meaning – but note that the action is all about monitoring and control, not yet about the wider function of coordination (or, taking 'integration' to cover both coordination and control, of integration).

Project coordination

Formal project coordination, carried out often, though by no means always, through the role of project engineers, became increasingly common in the

1920s, for example, due to the quasi-bespoke nature of much aircraft production at the time, in the US aircraft industry. Goldman, in 1929, quotes
examples of project engineers acting as coordinators both in line and staff
positions[30]. The aircraft manufacturer Wright Field assigned project engineers to monitor design and development from the 1920s and by the late
1930s had a 'project officer' assigned to each aircraft supported by a project
engineer. Project officers focused on safety and looked for design weaknesses[31]. Wright Field stressed the importance of the role: "the success or
failure of a job, particularly financially, is usually measured by the managerial and executive flexibility of the project engineer"[32], not least because of
the disruptive impact of change orders – a line of thinking later pursued by
Jay Galbraith in the 1960s, with his research on forms of integration at
Boeing[33]. A particular need was for coordination between design and production[34], an issue raised 40 years earlier, as we saw earlier, with the British
architects' divorce from the act of building.

In the mid-1930s, Luther Gulick, following Henri Fayol and others such
as Harrington Emerson and Russell Robb, sought to generalise the then
thinking on organisational coordination. (Fayol was one of the earliest
writers [around 1900–1918] to outline the principles of general management; Emerson developed the line and staff distinction; and Robb emphasised the importance of seeing the organisation as a whole[35].) Gulick
proposed that an organisation's primary task was to provide coordination
and, to do this effectively, organisations could be structured around function
(i.e. purpose), process (i.e. discipline), resources (people or things), or
place[36]. Crucially, groupings could occur around different bases at different
levels. In effect, Gulick was prefiguring the matrix organisation, where one
axis of organisation is project (purpose) and another is functional (discipline). The first real matrix organisation wasn't reported until around
1952–1953 in the Martin aerospace company, and we would have to wait
till the 1960s and 1970s for a fuller, more rigorous academic exposition – as
we shall see in Chapter 4.

It may be worth reminding ourselves that with these aircraft examples
we are dealing with the coordination and control of relatively small engineering production orders. This is a long way from the management of large
projects, of which there were many in this time period: from huge construction undertakings like the Hoover Dam (indeed the whole Colorado River
management/California irrigation system), completed for the Bureau of
Land Reclamation in 1936 two years ahead of schedule and under budget[37],
to the building of public transportation systems – New York's Robert Moses
stands high in the pantheon of devastatingly ruthless and effective public
works' promoters[38]. Projects certainly existed, but it was still hard to see
much more than the beginnings of a formal project management discipline
by the time World War II broke out.

World War II and the Manhattan Project

The Second World War saw thousands of projects and programs but no
evidence of a formal management approach called project management, or

project coordination or anything similar. Clearly D-Day, to pick one of the most obvious examples, was a massive 'program of programs of projects', but the project management terms weren't used. No-one talked about project management, or program management. While schedules and coordinators clearly existed in their thousands, there was no standard p.m. approach, no acknowledgement that there might be a body of knowledge or a discipline for managing projects.

Operations Research (OR), which is often suggested as contributing to the evolution of the discipline, certainly was effective during the war in helping, for example, to improve bombing accuracy, anti-aircraft firing patterns, convoy tactics, and air intercepts. OR did influence the development of systems analysis[39], which, as we are just about to see, was in turn very influential in the early 1950s on USAF (US Air Force) thinking that led to the formal introduction of project management, but there is no evidence of OR stimulating the evolution of project management, as there is for example of scientific management.

Many would argue that the Manhattan Project – the US program to develop the atom bomb – is the exception[40]. As I wrote in *The Management of Projects* (1994): "The Manhattan Project certainly displayed the principles of organization, planning and direction that typify the modern management of projects"[41] – but it didn't formally apply any of the tools, techniques, or arguably language or concepts of the discipline as it became articulated post the early to mid-1950s. That's why it's a fundamental mistake, I believe, to try and derive lessons for project management today from the Manhattan Project[42] (Figure 2.4). It simply wasn't a project – program – that was managed by the discipline of project or program management. The discipline wasn't known; the rules of good project and program management were, in several instances, flouted; the operational constraints were abnormal: no Front-End to speak of; no resource constraints; the schedule was absolutely dominant. Except, and it's a very important exception, there is the lesson of a dedicated project-oriented organisation (although it could hardly be claimed that this had never been tried before). J. Robert Oppenheimer, who headed the teams that were developing the bombs at Los Alamos, finding progress was too slow, decided to increase focus and pressure by turning the organisation of the site 90° from a functional orientation that reflected the primary scientific disciplines to one structured on "divisions organized around the end-product – a project organization"[43], thereby improving communication, integration and the sense of urgency. But would Leslie Groves, of the US Army Corps of Engineers, who was in overall charge of the program, or Oppenheimer have called themselves project or program managers? Not really; the language hadn't yet been invented. Nevertheless, General Groves, who had come to the project having just finished building the Pentagon, had as his top two factors responsible for the program's success: first, "a clearly defined unmistakeable, specific objective. Second, each part of the project had a specific task"[44]. Resonant but not sufficient to justify the assertion that project management began with Manhattan.

If formal project management has been prefigured so far largely in a hesitant, informal mode, this was now about to change. Big, urgent, superimportant projects and programs, involving brand new, interdependent technologies,

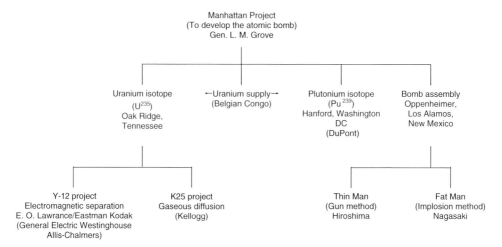

Figure 2.4   Work Breakdown Structure of the Manhattan Project.
Source: Morris, P. W. G. (1994), *The Management of Projects*, Thomas Telford: London, p. 16.

were from the early 1950s to stimulate the invention of new engineering and management disciplines, specifically systems engineering and project management. Energised by the US missile development programs, concepts, language, tools and techniques were to emerge that professionals today would recognise as constituting the discipline of project management.

References and Endnotes

1   *Oxford English Dictionary.*
2   Lewis-Williams, D. (2002), *The mind in the cave*, Thames and Hudson: London.
3   Mendelssohn, K. (1972), *The riddle of the pyramids*, Praeger: New York.
4   Atkinson, R. (1979), *Stonehenge*, Penguin: Harmondsworth.
5   Burl, A. (1976), *The stone circles of the British Isles*, Yale: New Haven, CT, pp. 310–311.
6   Burl, A. (1987), *The Stonehenge people*, Barrie and Jenkins: London; Renfrew, C. (1973), Social organization in Wessex, In: Renfrew, C. (ed.) *The explanation of culture change models in prehistory*, Duckworth: London, pp. 539–558.
7   Atkinson, R. (1961), *Neolithic Engineering*, 35, pp. 292–299.
8   Wren, D. A. (2005), *The history of management thought*, Wiley: Hoboken, NJ.
9   Straub, H. (1952), *A history of civil engineering*, MIT Press: Cambridge, MA.
10   *Ibid.*
11   Whinney, M. (1971), *Wren*, Thames and Hudson: London.
12   Cooper, M. (2003), *A more beautiful city: Robert Hooke and the rebuilding of London after the Great Fire*, Sutton: Stroud.
13   Bayeux Tapestry; Bates, D. (2001), *William the Conqueror*, Tempus: Stroud.
14   Chandler, D. (1973), *Marlborough as military commander*, B T Batsford: London, p. 130.
15   Straub, H. (1952), *op. cit.*: 9.

[16] See for example the engineering biographies by Rolt, L. T. C. (1957), *Isambard Kingdom Brunel: A biography*; (1958), *Thomas Telford*; (1960); *George and Robert Stephenson: The railway revolution*, all Longmans Green: London; and (1962), *James Watt*, Batsford: London.

[17] Walker, C. (1969), *Thomas Brassey: Railway builder*, Cox and Wyman: London.

[18] Bowley, M. (1966), *The British building industry*, Cambridge University Press: Cambridge, p. 342.

[19] Foxhall, W. B. (1972), *Professional construction management and project administration*, The American Institute of Architects and Architectural Record: New York.

[20] Buchanan, R. A. (2006), *Brunel: The life and times of Isambard Kingdom Brunel*, Hambledon and London: London, p. 68.

[21] Linder, M. (1994), *Projecting capitalism: A history of the internationalization of the construction industry*, Greenword Press: Westport, CT, p. 33.

[22] Wilson, A. T. (2007), *The Suez Canal: Its past, present and future*, The Looney Press: London.

[23] Linder, M. (1994), *op. cit.*: 21.

[24] Hughes, T. P. (1983), *Networks of power: Electrification in Western society: 1880–1930*, The Johns Hopkins University Press: Baltimore, MD.

[25] Pinney, B. W. (2001), *Projects, management, and protean times: Engineering enterprise in the United States, 1870–1960*, PhD thesis, Massachusetts Institute of Technology: Cambridge MA.

[26] Marsh, E. R. (1975), The harmonogram of Karol Adamiecki, *The Academy of Management Journal*, 18, 2, pp. 358–364. (Adamiecki used both the terms Harmonygram and Harmonygraph.)

[27] Pinney, B. W. (2001), *op. cit.*: 25.

[28] *Ibid.*

[29] *Ibid.*

[30] Goldman, S. H. (1929), Engineering organization, *Aviation Engineering*, pp. 15–16.

[31] Pinney, B.W. (2001) op. cit.: 25.

[32] Wright, C. (1938), *Aviation Engineering*, p. 388.

[33] Galbraith, R. (1970), Achieving integration through information systems, *Working Paper No. 361–368*, Alfred P. Sloan School of Management, Cambridge MA: Massachusetts Institute of Technology.

[34] Hardecker, J. F. (1929), Organization and supervision of the design department, *American Machinist*, November, p. 29.

[35] Wren, D. A. (2005), *op. cit.*: 8.

[36] Gulick, L. (1937), Notes on the theory of organization, In: Urwick, L. (ed.) *Papers on the science of administration*, Columbia University Press: New York, pp. 1–46.

[37] Hitzik, M.A. (2010) *Colossus: Hoover Dam and the making of the American Century*, Free Press: New York, pp. 323–324.

[38] Caro, R. (1974), *The power broker*, Vintage Books: New York.

[39] Johnson, S. B. (1997), Three approaches to big technology: Operations research, systems engineering, and project management, *Technology and Culture*, 38, 4, pp. 891–919.

[40] Rhodes, R. (1988), *The making of the atomic bomb*, Penguin: Harmondsworth.

[41] Morris, P. W. G. (1994), *The management of projects*, Thomas Telford: London, p. 18.

[42] Lenfle, S. (2011), The strategy of parallel approaches in projects with unforeseeable uncertainty: the Manhattan case in retrospect, *International Journal of*

Project Management, 29, 4, pp. 359–272; Lenfle, S. and Loch, C. (2010), Lost roots: How project management came to emphasise control over flexibility and novelty. *California Management Review*, 53, 1, pp. 32–55.

[43] Johnson, S. B. (2002), *op. cit.*: 31; Rhodes, R. (1988), *op. cit.*: 40: pp. 454–255.

[44] Groves, L. G. (1962/1983), *Now it can be told*, De Cap Press: Harper, p. 414.

Systems Project Management

Project management as a term seems to first appear in 1952–1953, in the US defence–aerospace sector, in three or four different places. It was followed towards the end of the decade in the engineering construction industries.

First, around 1952–1953 in the US Air Force (USAF), the project office idea got strengthened. As aircraft became more technically sophisticated after World War II, the challenges of integrating component development with airframe development became significantly harder. Difficulties were experienced particularly due to unstable, changing requirements and a failure to focus on the 'complete weapon'. (*Plus ça change*!)

Meanwhile, the United States was faced with an immense technical challenge: to develop a brand new set of technologies – intercontinental ballistic missiles, thermonuclear warheads, advanced guidance systems – in an environment of extreme urgency, for the Russians seemed to be on the verge of deploying their version of this weaponry before the Americans had a counter weapon. It was to organise and manage this effort that the project manager role and the matrix organisation were invented around 1952–1953. Brigadier Bernard Schriever, the father of modern project management, formalised the role on Atlas and Admiral Raborn followed on Polaris. By 1957, PERT and CPM had been invented. Project management practices were promulgated by the Department of Defense in the early 1960s, and by 1969 project management had landed man safely on the Moon. Apparently fit for anything, 'project systems management' soon thereafter found itself bumping into limits set by environmentalist opposition and other 'external' factors.

USAF Integration: The Formal Recognition of Project Management

The Cold War was technologically and managerially quite different from World War II, and the challenges of coordination were central to addressing

Reconstructing Project Management, First Edition. Peter W.G. Morris.
© 2013 John Wiley & Sons, Ltd. Published 2013 by John Wiley & Sons, Ltd.

the new reality. World War II had been "a weapon production race while the cold war was a weapons development race, where technological performance mattered more than numbers"[1]. The primary reason was the move from aeroplanes to missiles.

Traditionally, weapon development had been carried out by Army and Navy laboratories and arsenals – the USAF had only become independent of the Army in 1947 – by functionally organised groups. However, the new science and technologies now being required (rocket engines, communications, guidance and control, high-speed aerodynamics, etc.) were proving too often to be beyond the capacity of many of these in-house resources and delivery progress was too frequently inadequate.

As a first step to address this, in what would be many, new USAF regulations were issued in 1951, tentatively and with some linguistic obfuscation, tightening coordination between R&D and production – between the Air Force Research and Development (AFRDC) and Air Force Material Commands (AFMC) – which mandated that every major project should have a Weapons Systems Project Office linking the work of these two arms. The Offices were to be headed by 'team captains' (one from AFRDC and one from AFMC) who would have full responsibility for the entire 'weapon system' and who would coordinate the activities across the project. (Project Offices were also set-up for aircraft programs as well, e.g. the B-47 and B-52)

A shade later, around 1953–1954, another milestone in the creation of project management emerged with the Martin (Marietta) and McDonnell Aircraft companies formally creating the project manager position. McDonnell began using the title of project manager in 1953, the project manager's prime responsibility being organisation and staffing[2]. More significantly perhaps, Martin has a claim to have established the first matrix organisation, creating in 1953–1954 "a number of miniature companies, each concerned with but a single project. The project manager exercises overall product control"[3]. All functions, from design to manufacturing and distribution, were covered: systems analysis being applied to determine requirements, systems engineering on design, and systems management on integration[4]. (The systems approach is outlined below.)

But of all the forces influencing the emergence of project management, the most significant was the USAF's management of Atlas, America's first ICBM (intercontinental ballistic missile) (Figure 3.1). For the development pressure wasn't just technology push. There was an enormous threat: the Soviet Union was developing ICBMs too and intelligence suggested that Soviet ICBMs would be operational before the United States'. Without a counter-deterrent, the USA would be vulnerable not just to a Soviet first-strike capability onto its mainland – and America's mainland had never yet been bombed from the air – but would be so from a thermonuclear bomb. Urgency to develop the US ICBMs was therefore huge. The Air Force (and Navy and Army) thus looked very seriously at how the development of their missiles could be accelerated.

Figure 3.1   Atlas.
Source: © US Air Force.

Schriever and the Atlas Program

Each of the service branches was anxious to promote itself as a provider of the United States' ICBM capability. Following the Hydrogen Bomb tests of 1952 which showed that nuclear warheads would soon be small enough to be carried on an ICBM, President Eisenhower ordered the USAF to accelerate its ICBM development program. Eisenhower also requested the government's Nuclear Weapons Panel to review the overall status of the three services' missile programs. The committee reported in February 1954. It believed Atlas could be operational by 1962 but only if a new development group were given direct responsibility for the entire project; in particular, that a civilian organisation coordinate systems integration over and above Convair, the incumbent Atlas prime contractor[5]. The Air Force responded by endorsing the idea of a Manhattan-type organisation (*q.v.*), recommending "that overall technical direction be in the hands of an unusually competent group of scientists and engineers capable of making systems analyses, supervising the research phases, and completely controlling the experimental and hardware phases of the program"[6]. This recommendation was agreed and in May 1954 the USAF made Atlas its highest R&D priority.

Air Force development projects were managed by AFRDC from its base in California – the 'Western Development Division' (WDD). (The majority of personnel working on the US missile programs were based in California.) Brigadier Bernard Schriever, an Air Force officer with a strong interest in science and the newly emerging systems approaches to planning and engineering, was appointed to head WDD in July 1954 (Figure 3.2).

Schriever felt that the complexity of the program required an organisation that would be able to call on scientific and technological expertise in a more flexible and responsive way than would be possible either through a dedicated Manhattan-style organisation, or by just relying on Convair. He thus proposed instead a 'systems integrator' function that would sit between the USAF and Convair performing systems engineering and providing technical support to the program. This would entail, *inter alia*, preparing the overall program strategy, drawing up contractual specifications, coordinating and monitoring contractor performance, and performing evaluation testing[7]. In September 1954, the newly formed Ramo-Wooldridge Corporation was appointed as Atlas' systems integrator (later renamed Space Technologies Inc., as a subdivision of TRW)[8].

Schriever was also anxious to minimise delays caused by program funding procedures. Having shown President Eisenhower how current arrangements were delaying development progress, new procedures were authorised and put in place in 1955. The result was effectively to make Schriever autonomous in his command of the program. "These new procedures represented the first full application of project management in the air force, where the project manager had both technical and budget authority for the project"[9].

Systems thinking was a powerful force in the shaping of these organisational and management structures. (See box[10].) From the systems approach

Figure 3.2   Bernard Schriever.
Source: © US Air Force.

comes such practices as interface management (a natural locus for project management[11]), emphasis on the 'whole system' (as, e.g. in integrating R&D with supply), and design reviews – and in fact the whole philosophy of systems management which first the Air Force and later the Department of Defense (DoD) were to adopt in the 1960s. Schriever's natural turn of mind moved quickly towards these concepts, emphasising the value of looking at the 'whole' weapons system; of comprehensive, forward-looking reporting; and of high-level (systems) engineering and testing. Responsibility for all USAF major development programs was transferred to Air Force Systems Command (AFSC) in 1961 – a new Command headed by Schriever.

The Systems Approach

Overarching all the different branches of systems thinking was – and still is – General Systems Thinking (GST). Developed and promoted by Ludwig von Bertalanffy from the mid-1930s, this highly ambitious schema emphasises principles such as holism, emergence and hierarchy, regulation, homeostasis, boundaries and interfaces. Cybernetics, the science of communication and control, is often associated with GST. Branches of general systems theory include systems analysis, systems engineering, systems integration, and systems management.

Concurrency and configuration management

Having just criticised attempts to extract from Manhattan the suggestion that parallel development is a desirable project management strategy, we now see it displayed by Schriever on Atlas. The same admonition holds. Schriever's place in the pantheon of project management is that it was under his leadership that the discipline emerged as a named function. Schriever was not interested in writing a best-practice definition of project management: what he was committed to was having an ICBM capability in place in time to offer a realistic deterrent to the Soviets.

The challenge for Atlas and the ICBMs was urgency coupled with high technical uncertainty. To address this, Schriever deliberately encouraged parallel development of system elements, regardless of their readiness – which in this case was defined overwhelmingly by the level of technical maturity. That practice that came to be known as 'concurrency', that is, beginning one phase before the previous one is stable. Concurrency inevitably leads to changes being required with concomitant rework, added cost, and delay.

Ballistic missiles had new, unknown technologies in abundance, not just in individual fields but in the way they interacted. Accordingly, design changes were frequently required and made, but for a while the resulting decisions were inadequately documented. When reliability problems arose during Atlas tests in 1956–1957, this lack of adequate documentation meant that it was often hard to identify the causes of the problems. As a result, new, more formal design and configuration management practices were introduced. Design reviews now comprised three stages: Preliminary Design Reviews, Critical Design Reviews, and Flight Readiness Design Reviews, each involving outside expert input and special attention to interfaces. Under the new practice of Configuration Management developed by Boeing, proposed changes had to be reviewed and signed-off by a Change Control Board, and the revised documentation formally communicated to all affected parties*. Quality management became much tighter. These practices were to prove central to the management of future missile programs such as Minuteman, Titan and Apollo.

---

*The following describes Configuration Management well, although the text actually refers to the Apollo program. "Engineers used a variety of formal documentation to record and track information, and a 'change board' ensured that any department affected by the proposed engineering and manufacturing changes reviewed the proposals and committed appropriate resources to implement them. [The] process included identification, change, and accountability control over all process designs and corresponding hardware. The process centralized decision-making, in that all departments had to provide technical and resource estimates for the project manager's decision. But departments retained significant authority, as they specified the technical methods and the resources required to do a job." Johnson, S. (2001), Samuel Phillips and the Taming of Apollo, *Technology and Culture*, 42, p. 694.

Polaris

Meanwhile, the Navy and the Army had been working together using the Army's Jupiter rocket as the basis for a possible Fleet Ballistic Missile (FBM). With both the Navy's Bureaus of Ordnance and Aeronautics vying for command of the program, it was decided in 1955 to create a dedicated 'Special Projects Office' (SPO) headed by Vice Admiral William Raborn, with equivalent status of a Bureau, to manage the Navy's portion of the program (Figure 3.3). Two or three years previously, the term project or program manager hadn't even existed; now we have an Admiral heading up a Project Office! By mid-1956, the Navy had realised that technical advances were such that the rate at which the warhead, and hence the rocket, was getting smaller and lighter was such that it should be possible, by the time they went into service, to have a solid fuel-based rocket deployed on submarines – Polaris. The decision to abandon Jupiter and replace it with a new Polaris missile system as the FBM was sanctioned in December 1956. The SPO remained in place to manage the overall program[12].

Like the Air Force, a systems management approach was adopted but, unlike the Air Force, most of the systems design and integration – on the

Figure 3.3   William Raborn.
Source: © US Navy.

missile, submarine, missile guidance and fire control, launcher, submarine navigation, bases, and communication systems – was performed in-house under the SPO. Only the nuclear reactors and missile warheads lay outside the SPO's responsibility (with Admiral Rickover).

Polaris was a very real success. In terms of this history of project management, however, it is notable probably for two things in particular: Raborn's attention to people (including what we would today describe as stakeholders) and the invention of the scheduling system, PERT, the Planning and Evaluation Review Technique.

Like Schriever, Raborn emphasised quality of people and team morale. SPO officers worked in dress uniform, flags flew from SPO offices, and SPO members were told they were the cream of the cream. But what remains in the project management world's mind most notably about Polaris is almost certainly PERT.

PERT and CPM

Raborn wanted a 'forward-looking' reporting capability, one that could tell him what the likely completion dates were going to be on activities, not just where he'd got to so far. So in 1957–1958, Polaris's project controls manager, Gordon Peterson, working with a team drawn from Lockheed and management consultants Booz Allen & Hamilton, developed an event-oriented network scheduling system, in which a three-point estimate was made of activities' completion dates – likely, less likely, more likely (mean ± one standard deviation) – plotted as a β distribution. PERT-Cost was introduced in 1961–1962 (Figure 3.4).

As it happened, at exactly the same time, 1957–1959, DuPont (using a small team of DuPont engineers, led by Morgan J. Walker, and Remington Rand Univac engineers, led by James E. Kelley) invented the Critical Path Method (CPM), a similar network scheduling tool. DuPont wanted the tool for planning and controlling its construction activities. Unlike R&D programs such as Polaris, construction lends itself to calculation of activities' rate of progress on a linear basis – cubic metres of concrete poured per day, bricks laid per hour, etc. – and so CPM was activity- rather than end-event oriented*. The computer program was trialled on plant shutdowns in 1957 and the first paper on CPM was published in March 1959[13] (Figure 4.1).

PERT never quite fulfilled its promise on Polaris but Raborn, cleverly and presciently, used it "as a tool to manage his external environment"[14], lauding it to Congress and the press: "How can our program not be being well managed? Here", pointing to reams of print-out, "we are using the first management tool of the nuclear and computer age". Raborn actually paid great attention to his stakeholders – undertaking what we today call stake-

---

*In R&D one doesn't know as well what will happen between events, so the focus is on the probability of the event being accomplished as forecast.

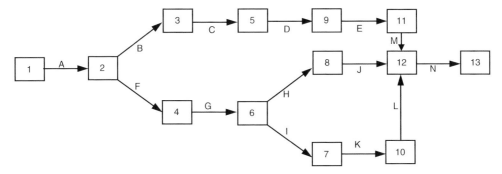

- Events are represented in numbered rectangles.
- Activities are represented on directional arrows that connect events.
- PERT focuses on the likelihood of events being accomplished by certain times. The duration of each activity is calculated in terms of a standard probability distribution. Mean, most optimistic and most pessimistic values are given (calculated by ± one standard deviation around the mean).

Figure 3.4   PERT.
Source: Author's own.

holder management – for example visiting the voter districts of congressmen whose support was important to the continuance of the program[15].

Both CPM and PERT became iconic exemplars for many, many years of what, for many people, project management was all about. (Both, of course, made use of the newly emerging, exciting power and possibilities of the computer.) Precedence scheduling ('Activity-on-Arrow') was invented in 1961 by John Fondahl at Stanford University[16]. Resource allocation was developed throughout the 1960s and early to mid-1970s. (See below.)

Construction

Less visibly, and more slowly than in defence–aerospace, project management began being taken up as an active role by the engineering-construction industries. Exxon had employed a 'project engineer' coordinating role since the 1930s but it was not until the 1950s that the more active deployment of project management began happening. Bechtel, an American architect-engineer firm, claims to have used the project manager role on the Trans Mountain Oil Pipeline in Canada (1951–1953), although the role related just to construction and it wasn't called project management. Nevertheless, according to Stephen Bechtel, Jr., "the approach and organisation was a forerunner of what was to come"[17]. (But this could have been said of Brassey's agents, and countless others for thousands of years: since the term was not yet in use, I don't count the emanation of the modern discipline yet to have happened here, which is why I award it to the Martin and McDonnell companies).

In Australia, Civil & Civic Pty Ltd was marketing its project management capability by 1958[18]. By the end of the 1950s, the idea of appointing a 'project manager' either as an individual or as an organisation to take full and undivided responsibility for achieving the construction project objectives had arrived and was starting to spread, but the field was engineering construction – process engineering – not building and civil engineering.

In the United Kingdom, the Operational Research Section of the UK Central Electricity Generating Board (CEGB) was also working on similar ideas to Kelley and Walker around 1955–1958. They developed the term the 'longest irreducible sequence of events' and applied their system to the shutdown and maintenance of Keadby Power Station in 1957, claiming a saving of 42% compared with the previous overall average time for similar shutdowns.

The *Harvard Business Review* Introduces the Project Manager!

The year 1959 saw a notable milestone in the development of project management with the publication of an article by Paul Gaddis in the *Harvard Business Review*: "The Project Manager"[19]. Project management was now beginning to receive serious general management attention. Interestingly, Gaddis emphasised the people-management aspects of the job rather than the planning and control tools and techniques which many later commentators assume were the dominant characteristics of project management at this time.

McNamara and the Bureaucracy of Systems

The broader, more developed systems approach to project and program management, having been pioneered on major missile programs, now spread rapidly in the 'high-tech' industries, becoming widely institutionalised in the world of defence–aerospace through the leadership of Robert McNamara, who became US Secretary of Defense in 1960 (Figure 3.5). It was then given added thrust by NASA (specifically Apollo). From there, it spread throughout the US aerospace and electronics industries, and then via NATO, into Europe. (In construction, meanwhile, it continued first with a strong OR bent, and second as a more *ad hoc* approach to integration, as we shall see shortly.)

McNamara was an OR enthusiast and a great centraliser. He used systems analysis extensively to organise and manage the Department of Defense, re-organising many aspects of operations, intelligence, communications, development and supply. Upon entering office for example he introduced the Program Planning and Budgeting System (PPBS) to help produce long-term, program-oriented budgets and he mandated several systems-based practices such as Life Cycle Costing, Integrated Logistics Support, Quality Assurance, Value Engineering, Configuration Management, and the Work Breakdown Structure (WBS).

These tools and techniques have since become core to project management as a discipline. The systems language of DoD forms the basis of the new discipline of project and program management and distinguishes it from any previous approach to the management of projects. (Though this

Figure 3.5   Robert McNamara.
Source: © US Department of Defense.

is not to say that they represent the core theoretical basis of project management: integration for example, as we shall see shortly, is a much more powerful organising theory.)

Apollo: Configuration Management and Project Leadership

In April 1961, the USSR launched Yuri Gagarin into space. The first man in space, put there by the Russians! American prestige was severely wounded. Russia appeared to be winning 'the space race'. President Kennedy responded in May, proposing that "this nation should commit itself to achieving the goal, before this decade is out, of landing a man on the Moon and returning him safely to Earth"[20]. The resulting Apollo program has several lessons for us in tracing the evolution of the discipline of project management: its strategy and targets, matrix management, configuration management, and the cost/quality relationship, but above all, Apollo was the great sales program for the new discipline of systems project management. The way 'Houston' addressed the astronauts and dealt with issues – the whole language – was both inspiring and inviting: project management was powerful, promising, and here to stay!

President Kennedy gave NASA five weeks before his famous speech to determine if getting to the Moon and back was technically feasible by the end of the decade, and if it were, to say how much it would cost. NASA proposed building a space station in Earth orbit from which a spacecraft would travel to and return from the Moon; the cost would be $13 bn. On hearing this, James Webb, the NASA Administrator (Head) slapped on $7 bn as contingency, on the basis of little more than hunch[21]. In the event, no space station was built (travel being straight from Earth to Lunar Orbit, and back again). Man landed on the Moon by the end of the decade, and the cost was approximately $21 bn! Not bad! But by 1963, it looked as though the budget, and possibly the schedule, would be blown.

D. Brainerd Holmes, head of the Office of Manned Space Flight, desperately sought more money but Webb sensed this would be politically unforthcoming. George Mueller (pronounced 'Miller'), formerly of TRW*, replaced Holmes and he in turn brought in USAF Brigadier General Samuel Phillips from Minuteman with the title and position, at Schriever's suggestion-cum-insistence, of Apollo Program Director (Figures 3.6 and 3.7). USAF practices

Figure 3.6   George E. Mueller.
Source: © NASA.

---

*Actually TRW's space systems integration company was known as Space Technology Laboratories.

Figure 3.7   Sam Phillips and Houston Control. Look at the control knobs!
Source: © NASA.

and personnel were introduced, several in key positions, to shake up NASA's more easy-going scientist culture.

Mueller was shocked when he learnt shortly after taking up his post that the chances of achieving a lunar landing by 1970 were less than 1 in 10! The only solution seemed to be to cut the number of proving flights. Instead of piece-by-piece, stage-by-stage prequalification testing, each part would be delivered thoroughly tested, ready for one single 'all-up' launch, in effect testing all the rocket stages simultaneously (rather, than had been the plan, via different launches for individual stages[22]). Mueller implemented a nested, boxed matrix structure across the program's four principal organisational levels, each one being a mirror of the others (Figure 3.8).

What we see here, crucially to the emerging discipline, is Mueller as, effectively, the program director making a huge strategic decision about the whole technical development strategy, based on a schedule risk analysis, with massive quality control and staffing implications, resulting in a restructured program management organisation. In one place – the head of the program – strategy, technology, schedule, risk, cost, organisation and personnel are brought together and integrated. Conceptually, Schriever, and

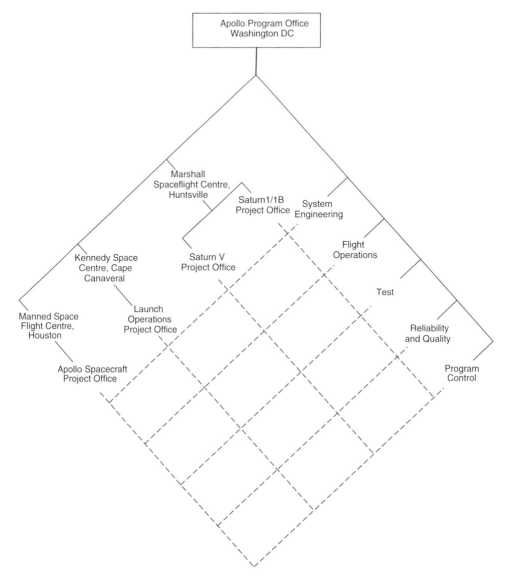

Figure 3.8   Apollo matrix organisation.
Source: Morris (1994), p. 55. (Morris, P. W. G. [1994], *The management of projects*, Thomas Telford: London.).

Raborn, occupied a similar role: the program (project) director. It was a role that was to prove curiously elusive in much project management that was to follow over the years ahead*.

It took Mueller and Phillips much effort, will and determination to bring program management-driven executive discipline to the organisation.

_____

*Because while in theory it is hard to argue against, particularly if the project is complex, urgent or otherwise demanding, in practice it is expensive and often difficult to fill with someone of the appropriate competence.

For example, it was not until January 1965 that Phillips was able to issue Apollo's master program control document: the Apollo Program Development Plan. This specified "how the Apollo objectives would be reached, how performance and proposed changes would be evaluated, and how these changes, after approval, would be implemented. . . . Other sections dealt with. . . . scheduling, procurement, data management, configuration management, logistics, facilities, funds and manpower, and systems engineering"[23].

Configuration Management was the bedrock of Phillips' control, along with design reviews, clear work package accountability and authority, and a program-wide information system providing comprehensive project control data, rolling up from contract work package, through stage/system level, to field centre, and finally to overall program level. But its formality and the rigour with which it was imposed by Phillips met with considerable opposition from existing NASA personnel (not least Werner von Braun*) and some of its contractors. The trouble was, NASA's people had been behaving less as managers than as scientists. Most of Apollo's contracts, for example, had been awarded, and many components already designed, manufactured and tested, before specifications had even been agreed[24]! USAF configuration management stamped on this, but against real opposition.

Neil Armstrong and Buzz Aldrin landed on the Moon on July 20, 1969. The program had met its goals, a triumph of systems project management (Figure 3.9). But manned spaceflight had become an embarrassment. Apollo was costing $3 bn a year. Post-Apollo planning had begun in 1964 but had been lacklustre. In September 1969, President-elect Richard Nixon set up a Space Task Group chaired by his Vice-elect, Spiro Agnew. It speaks volumes that long-term strategy should only have become a serious issue once the immediate goals had been achieved. NASA has had to live with the consequences of this uncertainty ever since. (The Shuttle was underfunded; the Space Station launched on confused and conflicting objectives[25]) Committing this nation to 'landing an American on the Moon before the decade was out' was an expensive riposte to Yuri Gagarin but, by a combination of science, technology and rigorous, driving, project management, the objective was accomplished! It was an immense achievement: now, many felt, surely we could tackle anything using this new management approach! But it wasn't going to be that easy.

DoD Bureaucratisation

Apollo broadcast systems project management to the world; DoD procedurised it, almost to death. Work Breakdown Structures had been promoted

---

*Once of the Nazi V2 program, then the US Army before joining NASA in 1960 as chief architect of the giant Saturn V booster used for Apollo.

Figure 3.9   Earth rise and Lunar Excursion Module (LEM).
Source: © Getty images.

in a joint DoD/NASA 1962 guide: *PERT/Cost Systems Design*. By 1964 DoD had up to 10 versions of PERT/Cost that contractors had to be familiar with. Industry objected. DoD responded in 1964 by offering a more adaptable version of PERT/Cost that had been developed by Phillips on Minuteman known as Earned Value Analysis (EVA). Then, in 1966 DoD issued a two-part 'Contractor Performance Measurement System': part 1 required contractors to have cost management systems that met the government's *Cost and Schedule Control Systems Criteria* (C/SCSC); part 2 required the contractor to report 'budgeted value of work performed'. Project control had become bureaucratised, so much so that by May 1966 the Aerospace Industries Association issued "an appeal for disengagement"[26]. DoD then revised its 375 procedures in November and NASA followed with its revised 500 version[27].

Unimpeded by such obstacles, project management was nevertheless now sweeping through the defence–aerospace industries, and was becoming embedded in Europe as well as America.

As it spread, many executives and service personnel found themselves pitched into project management roles for the first time in their lives, many in matrix organisations where, though called project managers, they lacked the big picture overview and authority of a Mueller, Raborn, Phillips or Schriever, even though they were still responsible for 'on time, in budget' delivery. Daunting as a job in itself, this seemed even more of a challenge when the specialised demands of the matrix and DoD, NASA or similar project management procedures were added.

Techniques such as PERT, WBS, EVA*, and practices such as Configuration Management, were genuinely new and difficult; many needed the new computer power now becoming available to help implement them. Conferences and seminars on how to do all this – but tending specifically to focus on project controls and planning and scheduling – proliferated. (Most of the literature throughout the 1960s and into the early 1970s was similarly focussed on planning and control tools and techniques, rather than the broader subject of project or program management[28]) These events spawned a phenomenon that was to prove highly significant: the project management professional associations. The US Project Management Institute (PMI) was founded in 1969; the International Management Systems Association (then also called INTERNET, now the International Project Management Association – IPMA) in 1967 with various European project management associations being formed contemporaneously. Initially these associations were communications fora. Later they were to have much greater normalising, and proselytising, roles, as we shall see.

Externalities

In fact, successful though Apollo had been, the limits of systems project management had already been reached. Initially, as NASA landed men on the Moon, it was being heralded as the management approach of the future. James Webb, NASA's Administrator (Mueller's boss), gave a series of lectures in 1968, published in 1969 as *Space age management: the large-scale approach*, in which he emphasised leadership and management in changing conditions, the virtues of flexible and adaptive organisational structures and the power of government, proposing the same systems approaches that had got man to the Moon – employing "adaptive, problem-solving, temporary systems of diverse specialists, linked together by coordinating executives"[29] – as the means to tackle society's broader challenges†.

But it was not going to be so easy. Apollo had succeeded because of the clarity of its goals, its political championing, and the stability that allowed Phillips determinedly, through configuration management, to manage scope and schedule (and cost). Where such stability failed to exist, which was most everywhere it sometimes appeared, systems project management rarely worked so effectively[30]. As Leonard Sayles and Margaret Chandler, two leading academics, pointed out in 1971: "NASA was a closed loop – it set its own schedule, designed its own hardware. . . . As one moves into the (more political) socio-technical area, this luxury disappears"[31]. Which is just what happened.

---

*Work Breakdown Structures; Earned Value Analysis – see Chapter 9.
†A much wider scope than what later came for many to be standard project or program management.

The US Super-Sonic Transport and Concorde

The development and ultimate failure of the US SST aeroplane – the Super-Sonic Transport, Concorde's US rival – epitomised the danger. Two Air Force (AFSC) officers were appointed to run the SST program: William McKee and Jewell Maxwell. The program demanded extraordinary technological advances. Investigating a swing-wing option ended up creating a two-year schedule slippage. In hindsight this proved fatal in that it allowed time for environmentalist opposition to strengthen while political support diminished, not least in the White House. In the end, the US Congress voted in 1971 to cease its R&D funding, reflecting nationwide opposition to the sonic boom. This opposition was coordinated by William Shurcliff, a 59-year-old semi-retired scientist, working out of a spare room in his Boston home. Despite ranting at Shurcliff, neither the US SST management nor the Federal Aviation Authority had an effective method of managing such stakeholder opposition. The SST's management, while "excellent project managers. . . . did not comprehend until it was too late that important societal changes could severely undermine their program"[32].

Concorde, the US SST's rival, like the US SST (Figure 3.10), was an enormous technology-push project[33]. Unlike the US systems management projects, however, there was no project management applied to the project at all! There was no one in authority having overall management control over the project. In reality, there was organisational duplication everywhere. PERT was introduced in 1964–1965 but abandoned in 1972 as unworkable. (Look at Figure 3.10 again!) When the plane was awarded its Certificate of Airworthiness in 1969, it was six years late and its development cost had risen to £1.129 bn against its budget of £135 m.

Concorde only survived because of the 1962 irrevocable treaty between France and Great Britain which governed it and which ensured it received vital political support*. France and Britain each wanted, on two or three occasions, to cancel the project but crucially never at the same time and were prevented from cancelling unilaterally by the terms of the treaty.

Second-generation systems project management

Concorde was only one of several UK aircraft project disasters of this time – for example TSR2, a beautiful, original 'tactical strike reconnaissance aircraft', or the RB-211, a jet engine development project which bankrupt Rolls-Royce – though things improved on the back of lessons painfully learned. The Downey report of 1966 concluded that "not enough time was being spent in front-end definition and preparation" and there was

---

*Concorde was wanted by the French for generic technology development reasons and by the British as an entrance ticket to the European Common Market, the forerunner to the European Union.

Figure 3.10   Concorde. Note the advanced manufacturing conditions!
Source: © Hulton-Deutsch Collection/CORBIS.

"inadequate control over design changes"[34]. Management of the Tornado program, on the other hand (begun in 1968), a joint German, Italian and British program undertaken via a company, Panavia, especially established for the project, was outstanding: first-rate leadership, unity of command, clear client requirements and organisation, with the whole panoply of project control techniques effectively deployed.

The USA, too, was experiencing problems in its defence sector. Despite all the initiatives of McNamara, weapons development projects were still suffering major performance and cost growth problems[35]. David Packard, who became US Deputy Secretary of Defense in May 1967, believed that having "competent people [with] clearly defined priorities" was central to getting any kind of improvements[36]. The Defense Systems Management College was established in 1971. The Services' habit of rotating personnel through different jobs, so that a program manager might only be in post a couple of years before moving on to a completely different assignment, significantly hampered this initiative, however[37]. As a result, program and project management was, to repeat the point made above with respect to the matrix, not a really major force, as it had been under the leadership of project/program directors like Schriever, Raborn, Mueller etc. Instead it had become institutionalised as a middle management activity carried out through specialist bureaucratic processes and techniques.

Packard, however, had things to say about these too. Greater emphasis needed to be given to achieving better front-end definition (in fact, in 1977 a new milestone was introduced into the acquisition process: Milestone 0 – approval of the 'mission element need statement'). Development, Packard said, should be 'heel-to-toe' and proposed proscribing concurrency but in the end resigned before doing so. (Packard's successor, William Perry, relaxed this in 1979, allowing concurrency where the threat justified the inefficiencies[38].) Cost estimating had to be improved; areas of high technical risk needed to be identified; project definition had to be agreed on before letting development contracts; milestone (Gate) reviews were strengthened; and change control and configuration management had to be improved. 'Design-to-Cost' and 'Life-cycle costing' were introduced.

Energy and Commodities Projects

Project management had meanwhile begun to be adopted by other industries. The process engineering sectors – oil, gas, chemicals, paper and pulp, power, water, etc. – for example were early adopters, along with mining and telecoms. (Software was a late adopter, struggling for years.) Unlike building and civil engineering, many of whose sponsoring owners were only occasional 'clients' (owners), the process engineering industries were generally younger and less encumbered by professional rules (for example, the RIBA's constraints on design–construction integration discussed in Chapter 2) and their projects were typically sponsored by more experienced, managerially sophisticated owner organisations. Nevertheless, their track record wasn't necessarily that good either, as the IMEC study found in 2000 (see Chapter 5, pp. 75–6).

Unlike weapons systems which were at least shielded from environmentalist action, many of these projects, in common with major infrastructure projects, faced a world not just of technical challenges but often of strong 'environmentalist' opposition. This new set of challenges was not always tackled well. Take the huge nuclear power industry which sprang up in the 1970s.

Nuclear Power

There was enormous government push to develop substantial nuclear power generation capacity in the USA, France and Britain in the late 1960s – a huge expansion of demand with little serious thought seemingly given to the industry's ability to service this demand. (Where were the engineers and scientists to come from? Was this known technology?) FERC, the US Federal Energy Regulatory Commission, regulated; it didn't manage. Thus the USA went from having no nuclear power plant ordered as of 1965 to 22 units in 1966; 32 in 1967; 45 between 1968 and 1970; and 105 between 1971 and 1974. Britain ordered eight (magnox – graphite moderated) plants between 1957 and 1964, then switched to a new, unproven, risky 'Advanced

Figure 3.11   Dungeness B nuclear power station.
Source: Public domain.

Gas-cooled Reactor' (AGR) technology, placing five orders in the late 1960s/
early 1970s.

To describe the US and UK programs as problematic would be an under-
statement: they were dreadful. Dungeness B, the UK's first AGR, was completed in 1983 13 years late and £140 m over its sanctioned budget of
£106 m (Figure 3.11). The British were essentially building 330 mW prototype plants as commercial generating stations, learning as they went. Scientists explored. Concurrency ruled. The emerging technical understanding
led to new regulations being issued ('regulatory ratcheting') which required
stripping-out and re-work – concurrency again! Management of the program
lacked political and technological clarity of purpose. And whilst there was
project management in name, there was really no effective risk management.
France, on the other hand, having built its first eight (gas-cooled) reactors
in the 1960s, switched in the 1970s to Pressurised Water Reactor (PWR)
technology. All were developed by Framatome from the initial Westinghouse
technology but on a highly standardised design. Fifty-eight plants were built
between the early 1970s and 2000, virtually on a production basis, certainly
as a program*.

To make matters worse, at the plant level the Americans experienced an
extraordinary series of confidence-sapping mistakes: an electrician's candle

---

*Three programs in fact: 900 MWe, 1300 MWe and 1450 MWe.

caused 15% of the Tennessee Valley Authority's production capacity to be shut down; Diablo Canyon was sited on an earthquake fault line; San Onofre was built back to front; piping was installed incorrectly on Comanche Peak and South Texas. Add to these the experience of Three Mile Island when, in March 1979, a stuck valve brought the plant to within 60 minutes of meltdown (plus the fatal disaster of Chernobyl in April 1986) and, taken together, it becomes little wonder that a population which had already seen off the US SST now rose in hostility to this threatening technology, being so incompetently developed and built. The 1970s saw environmentalists clashing with the authorities in Germany, France, the UK, Japan, the USA and elsewhere. Little attempt was made to influence public thinking. Stakeholder management was almost non-existent. There was no attempt to manage environmentalists professionally. Instead they were considered as quasi-criminals, not as people with a valid if different point of view which required negotiation and adjustment. There was no real management of the industry's institutional context.

The Extractive Industries

In the oil and gas sector meanwhile, as the price of oil rose rapidly from $3 per barrel in 1971 to over $30 per barrel by the end of the decade, developments began in new, tougher environments like the North Sea and Alaska. Not only did the North Sea entail "greater water depths, rougher and colder seas, harsher weather conditions and greater distances from land"[39] but cost inflation was huge, running around 100% p.a. in the early 1970s. Not surprisingly, projects were underestimated, in places cavalierly: the TransAlaskan Pipeline for example was estimated based on 'Lower 48'* pipe-laying prices ($1 million per mile), ignoring the fact that the territory was uninhabited wilderness with three huge mountain ranges, seismic activity, permafrost and a host of special flora and fauna. (The project ended up $7.6 bn. over its initial $900 m budget)[40] (Figure 3.12).

Not every project of this era was a failure but certainly the trend, fed by underestimation of cost inflation, poor management of risky technology, and inexperienced project staff and stakeholder opposition, was in this direction. Thus Sir Alistair Frame, Chairman of the mining company RTZ, Rio Tinto Zinc, one of the very largest mining companies in the world, presented an alternative, positive view, emphasising that "the most important part of project management is the people . . . [and] the most important factor in selecting the top man is his qualities of leadership"[41]. Frame's formula also required that the sponsor must be strong, as must the project engineering team; the project procedures and authorities should be unambiguously clear; the project director should be of the highest calibre and have his responsibilities clearly designated; and the design/scope, budget and schedule must be signed off before work begins (on site). The scope should clearly reflect the client's requirements and proposed method of construc-

*That is, the US states below the 49th parallel: those excluding Alaska and Hawaii.

Figure 3.12   The TransAlaskan Pipeline.
Source: © BP p.l.c.

tion. There should be alignment of the project's logistics, personnel, basis of contracting, financial reporting, and communications – the whole kerboodle, in short!

This said, in general, management in the world of projects was clearly not going well. The data showed that, just as project management was emerging formally onto the world stage, the record of project development and delivery seemed to be in decline – severely! A number of studies sought to find out why and what the lessons might be. DoD had commissioned several, as had the World Bank, the US General Accounting Office and the UK National Audit Office and others. (See Appendix 1) Their collective findings began to suggest that the causes of project failure lay in areas such as technological uncertainty, scope changes, concurrency, and contractor engagement. In other words, in areas other than project execution. Yet this was precisely where PMI, the US-based Project Management Institute, focussed its efforts in trying to delineate the knowledge that, in their opinion, was 'unique' to project management, as we shall now see.

References and Endnotes

[1] Sapolsky, H. (2003), Inventing systems integration, in: Prencipe, A., Davies, A. and Hobday, M. (eds.) *The business of systems integration*, Oxford University Press: Oxford, p. 19.
[2] Bergen, W. B. (1954), New management approach at Martin, *Aviation Age*, 20, 6, pp. 39–47.

[3] Bugos, G. E. (1954), Manufacturing certainty: Testing and program management for the F-4 phantom II, *Social Studies of Science*, 23, pp. 271–275.

[4] Lanier, F. (1956), Organizing for large engineering projects, *Machine Design*, 27, p. 54.

[5] Beard, E. (1976), *Developing the ICBM*, Columbia University Press: New York, p. 161.

[6] "Recommendations of the Teapot Committee" 1 February,1954, in: Neufeld, J. (1990), *Ballistic missiles in the United States Air Force 1945–60*, Office of Air Force History, United States Air Force: Washington, DC, pp. 260–261.

[7] Hughes, T. P. (1998), *Rescuing Prometheus*, Vintage: New York; Neufeld, J. (1990) *op. cit.*: 6, pp. 102–105, 111.

[8] McKenzie, D. (1990), *Inventing accuracy: A historical sociology of nuclear missile guidance*, MIT Press: Cambridge.

[9] Johnson, S. B. (1997), Three approaches to big technology: Operations research, systems engineering, and project management, *Technology and Culture*, 38, 4, pp. 891–919.

[10] Checkland, P. (1999), *Systems thinking, systems practice*, John Wiley & Sons: Chichester; Beer, S. (1959), *Cybernetics and management*, Unibooks: London.

[11] Ancona, D. G. and Caldwell, D. (1990), Beyond boundary spanning: Managing external dependence in product development teams, *Journal of High Technology Management*, 1, pp. 119–135.

[12] Spinardi, G (1994), *From Polaris to Trident: The development of US Fleet Ballistic Missile Technology*, Cambridge University Press: Cambridge.

[13] Kelley E. J. and Walker M. R. (1959), Critical path planning and scheduling, Proceedings, Eastern Joint IRE-AIEE-ACM Computer Conference, Boston, MA, December 1–3, pp. 160–173.

[14] Morris, P. W. G. (1994), *The management of projects*, Thomas Telford: London, p. 18.

[15] Sapolsky, H. (1972), *The Polaris system development: Bureaucratic and programmatic success in government*, Harvard University Press: Cambridge, MA, p. 19; Spinardi, G. (1994) *op. cit.*: 12.

[16] Fondahl, J. (1987), The history of modern project management – Precedence diagramming methods: Origins and early development, *Project Management Journal*, 18, 2, pp. 33–36.

[17] Bechtel, S. D. (1989), Project management – Yesterday, today and tomorrow, *PM Network*, III, 1, pp. 6–8.

[18] Stretton, A. (2011), Forward through the Fifties, *The Project Manager*, 30, 2, pp. 8–10.

[19] Gaddis, P.O. (1959), The project manager, *Harvard Business Review*, May–June, pp. 89–97.

[20] Logsdon, J. (2011), *John F. Kennedy and the race to the moon*, Palgrave Macmillan: London.

[21] Personal communication to the author by Dr. Albert J. Kelley of NASA who was present at the time.

[22] Brooks, C. G., Grimwood, J. M. and Swenson. L. S. (1979), *Chariots for Apollo: A history of manned lunar spacecraft*, NASA: Washington, DC, p. 1303.

[23] *Ibid.*

[24] Johnson, S. B. (2002) op. cit.: 9.

[25] Morris, P. W. G. (1994) op. cit.: 14.

[26] Acker, D. D. (1980), The maturing of the DoD acquisition process, *Defense Systems Management Review*, 3, 3, p. 26.

[27] Morrison, E. J. (1967), Defense systems management: The 375 series, *California Management Review*, 9, 4.

[28] Moder, J. J. and Phillips, C. R. (1964), *Project management with CPM and PERT*, Van Nostrand Reinhold Company: New York; Antill, J. M. and Woodhead, R. W. (1965), *Critical path methods in construction practice*, John Wiley & Sons: New York; Archibald, R. D. and Villoria, R. L. (1967), *Network-based management systems (PERT/CPM)*, John Wiley & Sons: New York.

[29] Webb, J.E. (1969), *Space age management: The large-scale approach*, McGraw-Hill: New York, p. 23.

[30] Johnson, S. (2001), Samuel Phillips and the taming of Apollo, *Technology and Culture*, 42, 4, pp. 685–709.

[31] Sayles, L. R. and Chandler, M. K. (1971), *Managing large systems: Organizations for the future*, Harper and Rowe: New York, p. 160.

[32] Horwitch, M. (1982), *Clipped wings: The American SST conflict*, MIT Press: Cambridge, MA, p. 344.

[33] Morris, P. W. G. and Hough, G. H. (1987), *The anatomy of major projects*, Wiley and Sons: Chichester.

[34] Ministry of Technology (1969), *Report of the Steering group on development cost estimating*, TSO: London.

[35] Drezner, J.A. and Smith, G.K. (1990), *An analysis of weapons acquisition schedules* Rand Corporation, RM-3927-ACQ Santa Monica, CA

[36] Department of Defense (1971), Acquisition of major defense systems, DoD Directive 5000.1 Washington, DC.

[37] Fox, J. R. (1974), *Arming America: How the US buys weapons*, Harvard University, Boston; Fox, J. R. (1984), Evaluating management of large, complex projects: A framework for analysis, *Technology in Society*, 6, pp.129–139.

[38] Harvey, T.E. (1980), Concurrency today in acquisition management, *Defence Systems Management Review*, 3, 1, pp. 50–56.

[39] Department of Energy (1975), *North Sea costs escalation study*, TSO: London, p. 2.

[40] Morris. P.W.G. (1994) op. cit.: 14: pp. 98–114

[41] Frame, A. (1988) Project management: A client's view, In: Burbridge, R. N. G. (eds.) *Perspectives on project management*, Institution of Electrical Engineers/Peter Peregrinus: London.

The Project Management Knowledge Base

As project management spread across the high-tech industries, first in America and, on its heels, Europe, so practitioners began to meet at seminars and conferences. Initially, as we've seen, attention was centred mostly on project management's specialist planning tools but also on the challenges of working in matrix organisations. From such meetings grew the project management professional associations. By the early 1980s the question of formalising project management's core knowledge base had arisen. The US headquartered Project Management Institute published its *Project Management Body of Knowledge* in 1983/1987 but had as the criterion for selecting the knowledge to be included 'knowledge which was unique to project management'. Meanwhile, studies (for example Morris and Hough, 1987, and Wheelwright and Clark, 1992) were showing that a much broader range of factors often had to be managed if the project was not to fail. Other researchers (predominantly the so-called Scandinavian School) began examining what project personnel actually do rather than what they maybe ought to be doing and began studying projects as organisational phenomena – as forms of temporary organisation.

The PMBOK® Guide

The 1980s and 1990s saw the project management professional associations begin formalising what they considered to be the intellectual 'scope', or definition, of the subject in their 'bodies of knowledge' (BoKs). This stream of work was to prove – and continues to prove – particularly influential in the establishment of project management not just as a profession but as a domain and as a discipline. Influential in that the BoKs define for hundreds of thousands of people what the subject's scope and knowledge content is about, even if only in outline; and influential too in that from now onwards discussion of the evolution – the 'construction' – of project management

Reconstructing Project Management, First Edition. Peter W.G. Morris.
© 2013 John Wiley & Sons, Ltd. Published 2013 by John Wiley & Sons, Ltd.

becomes more involved with theory and with concepts. First came PMI's BoK in the early to mid-1980s.

The drive behind the development of a project management Body of Knowledge was the idea first mooted by PMI in the mid-1970s that if project management was to be a profession surely there ought to be some form of certification of competence[1], for one of the attributes of professionals* is evidence of the mastery of a body of knowledge distinctive to it, leading to a 'license to practise'[2]. This obviously implies some definition of the knowledge area. As a result, in 1983 a pilot *Project Management Body of Knowledge* (PMBOK) was published. The BOK identified six knowledge areas that were considered "unique to the project management field"[3]: scope, time, cost, quality, human resources, and communications. In 1987, PMI formally published the PMBOK and in doing so added risk and contract/procurement. A 1996 revision added integration and importantly changed the document's name to *A Guide to the Project Management Body of Knowledge* – also known as the *PMBOK® Guide*. There have since been several more updates, most relatively minor, although the 2013 Fifth Edition introduced a tenth knowledge area: project stakeholder management[†].

In creating the PMBOK, PMI wrestled hard with trying to decide what was 'unique' to project management, which is what they felt they needed to define. The distinctions were drawn between project management knowledge and more general knowledge. The latter, they said, comprised that lying either:

- in general management (for example, business strategy, organisational behaviour, staffing, marketing and sales, decision-making);
- in supporting disciplines such as configuration management, quality management, contracting and procurement (there is some confusion here

---

*Hodgson and Muzio, writing on professionalism and project management, say "professions seek to control entry to and competition within labour markets, while ensuring some degree of 'institutional autonomy' to regulate their own affairs. Crucial to many such professional projects is the ability to monopolise specific jurisdictions of work, creating and defining core knowledge": Hodgson, D. and Muzio, D. (2010), Prospects for professionalism, in: Morris, P. W. G., Pinto, J. K. and Söderlund, J. (eds.) *The Oxford handbook of project management*, Oxford University Press: Oxford, p. 109.

†The 2008 revision introduced two new processes, Collect Requirements and Identify Stakeholders. The 2013 edition expanded this latter by splitting out Communications Management and creating a new knowledge area, Project Stakeholder Management. It also added new processes for Plan Scope Management and ditto [i.e. Plan] for Cost, Schedule, and Stakeholder Management. The treatment of both requirements and 'plan' can be criticised as not going far enough (see Chapter 12, p. 169, and Chapter 11, pp. 162–3 respectively, and Chapter 20). They read like 'execution' management trying to reach into the front-end rather than the front-end flowing into execution; which of course is what they are.

in that quality management was already, and contracting and procurement were just about to be, included in the *PMBOK*);
- or in industry-specific ('technical') knowledge*.

Thus "much of the general management body of knowledge should be recognized as a given or prerequisite for project management and not included in the *PMBOK® Guide* unless it was considered that aspects of this knowledge are an integral part of the project management process"[4].

This unfortunately created a fundamental shortcoming: *the PMBOK® Guide did not, and still does not, represent the knowledge that is necessary for managing projects successfully* but only that which was considered truly unique to project management. PMI had asked the wrong question. A professional body of knowledge, like that of a medic for example, covers that knowledge which is necessary in order (to be licensed) to practise one's profession. The *PMBOK® Guide* did not – does not – do this. It sought to describe only the knowledge relevant to project management that was not, it was claimed, available in other fields. (Yes: hindsight is a wonderful thing!)

To define the knowledge that is supposedly unique to project management, one had to be able to describe project management. On what basis do you do this? How do you characterise the thing, project management, in order to characterise the knowledge about it? After all, project management is a social construct. Projects are 'invented not found'. One needs a mental model as a starting point – a framework to enable discussion to progress[†]. PMI's construct was a process – not primarily the project/product development process (the project life-cycle), despite its fundamental importance, although it is given passing acknowledgement – but a simple 'initiate → plan → execute → monitor and control → close' set of process groups[‡]. Thus, according to the *Guide* (fifth edition, p. 5), "project management is the application of knowledge, skills, tools, and techniques to project activities to meet project requirements. Project management is accomplished through the appropriate application and integration of the [now] 47 logically grouped project management processes comprising the 5 Process Groups. The 5 Process Groups are Initiating, Planning, Executing, Monitoring and Controlling, Closing". An extraordinarily disembodied and inadequate definition of the thing that PMI is the profession for!

---

*This differentiation, or at least one like it, was retained up until the publication of the fourth edition in 2008, when it was dropped and replaced by discussion of program management and portfolio management.

[†]This, for those that are interested, is surely an example of what Immanuel Kant famously meant by *a priori* judgements in *The Critique of Pure Reason* of 1787: that all scientific and moral judgements are imposed on the world by the mind. Without our own conceptions of such things we would not be able to envision or make sense of them.

[‡]Monitoring, in cybernetics, is a part of the control loop; or put another way, the cybernetic control cycle includes first planning and then monitoring (and correcting/re-planning) – see Chapter 3. However, the distinction between monitoring and control is a fair one. Monitoring may tell you you're late; control does something about it. You can have monitoring but no control!

This process basis leads to a number of problems.

First, it failed, and still fails, to reflect the distinctive characteristics and challenges to management of the different project/product development stages – stages which fundamentally are common to all projects as we saw in Chapter 2. The PMI process is firmly rooted in the Deming 'Plan–Do–Check–Act' cycle. Nothing wrong with that except that, being so generic, it fails to bring out the characteristics of the product development sequence – the project life-cycle as it is more usually termed – and its stages, and the impacts of these on project management. Change control is different in concept design than in detailed design for example.

Second, insufficient weight was and still is given to managing the front-end definition development stages and to the establishing of cost and schedule targets and the definition of scope. The spirit of the *Guide* – its ontology – has been and is essentially an execution one: post 'requirements "collection"' with no mention of developing or shaping the project. (Though the fourth and fifth editions have moved some way to diminish this execution emphasis.)

Third, even the knowledge that was proposed as unique to project management was, as such, questionable, for non-project undertakings also involve many of the things identified in *The PMBOK® Guide* under cost, risk, procurement, quality, human resources and communications (Figure 4.1). The text referring to unique knowledge changed a little in the fourth edition* but the contents of the *Guide* did not.

Fourth, by focussing only on that knowledge which was supposedly unique to managing a simple generic process, many of the things which we've seen to be important in the effective management of projects ended up, despite the broadening of recent editions, being excluded (or by default, failed to be included): technology, strategy, etc. Thus for example, leadership and people, considered by Sir Alistair Frame as *the* most important area, was initially excluded more or less totally since it is not unique to project management. It now rests largely in Human Resource Management and Appendix X3 – with virtually no reference to any of the huge body of theory that is relevant to these topics[†].

Surely it would have been better to define the BOK around the knowledge needed to manage projects successfully? And this, as we shall see shortly, is what most of the other professional project management associations subsequently did.

Theoretical Underpinnings

In fairness, there wasn't much data of quality around at the time of drafting the *Guide* on the knowledge needed to manage projects successfully. What texts there were tended to be 'how-to' prescriptive books such as Kerzner's

---

*The fourth edition, 2008 text "Knowledge unique to the project management field" is replaced by "This standard is unique to the project management field" – p. 13; p. 18 in the 2013 fifth edition.

[†]In fact such is the emphasis on process and the lack of theory that one might seriously question whether this is a body of knowledge at all but rather a 'how to' process manual.

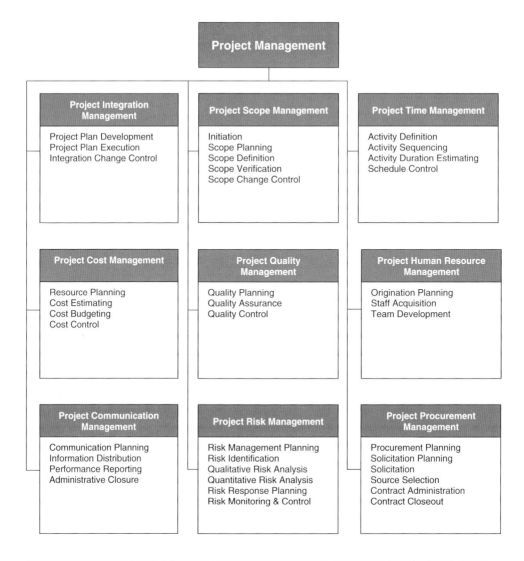

Source: PMI (1996) [PMI (1996), A Guide to the Project Management Body of Knowledge (PMBOK Guide), Project Management Institute, Newtown Square, Pennsylvania, USA.]

Figure 4.1   Knowledge areas and processes of the *PMBOK® Guide*.
Source: © PMI (1996).

Project Management: A Systems Approach (1979), Cleland and King's *Systems Analysis and Project Management* (1968) or Johnson, Kast and Rosenzweig's broader *The Theory and Management of Systems* (1973)[5]. There was some theory but not much; little critical examination of the aims, the ontology or the epistemology of the subject – what 'it was really all about' and 'how we know what we purport to know about it'. But equally PMI was very self-referential in drawing up the *Guide*. There was an almost total lack of reference to any scholarship in the undertaking, which is strange in a document all about knowledge; the Terms of Reference for PMI

'Project 121', the 1986 BOK upgrade, for example, limited the upgrade to capturing "the knowledge applied to project management by PMI members"[6]. Theory was now beginning to emerge, however, from a number of directions. From now on in this account of the 'construction of project management' theory and practice begin a richer interplay.

Operations research

Theory's contribution to the subject had begun, as we've seen, with Operations Research (OR). OR essentially is, in this sense, the outgrowth of Taylor's 'Scientific Management' and includes work planning (WBS, etc.); activity scheduling – James Kelley and Morgan Walker on CPM, Gordon Peterson *et al.* with PERT, and John Fondahl on Precedence; and in France, Bernard Roy in developing MPM, the Metra Potential Method, and in Germany the Schleip brothers, Walter and Rainer, developing RPS[7]. Then later, quite specific resource allocation modelling work[8] of the 1970s and the performance measurement tools (EVA) ushered in by the DoD. By the late 1970s, however, interest in OR as an approach to project management had begun to wane, quite substantially.

Integration

Just as modern project management had originated, as we've seen, with an interest in scheduling and cost control techniques but then developed as a means of coordination, so it was coordination that now began to receive theoretical attention. Around 1965–1967 writers from the United Kingdom's Tavistock Institute, building off systems theory, pointed up the significance of managing system boundaries (or interfaces between interacting system boundaries): Miller and Rice showed in 1967 that boundaries occur where there are discontinuities in time, technology or territory (and I'd add, in organisation[9]), thereby suggesting where project management attention would be most required[10]. Also in 1967, James Thompson, at the University of Indiana, showed that the amount, and type, of integration – that is, coordination and control – required is greater where the parties are in reciprocal interdependence rather than sequential, or pooled[11]. (See Figure 4.2).

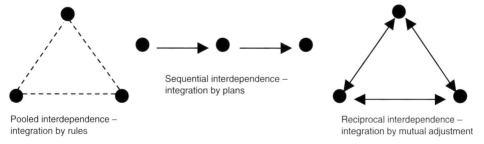

Pooled interdependence –
integration by rules

Sequential interdependence –
integration by plans

Reciprocal interdependence –
integration by mutual adjustment

Figure 4.2   Thompson's classification of types of interdependence and concomitant integration types.
Source: Author's own.

In 1967–1968 Paul Lawrence and Jay Lorsch of Harvard showed how integration varied both with the need for integration and with the nature of the differentiation between the organisational elements needing integrating and the stability of the environment being worked in. (Unstable environments need a more decentralised integration.) This resulted, they concluded, in a "new management job: the integrator"[12] – a project manager in all but name. (Russ Archibald in 1976 described the project manager, classically and pithily, if slightly inaccurately in today's terms, as "the single point of integrative responsibility" – 'accountability' would be more accurate in current terminology[13].) In 1968, Jay Galbraith at MIT showed, by studying the information that needed processing, that the amount of integration needed increased as the environmental, technical and organisational rate-of-change and complexity increased[14]. My own work at Manchester in 1972 suggested that the need for integration increased as the project became larger, more urgent, or more complex[15]. (Jumping ahead for a moment, Aaron Shenhar and Dov Dvir in 2005 updated this line of work and proposed that project management varies according to the project's novelty, complexity, technology and 'pace' – i.e. speed, urgency – of the project[16]. See Chapter 18, p. 241, for a note on their methodology, however.)

In 1973, Galbraith showed further that there is a range of integrating mechanisms available going from liaison and expeditor roles, through special teams, project coordinators, to project managers, and the matrix, each providing more powerful integration than the preceding mechanism[17] (Figure 4.3). Galbraith's sequence was used by Henry Mintzberg in 1979 in his general review of structural options in organising and, in particular, on coordinating mechanisms[18].

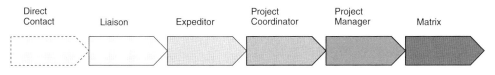

Sequence of increasingly powerful means of effecting lateral integration

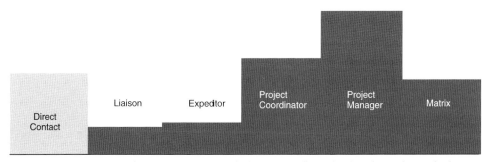

Indicative potential amount of project management authority corresponding to the above integration mechanisms

Figure 4.3   Galbraith's range of integrating mechanisms.
Source: Author's own.

Other research at this time looked at related topics such as the project manager's authority[19], conflict management[20], and the matrix[21].

Critical success factor studies

A third line of analysis, meanwhile, was beginning that was to have an important impact on our thinking as to what constitutes the discipline and hence at least some of the BOKs: a growing interest in looking at projects as 'whole entities' – organisations in their own right – and studying how successfully they were being managed – what are the 'critical success factors' to doing this, as Kam Jugdev and Ralf Müller put it in 2005[22]. (Indeed, how we measure success: isn't it sometimes more than simply 'on time, to schedule, to specification'? A subject we shall be returning to again.)

We have already noted that DoD had commissioned a number of studies on the apparent failure of many of their projects, despite the enormous amount of horsepower (to say nothing of technique, process and intellect – not least inventing the whole discipline) which it had put in to try to develop and deliver projects successfully[23]. Several other institutions were now looking at the factors influencing project success and failure[24]. Appendix 1 summarises many of those studies. Figure 4.4 selects and summarises five of the most notable CSF studies, all of which we shall note in this historical account of the development of the discipline (Figure 4.4).

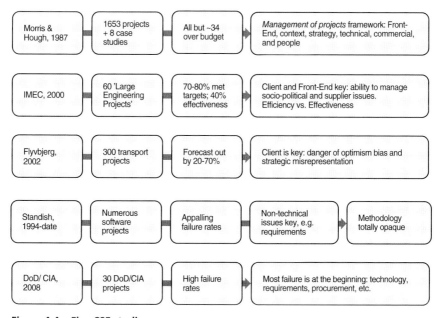

Figure 4.4   Five CSF studies.
Source: Author's own.

'The Management of Projects'

I myself, in collaboration with George Hough, addressed the issue of project success and failure in our 1987 study *The Anatomy of Major Projects*[25]. After reviewing reports on 1,653 projects (with another 1,600 excluded on quality grounds – see end note 24), conducting interviews with dozens of project executives, and preparing and analysing eight case studies of major projects, we found that typical sources of difficulty were such things as unclear success criteria, changing sponsor strategy, poor project definition, technology (fascination with; uncertainty of; design management), concurrency, poor quality assurance, poor linkage with sales and marketing, inappropriate contracting strategy, unsupportive political environment, lack of top management support, inflation, funding difficulties, poor control, inadequate manpower, and adverse geophysical conditions. Most, though not all, of these factors fell outside the standard project management rubric of the time, as expressed in the textbooks that were now emerging and on conference hall floors, and as formalised by the *PMBOK® Guide*. In classical project management terms, this was a failing discipline: only 34 projects came in on or under budget; overruns of 40 to 200% were common (see Table 1.1).

In short, what '*The Anatomy*' emphasised – and we shall see subsequent studies of project success and failure in Chapters 5, 19 and 20 and Appendix 1 repeating the emphasis – was the importance of managing the front-end project definition stages of a project, the pivotal role of the owner (or sponsor), and the need to manage, or influence, in some way the project 'externalities' – its context.

Yet none of these points were even alluded to in the *PMBOK® Guide*, at least until the fourth and fifth editions. In fact, the *Guide* underplays to the point of missing almost completely management's role in the development of the project front-end: the establishment of the project definition and targets, precisely the area where evidence shows management needs to concentrate*.

The model of project management represented by the *PMBOK® Guide* is one essentially of delivery execution: one where the requirements have at most to be 'collected'[26]; where the cost, schedule, scope and other targets have already been set. The ethos of the discipline is then to 'monitor and control', not to actively shape and drive solutions, a topic we shall return to in Chapter 20.

---

*IPA, the oil, gas and minerals project benchmarking company, coined the useful term 'Front-End Loading': "Front-End Loading (Business FEL) is a tool for determining which is the "right" project to meet the needs of business. The FEL tool assesses the level of definition of a number of critical items that are used to determine what, if any, asset should be built to meet a particular business need. The Business FEL Index is made up of three components: the business case, the business/engineering interface, and the conceptual engineering and facilities planning factor." (http://www.ipaglobal.com)

'The Management of Projects' Paradigm versus 'Execution Delivery'

The critical success studies signposted a growing bifurcation in the way project management was, and still is, perceived, with many taking the predominantly middle management, execution/delivery-oriented perspective, others responding to the success and failure data and taking a broader, more holistic view where the focus is on managing projects – in their environment, for stakeholder success. In the latter, the unit of analysis moves from execution management to the project as an organisational entity which has to be managed successfully, both in its development and in the execution of its delivery (Figure 4.5). Both paradigms involve managing multiple elements but the 'management of projects' paradigm, as I called the broader approach in 1994[27], is a much richer, more complex domain than execution management. This intellectual contrast was marked clearly in the Association for Project Management's Body of Knowledge (the APM BOK) which was developed around 1990.

The APM, IPMA, and Japanese BOKs

The APM based its Body of Knowledge not on the knowledge that is 'unique to project management' but on what you need to know in order to manage projects successfully. In practical terms, it considered the *PMBOK®* *Guide* misguided in its omission of the front end and too narrow in its definition of the subject. APM thus produced a broader document which followed the 'management of projects' model, recognising topics such as objectives, strategy, technology, environment, people, business and commercial issues, and so on. The APM BOK has gone to six revisions, with versions 3 and 5 being based on specially commissioned research (Figure 4.6)[28].

In 1998, the IPMA (the International Association of Project Management – an association of national project management associations which number over 50, representing more than 50,000 members) published its International Competence Baseline document to support its certification program[29]. In doing so, it adopted the APM BOK almost wholly as its model of project management. In 2002, the Japanese project management associations, ENAA and EPMF, produced a similarly broadly based BoK: P2M (Project and Program Management).

The Japanese BOK is really interesting with its emphasis on innovation (*Kakusin*), development (*Kaihatsu*) and improvement (*Kaizen*), which together make up innovation reform (*Kaikaku*). Linkage with mission and strategy are critical; technology management and value creation are very important. All these are concepts or approaches advocated in Part 3 of this book but which until now are too often ignored in the face of the planning and monitoring, iron triangle, efficiency paradigm pushed by Western standards-setting organisations (Figure 4.7)[30].

'Project Management', as represented by PMBOK, is pre-eminently about "on time, in budget, to scope" execution/delivery

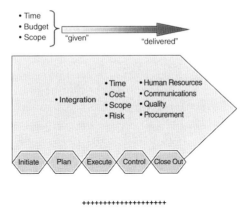

++++++++++++++++++++

The 'Management of Projects' conception, on the other hand, extends the PMBOK view to include the front-end definitional stages

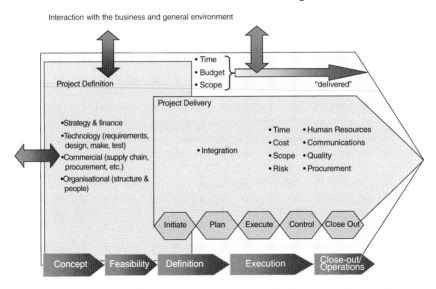

The 'Management of Projects' involves managing the definition and delivery of the project for stakeholder success. The focus is on the project in its context.

Figure 4.5   'MOP' – 'the management of projects' – compared with *PMBOK® Guide*.
Source: Author's own.

	General			
	Project Management Programme Management Portfolio Management		Project Context Project Sponsorship Project Office	
	Strategic			
	Project Success Criteria and Benefits Management Stakeholder Management Value Management		Project Management Plan Risk Management Quality Management Safety, Health & Environment	
Executing the strategy	**Techniques**	**Business and Commercial**	**Organization and Governance**	**People and the Profession**
Scope Management Scheduling Resource Management Budgeting & Cost Management Change Control Earned Value Management Information Management and reporting Issue Management	Requirements Management Development Management Estimating Technology Management Value Engineering Modelling & Testing Configuration Management	Business Case Marketing & Sales Financial Management Procurement Legal Awareness	Project Life Cycles Concept Project Reviews Organization Structure Organizational Roles Methods and procedures Governance	Communication Teamwork Leadership Conflict Management Negotiation Human Resource Management Behavioural Characteristics Learning and Development Professionalism and Ethics

Figure 4.6   The APM Body of Knowledge.
Source: © APM (2006), *APM Body of Knowledge*, 5th edition, Association for Project Management: London.

'Management by projects'

If 'the management of projects' was a broad conceptualisation of what one needed to know not "just on the processes and practices needed to deliver projects 'to scope, in budget, on schedule' but rather on how we set-up and define the project to deliver stakeholder success"[31], an even broader conceptualisation – the theme of the 1990 IPMA World Congress in Vienna, Austria – was that of 'management by projects' developed by Roland Gareis[32]. (Albert Hamilton of Limerick University has a book with this as its title[33] but the text is hardly polemical in the sense that Gareis' work is.) Gareis' 'Management-by-Projects' model seeks to characterise the collected attributes of the 'Project-Oriented Company'. Conceptually attractive and logically valid, it seems to have lacked the intellectual hooks which would have made it compelling to the researcher or valuable to the practitioner. It was, to invert Kuhn, a paradigm in search of distinctive evidence*.

Not only was the 'scope' of project management changing, its environment was becoming more managerially complex with new concerns and concepts enriching the way we approached the subject. In the 1980s, Total Quality Management, for example, together with Lean Management, had revolutionised the way the Japanese built automobiles, with crushing effect on American and European competition. This led to a swathe of new practices in New Product Development which collectively have had an incredibly important influence on project management.

---

*The philosopher Thomas Kuhn coined the term 'paradigm shift' to describe how scientific revolutions occur in the way a field is perceived: namely when evidence accumulates to the point that new models replace the previously prevailing 'exemplar piece of research': Kuhn, T. (1962) *The structure of scientific revolutions*. Chicago: The University of Chicago Press.

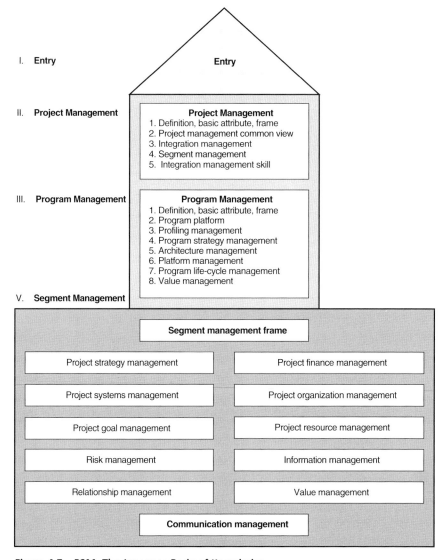

Figure 4.7   P2M: The Japanese Body of Knowledge.
Source: © Ohara, S. (2009), Framework of contemporary Japanese Project management (1): Project management paradigm-interpretation, application and evolution to KPM, In: Ohara, S. and Asada, T. (eds.) *Japanese Project Management*, World Scientific Publishing: Singapore, p. 14.

Quality Management

For many in project management, quality had been seen as a technology measure: the iron triangle of 'time, cost and quality'. But, as the Quality gurus – Deming, Crosby, Juran, Ishikawa[34] – of the 1970s and 1980s insisted, quality relates to the total work effort. It is about more than just technical performance.

The 1980s and 1990s saw a marked impact of quality thinking on project management. Quality Assurance became a standard management practice in many project industries (ISO 10006[35]). More fundamental was the increasing popularity of Total Quality Management with its emphasis on continuous improvement, putting the customer first, performance metrics and aligned, stable supplier relationships. This was to have many impacts on the practice of project management. We've just seen one with relation to the Japanese Body of Knowledge, but a particularly immediate one was to strengthen the philosophy of aligned supply chains – 'partnering', see Chapter 5 – which was to sweep in from Toyota and which was so different from the adversarial nature of much contracting and which had characterised the management of many projects, certainly in construction, for hundreds of years. But there was much more than this that came from the Japanese automobile industry.

New Product Development: Lessons from Toyota

The late 1980s and early 1990s saw some seminal articulation of much that is now considered mainstream good practice in the management of projects. It came from research in an area hardly touched before: new product development (NPD). The initial impetus was (again) studies of success and failure on new product development, notably those by Elko Kleinschmidt and Robert Cooper, the result of which was to recommend a staged approach to development, with strong scrutiny at stage-gates with a predisposition not to proceed unless assured of the investment and management health of the development process[36]. (This has become standard good practice in the management of projects, as witnessed for example by the UK's Office of Government Commerce (OGC), with its Gate Review process introduced in the late 1990s[37]. It springs from Schriever's design reviews and the USAF 375 series' milestone reviews, as previously discussed.)

The stage-gate ideas were absorbed into two research programs which were to have a strong influence on practice across many project-based sectors – pharmaceuticals and other R&D industries, manufacturing, oil and gas, utilities, systems development: one based at Harvard[38]; the other, the International Motor Vehicle Program (IMVP), centred at MIT[39]. Both drew heavily on Japanese auto manufacturers' practices, particularly Toyota's.

Wheelwright and Clark

Wheelwright and Clark's book, *Revolutionizing Product Development* (1992)[40], based on the Harvard work, is quite remarkable in the way it articulates so many of the practices that were to characterise project management over the next 10–20 years. They begin with the premise that different kinds of projects require different kinds of management, that projects' development strategies should reflect their sponsoring entity's business strategy – the 'aggregate plan' – agreed early in the project's development. The book illustrates the portfolio selection and management process (in relation to market demands and technology strategy and the pace of scheme development). The 'mouth' of the development funnel should be wide but the 'neck' narrow. Stage (gate) reviews manage the transition of the product through the funnel. Emerging products will offer different opportunities and, adding in their different characteristics, projects will therefore require different organisational team structures and leadership, going from lightweight, through middleweight to heavyweight (per Galbraith, 1973, as we saw previously). The heavyweight team, led, using a Japanese term, by the *Shusha*, the heavyweight project leader[41], is made up of all the functions critical to overall project success whose representatives are dedicated to it "for the duration of the project" (p. 205). (A key, 'revolutionary' idea.)

This leads, however, to some confusion over the place of project management: initially it is positioned in execution following approval of the project plan (Exhibit 2–3, p. 35) but later we read that project management "sets the scope of the development project, establishes the bounds of what is and is not included in it, and defines the business purposes and objectives of the project. Activities such as initial concept development, defining and scoping the project effort, obtaining both internal and external preliminary inputs, and selling the project" are part of its early responsibilities (p. 135). Sounds right, but the confusion reflects the problems of thinking about project management as only execution management rather than as the management of the overall project.

Concurrent engineering

Effective communication is, in Wheelwright and Clark's view, central to achieving integration, and integration is a central management preoccupation. Achieving downstream–upstream interaction is critical for them (building on Thompson, 1967), just as it was for Schriever. This leads to concurrent engineering and partnering.

Concurrency has been the wicked witch so far in this history of the development of project and program management. Crucial to Schriever (and the Manhattan Project, or other projects where urgency is the essence and where, given the extraordinary schedule urgency, good practice rules, like freezing the design before going to 'build', can be overridden) it is nevertheless bound to lead to problems – wasted work, re-work, and additional costs. In the automobile programs, however, concurrency (bad) became replaced by concurrent engineering (good). Concurrent engineering, as practised by Toyota, comprises parallel working where possible (simultaneous engineering); integrated teams drawing on all the functional skills needed to develop and deliver the total product (marketing, design, production – hence design-for-manufacturability, design-to-cost, etc.); integrated data modelling; and a propensity to delay decision-taking for as long as possible (Figure 4.8)[42].

Concurrency was often really part of the broader issue of how to manage technical innovation in a project environment. Various solutions began to emerge in the 1980s: prototyping off-line so that only proven technology is used in commercially sensitive projects (compare the nuclear power story with its 330 mW power plants – essentially prototypes); rapid prototyping where quick impressions could be gained by quasi mock-ups[43]; use of pre-planned product improvements (P[3]I), particularly on shared platforms – a form of program management (see below, Chapter 5).

The paradigm was shifting. We had moved well beyond the relatively closed world of systems project management; beyond analysing what the critical success factors are. We are in a multidisciplinary, managerially complex environment where behaviour and conceptual ability are as important as technical and commercial finesse.

And as it happens, just at this time the academic world of organisation theory began seriously to engage with projects and their management. For although there had been some theoretical work regarding projects since the 1960s, as we've seen, the amount had been relatively modest. As the century approached its end, this was changing. Organisation theorists began to be engaged more widely, developing interests in projects as novel organisational forms, with the people who worked in them, and how they did so. And as they theorised, so the discourse changed.

Academic Engagement

The tenor of the theoretical, conceptual work in the development of project management so far has been practitioner-focussed, functionalist, instrumental, and largely prescriptive and normative: what could, or should, be done to improve our ability to manage projects better. But now organisational theorists, led by, it is fair to say, a group of Scandinavian scholars – the 'Scandinavian School' – focussed instead on the 'actors' and the organisations working on projects and the reality of that work: "putting less energy into studying what is meant to happen, and more into what is actually happening" as one of their leading theorists, Johann Packendorff of the University of Umeå, put it in 1996[44]. There were three main deficiencies in the studies of project management to date, according to Packendorff: the assumed universality of project management theory, the lack of empirical studies of projects, and the lack of alternative representations of projects[45].

Thus, Kerstin Sahlin-Andersson of Uppsala University offered an alternative model of managing projects, contrasting those that are developed 'messily', allowing ambiguity but which may lead to more interesting realisations with those developed on more 'clear' grounds, which on the other hand typically allow better project control[46]. (Though to my mind, she is really talking about programs, not projects: her example is a large urban development comprising a number of separate buildings having different

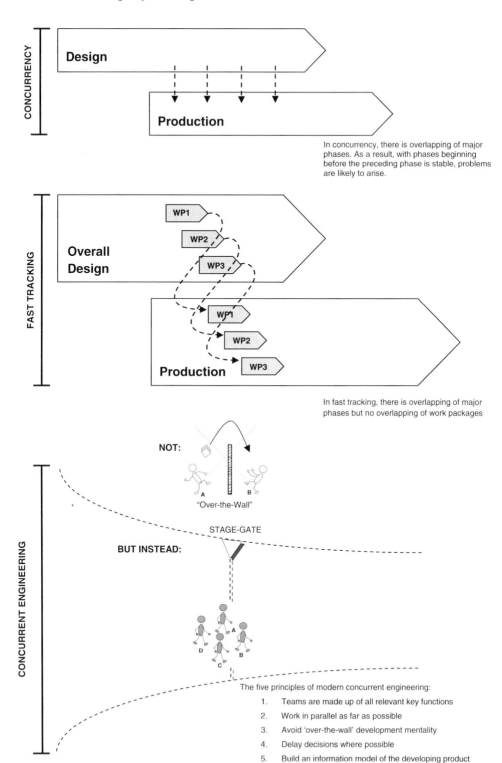

CONCURRENCY

In concurrency, there is overlapping of major phases. As a result, with phases beginning before the preceding phase is stable, problems are likely to arise.

FAST TRACKING

In fast tracking, there is overlapping of major phases but no overlapping of work packages

CONCURRENT ENGINEERING

The five principles of modern concurrent engineering:

1. Teams are made up of all relevant key functions
2. Work in parallel as far as possible
3. Avoid 'over-the-wall' development mentality
4. Delay decisions where possible
5. Build an information model of the developing product

Figure 4.8   Concurrent Engineering.
Source: Author's own.

needs and characteristics. Her description rings true as a possible program management scenario.)

Another feature of the Scandinavian School is to emphasise projects as temporary organisations. Thus Rolf Lundin of Jönköping and Anders Söderholm of Umeå[47], drawing on work by Richard Goodman[48] and Connie Gersick[49], argued that projects are predominantly time-defined and are less goal-oriented than permanent organisations – again, this is surely questionable: projects are typically very goal-oriented. (See Chapter 18 on teleology.)

Thirdly, the School emphasised the institutional context within which projects occur – 'no project is an island'* – and thus questioned the value of generic bodies-of-knowledge and critical success studies (Mats Engwalls[50]). (Does no level of normative or prescriptive abstraction have value?)

This, and other work such as that of Arthur Stinchcombe on contractual arrangements and administration[51] – though his work is not of the Scandinavian School proper but stems from the 'markets and hierarchies' perspective of Oliver Williamson[52] – should be seen as adding to the theoretical work referred to earlier on integration (Thompson; Lawrence and Lorsch; Galbraith; Mintzberg).

As the new century got underway the management of projects would be looked at from many other theoretical perspectives: governance (Müller, Turner), strategy (Artto, Loch, Morris and Jamieson), technology management (Davies, Hobday *et al.*), innovation (Gann), supply-chain management and networks (Cox, Pryke), organisational learning (De Fillipi, Sence, Grabher), critical management (Hodgson and Cicmil), institutional theory (Bresnen, Orr), and many others (concurrent engineering, complexity, culture, epistemology, ethics, funding, knowledge management, ICT, sustainability, trust, etc.)[53].

Many of these theoretical approaches offer insight into the practical challenges of managing projects and programs, but for the non-academic, their treatment can prove difficult to enter and engage with being written largely for other academics to read. Normative guidance is treated with suspicion. In the Scandinavian School's actor-centric mode of enquiry, for example, with the main source of information being individuals with their subjective realities, this brings about "a change in epistemology, in that there will not be any 'truth' beneath or beyond the narration of the project member"[54]. So while the analysis becomes richer, it's an observer's theoretical discussion that often results; too frequently failing to help companies, or people, know

---

*But compare the sociologists Jeffrey Pressman and Arnold Wildavsky 20–30 years previously: "As programs are altered by their environments and organizations are affected by their programs, mutual adaptation changes both the context and content of what is implemented." (Their definition of 'program' is not a p.m. one, though: "a program can be conceived of as a system in which each element is dependent on the other": p. xxiii): Pressman, J. L. and Wildavsky, A. (1984) *Implementation*, 3rd edition, University of California Press: Berkeley, p. xvii. We shall examine the interaction between context and the project in Chapter 19.

what to do to manage their projects better*. To quote Packendorff again: "theories can never in themselves enhance the practice of project management but they might be helpful to the people actually involved in the work"[55]. To the extent I understand this, I think I disagree: personally I have found the ideas on integration and contingency, for example, enormously useful – even essential – in helping me, as a consultant, to design and explain organisational structures. Similarly on much relative to behaviours: Myers-Briggs, leadership, team building, conflict management, emotional intelligence.

References and Endnotes

[1] Cook, D. L. (1977), Certification of project managers – Fantasy or reality? *Project Management Quarterly*, 8, 3, pp. 32–34.
[2] Hodgson, D. and Muzio, D. (2010), Prospects for professionalism, in: Morris, P. W. G., Pinto, J. K. and Söderlund, J. (eds.) *The Oxford handbook of project management*, Oxford University Press: Oxford.
[3] Hodgson, D. and Muzio, D. (2010), *ibid.*; Project Management Institute (2004), *A guide to the Project Management Body of Knowledge (PMBOK® Guide)*, 3rd edition, Hoboken, NJ, 1.5.1.
[4] Stuckenbruck, L. C. (1986), Project management framework, *Project Management Quarterly*, XVII, 2, pp. 27–28.
[5] Cleland, D. I. and King, W. R. (1968), *Systems analysis and project management*, McGraw-Hill: New York; Johnson, R. A., Kast, F. E. and Rosenzweig, J. E. (1973), *The theory and management of systems*, McGraw-Hill: New York; Kerzner, H. (1979), *Project management: A systems approach to planning, scheduling and controlling*, van Nostrand Reinhold: New York.
[6] Terms of Reference for PMI Project 210: see Wideman, R. M. (1986), The PMBOK report: PMI Body of Knowledge standards, *Project Management Quarterly*, XVII, 2, pp. 15–24.
[7] Roy, B. (1962), Graphs et ordonnancements, *Revue Française Opertionelle*, 6, No. 323; Schleip, W. and Schleip, R. (1972), *Planning and control in management: The German RPS system*, Peter Peregrinus: Düsseldorf.
[8] For example: Arisawa, S. and Elmaghraby, S. E. (1972), Optimal time-cost trade-offs in GERT networks, *Management Science*, 18, 11, pp. 589–599; Davis, E. W.

---

*This criticism is being made of actor-focussed enquiry and doesn't apply to all academic research into project management, of course. But this orientation has clearly influenced the academic community: project management conference papers (at IRNOP) for example have been shown to have a bias towards subjectivity, interpretivism, qualitative methods and case studies (Biedenbach, T. and Müller, R. [2011], Paradigms in project management research: Examples from 15 years of IRNOP conferences, *International Journal of Managing Projects in Business*, 4, 1, pp. 82–104). The influence is not universal, however. Analysis of a different set of publications – IJPM papers published in 2005 – showed positivism as the dominant epistemology followed by empiricism (Smyth, H. J. and Morris, P. W. G. (2007), An epistemological evaluation of research into projects and their management: Methodological issues, *International Journal of Project Management*, 25, 4, pp. 423–436.

(1973), Project Scheduling under resource constraints- historical review and categorization of procedures, *AIIE Transactions*, Dec.; Pritsker, A. A. B. (1966), *GERT: Graphical Evaluation and Review Technique*, Rand Corporation, RN-4973-NASA, Santa Monica.

[9] For those wanting all academic references, see Morris, P. W. G. (1972), *An organisational analysis of building projects*, PhD thesis, UMIST: Manchester.

[10] Miller, E. J. and Rice, A. K. (1967), *Systems of organization: The control of task and sentient boundaries*, Tavistock: London.

[11] Thompson, J. D. (1967/2003), *Organizations in action*, McGraw-Hill/New Brunswick: New York.

[12] Lawrence, P. R. and Lorsch, J. W. (1967), The new management job: The integrator, *Harvard Business Review*, Nov-Dec, pp. 142–151; Lawrence, P. R. and Lorsch, J. W. (1967), *Organisation and environment: Managing integration and differentiation*, Harvard University Press: Cambridge.

[13] Archibald, R. D. (1976, 1997, 2003), *Managing high-technology programs and projects*, Wiley: New York.

[14] Galbraith, J. R. (1968), *Achieving integration through information systems*, Working Paper No. 361-68, Alfred P. Sloan School of Management, Massachusetts Institute of Technology.

[15] Morris, P. W. G. (1973), An organizational analysis of project management in the building industry, *Build International*, 6, 6, pp. 595–616.

[16] Shenhar A. J. and Dvir, D. (2007), *Re-inventing project management*, Harvard Business School Press: Cambridge, MA.

[17] Galbraith, J. (1973), *Designing complex organizations*, Addison-Wesley: Reading, MA.

[18] Mintzberg, H. (1979), *The structuring of organizations*, Prentice-Hall: Englewood Cliffs, NJ.

[19] Goodman, R. A. (1967), Ambiguous authority definition of project management, *Academy of management Journal*, 10, 4, pp. 395–407; Wilemon, D. L. and Cicero, J. P. (1970), The project manager – Anomalies and ambiguities, *The Academy of Management Journal*, 13, 3, pp. 269–282.

[20] Butler, A. G. (Mar., 1973), Project management: A study in organizational conflict, *The Academy of Management Journal*, 16, 1, pp. 84–101; Gemmill, G. R. and Wilemon, D. L. (1970), The power spectrum in Project Management, *Sloan Management Review*, 12, 4, pp. 12–25; Wilemon, D. L. (1971), Project management conflict: A view from Apollo, *Proceedings of 3rd Annual Symposium*, Project Management Institute, Drexel Hill, PA; Wilemon, D. L. and Gemmill, G. R. (1971), Interpersonal power in temporary management systems, *Journal of Management Studies*, 8, Oct., pp. 315–328; Thamhain, H. J. and Wilemon, D. L. (1974), Conflict management in project-oriented work environments, *Proceedings of 6th Annual International Meeting of the Project Management Institute*, Project Management Institute, Drexel Hill; Thamhain, H. J. and Wilemon, D. L. (1977), Leadership effectiveness in program management, *IEEE Transactions on Engineering Management*, EM-24, pp. 102–108.

[21] Knight, K. (1976), Matrix organization: A review, *Journal of Management Studies*, 13, pp. 111–130; Davis, S. M. and Lawrence, P. R. (1977), *Matrix*, Addison Wesley: Reading, Mass.; Might, R. J. and Fischer, N. Z. (1985), The role of structural factors in determining project management success, *IEEE Transactions on Engineering Management*, 32, 2, pp. 71–77; Larson, E. W. and Gobeli, D. H. (1989), Significance of project management structure on development success, *IEEE Transactions on Engineering Management*, 36, 2, pp. 119–125.

22 Jugdev, K. and Müller, R. (2005), A retrospective look at our evolving understanding of project success, *Project Management Journal*, 36, 4, pp. 19–31.

23 Marshall, A. W. and Meckling, W. H. (1959), *Predictability of the costs, time
and success of development*, Rand Corporation: Santa Monica, CA; Summers,
R. (1965), *Cost estimates as predictors of actual weapons costs*, Rand Cor-
poration: Santa Monica, CA, RM-3061-PR; Perry, R. L., DiSalvo, D., Hall,
G. R., Harman, A. L., Levenson, G. S., Smith, G. K. and Stucker, J. P. (1969),
System acquisition experience, Rand Corporation: Santa Monica, CA, RM-6072-
PR; Peck, M. J. and Scherer, F. M. (1962), *The weapons acquisition process:
An economic analysis*, Harvard University Press: Cambridge MA; Large, J. P.
(1971), *Bias in initial cost estimates: How low estimates can increase the cost
of acquiring weapons systems*, Rand Corporation: Santa Monica, CA, R-1467-
PA&E.

24 Baum, W. C. and Tolbert, S. M. (1985), *Investing in development*, Oxford
University Press: Oxford; Hirschman, A. O. (1967), *Development projects
observed*, The Brookings Institute: Washington, DC. A potentially interesting
group should be that by Kathleen Murphy, who analysed McKinsey-originating
data on 1,600 'third world development' macro-projects. Unfortunately very little
of substance comes from the analysis, except possibly the emphasis on technological IPR (intellectual property rights) and the importance of people. Murphy, K.
(1983), *Macro-project development in the Third World*, Westview Press: Denver,
CO.

25 Morris, P. W. G. and Hough, G. H. (1987), *The anatomy of major projects*, John
Wiley and Sons: Chichester.

26 See Section 5.2 of the fifth edition where, amidst the emphasis on planning and
documenting, there is twice, welcomingly, reference to 'managing' requirements
'to meet project objectives': PMI (2013), *A guide to the Project Management
Body of Knowledge (PMBOK® Guide: fifth edition)*, Project Management Institute, Newtown Square, PA.

27 Morris, P. W. G. (1994), *The management of projects*, Thomas Telford:
London.

28 Morris, P. W. G., Patel, M. B. and Wearne, S. H. (2000), Research into revising
the APM Project Management Body of Knowledge, *International Journal of
Project Management*, 18, 3, pp. 155–164; Morris, P. W. G., Crawford, L.,
Hodgson, D., Shepherd, M. M. and Thomas, J. (2006), Exploring the role of
formal bodies of knowledge in defining a profession – The case of Project Management, *International Journal of Project Management*, 24, 8, pp. 710–721. Note
that in publishing the APM BOK the 'technology' section got 'lost in translation'
and has been printed as 'techniques' which is wrong. Techniques surface everywhere; technology is about systems engineering, requirements, innovation, design,
build, testing, etc.

29 Pannenbacker, K., Knopfel, H., Morris, P. W. G. and Caupin, G. (1998), *IPMA
and its validated four-level certification programmes*, Version 1.00, International
Project Management Association: Zurich.

30 ENAA (2002), *P2M: A guidebook of project & program management for enterprise innovation: Summary translation*, Project Management Professionals Certification Center (PMCC): Tokyo, Japan; Ohara, S. and Asada, T., (eds.) (2009),
Japanese Project Management, World Scientific Publishing: Singapore.

31 Morris, P. W. G. and Pinto, J. K. (2004), *The Wiley guide to managing projects*,
Wiley: Hoboken, NJ, p. xvi.

32 Gareis, R.(2004), Management of the project-oriented company, in: Morris,
P. W. G. and Pinto, J. K. (eds.), *ibid.*

[33] Hamilton, A. (1997), *Management by projects*, Thomas Telford: London.

[34] Deming, W. E. (1986), *Out of the crisis: Quality, productivity and competitive position*, Cambridge University Press: Cambridge; Crosby, P. (1979), *Quality is free: The art of making quality certain*, McGraw Hill: New York; Juran, J. J. (1982), *Upper management and quality*, 4th edition, Juran Institute Inc., New York; Ishikawa, K. (1986), *Guide to quality control*, 2nd edition, Kraus International Publications: White Plains, NY.

[35] ISO 10006 (2003), *Quality management systems: Guidelines for quality management in projects*, British Standards Institute: London.

[36] Cooper, R. G. (1986), *Winning at new products*, Addison-Wesley: Reading, MA.

[37] Office of Government Commerce (2004), *The OGC gateway process*, TSO: London.

[38] Clark, K. B. and Fujimoto, T. (1991), *Product development performance*, Harvard Business School Press: Cambridge, MA; Wheelwright, S. C. and Clark, K. B. (1992), *Revolutionizing product development*, Harvard Business School Press: Cambridge, MA; Clark, K. B. and Fujimoto, T. (1990), *Product development performance*, Harvard Business School Press: Cambridge, MA.

[39] Womack, J. R., Jones, D. T. and Roos, D. (1990), *The machine that changed the world*, Macmillan International: New York.

[40] Wheelwright, S. C. and Clark, K. B. (1992), op. cit.: 38.

[41] Clark, K. B., & Wheelwright, S. C. (1992). Organizing and leading 'Heavyweight' development teams. *California Management Review*, 34, 3, pp. 9–28.

[42] Prasad, B. (1996), *Concurrent engineering fundamentals*, Volume I and II, Prentice Hall: New York; Gerwin, D. and Susman, G. (1996), Special issue on concurrent engineering, *IEEE Transactions on Engineering Management*, 32, 2, pp. 118–123; Anuba, C., Kamara, J. M. and Cutting-Decelle, A. F. (2006), *Concurrent engineering in construction projects*, Routledge: London.

[43] Thomke, S. H. (2003), *Experimentation matters*, Harvard Business School Press: Cambridge, MA.

[44] Packendorff, J. (1996), Inquiring into the temporary organization: New directions for Project Management research, *Scandinavian Journal of Management*, 11, 4, pp. 319–334.

[45] *Ibid.*: 326.

[46] Sahlin-Andersson, K. (1992), The use of ambiguity – The organizing of an extraordinary project, In: Hagg, J. and Segelod, E. (Eds.) *Issues in empirical investment research*, Elsevier: Amsterdam, The Netherlands, pp. 143–158.

[47] Bakker, R. M. (2010), Taking stock of temporary organizational forms: A systematic review and research agenda. *International Journal of Management Reviews*, 12, 4, pp. 466–486; Lundin, R. A. and Söderholm, A. (1995), A theory of the temporary organization, *Scandinavian Journal of Management*, 11, 4, pp. 437- 455; Packendorff, J. (1995), *op. cit.*: 44.

[48] Goodman, L. P. and Goodman, R. A. (1972), Theater as a temporary system, *California Management Review*, 15, 2, pp. 103–108; Goodman, R. A., *op.cit.*: 19.

[49] Gersick, C. (1989), Marking time: Predictable transitions in Task Groups, *The Academy of Management Journal*, 32, 2, pp. 274–309.

[50] Engwall, M. (2003), No project is an island: Linking projects to history and context, *Research Policy*, 32, pp. 780–808.

[51] Stinchcombe, A. L. (1959): Bureaucratic and craft administration of production, *Administrative Science Quarterly*, 4, pp. 168–187.

[52] Williamson, O. E. (1975), *Markets and hierarchies, analysis and antitrust implications: A study in the economics of internal organization*, Free Press: New York.

[53] See for example: Müller, R. and Turner, J. R. (2005), The impact of principal-agent relationship and contract type on communication between project owner and manager, *International Journal of Project Management*, 23, 5, pp. 398–403; Turner, J. R. and Keegan, A. (2001), Mechanisms of governance in the project-based organization: Roles of the broker and steward, *European Management Journal*, 19, 3, pp. 254–267; Artto, K., Kujala, J., Dietrich, P. and Martinsuo, M. (2007), What is project strategy? *International Journal of Project Management*, 26, 1, pp. 4–12; Loch, C. (2000), Tailoring product development to strategy: Case of a European technology manufacturer, *European Management Journal*, 18, 3, pp. 246–258; Morris, P. W. G. and Jamieson, H. A. (2004), *Translating corporate strategy into project strategy*, Project Management Institute: Newtown Square, PA; Davies, A. and Hobday, M. (2005), *The business of projects*, Cambridge University Press, Cambridge; Grabher, G. (2002), Cool projects, boring institutions: Temporary collaboration in social context, *Regional Studies*, 36, 3, pp. 205–214; Grabher, G. (2004), Temporary architectures of learning: Knowledge governance in project ecologies, *Organization Studies*, 25, 9, pp. 1491–1514; Cox, A. and Ireland, P. (2006), Relationship management theories and tools in project procurement, in: Pryke, S. and Smyth, H. (eds.) *The management of complex projects: A relationship approach*, Blackwell Publishing: Oxford, pp. 251–281; Pryke, S. D. (2001), Analysing construction project coalitions: Exploring the application of social network analysis, *Construction Management and Economics*, 22, pp. 787–797; Hodgson, D., and Cicmil, S. (2006), *Making projects critical*, Palgrave Macmillan: Hampshire and New York; Orr, R. J., and Scott, W. R. (2008), Institutional exceptions on global projects: A process model, *Journal of International Business Studies*, 39, 4, pp. 562–588; Gann, D. M. and Salter, A. J. (2000), Innovation in project-based, service-enhanced firms: The construction of complex products and systems, *Research Policy*, 29, 7–8, pp. 955–972; DeFillippi, R. J. and Arthur, M. B. (1998), Paradox in project-based enterprise: The case of film making, *California Management Review*, 40, 2, pp. 125–139; Sense, A. J. and Antoni, M. (2003), Exploring the politics of project learning, *International Journal of Project Management*, 21, 7, pp. 487–494; Bresnen, M. (2007), Deconstructing partnering in project-based organisation: Seven pillars, seven paradoxes and seven deadly sins, *International Journal of Project Management*, 25, 4, pp. 365–374; Bresnen, M. and Marshall, N. (2000), Building partnerships: Case studies of client-contractor collaboration in the UK construction industry, *Construction Management and Economics*, 18, 7, pp. 819–32; Orr, R. J. and Scott, W. R. (2008), Institutional exceptions on global projects: A process model. *Journal of International Business Studies*, 39, pp. 562–588.

[54] Packendorff, J. (1996), *op. cit.*: 44.

[55] *Ibid.*

Developing Project Management

Project management now grew rapidly and substantially, both in concept and in practice. Health, Safety and Environment; risk, value, benefits; requirements, innovation and technology; strategy and governance – all had become mainstreamed by the early 21st century. Effectiveness was beginning to be as important as efficiency.

Partnering, as a new form of contracting and procurement, led to radically new ways of behaving and managing projects. Add to this new forms of acquisition centred on supplier financing and operation (Build–Own–Operate and the Private Finance Initiative [PFI]) which brought increased sensitivity to value, whole life costing and more sophisticated perspectives on risk, so by the late 1990s the project management field was very different from 20 or 30 years previously.

These changes were heralded by a number of seminal Critical Success Factor studies which we've already touched upon – Figure 4.4 – but which we should now look at in more detail, not least because many were to shape further our thinking on the management of projects and programs: principally, the IMEC program on large engineering projects, Flyvbjerg *et al.*'s studies of transport projects, the US DoD/CIA study of 'intelligence' projects, and the Standish Group's studies of software development projects, in addition to many others by scholars, and government or regulatory agencies[1] (see Appendix 1).

First the IMEC study.

IMEC: 'Large Engineering Projects'

Roger Miller and Don Lessard's 2000 book – Miller of Montreal University, Lessard of MIT – on the IMEC program, the International Program in the Management of Engineering and Construction, a multi-university international program of research, similar to the IMVP previously discussed[2] – examined 60 'Large Engineering Projects' (LEPs) of average cost $1 bn, in

Reconstructing Project Management, First Edition. Peter W.G. Morris.
© 2013 John Wiley & Sons, Ltd. Published 2013 by John Wiley & Sons, Ltd.

the energy and transportation sectors. This was not a successful group: "Close to 40% of them performed very badly; by any account, many are failures"[3]. 82% met their cost targets, 72% their schedule targets. These Miller and Lessard called their 'efficiency' measures. But then they introduced the notion of 'effectiveness' measures (defined as 'overall utility' – did the project investment essentially turn out as it was meant to?[4]). Of these effectiveness measures, only 45% met their sponsors' objectives, 18% without 'crises' while 17% needed restructuring after crises and 20% were abandoned or taken over*. The causes of failure were not particularly technical difficulties, social issues or size but turbulence – though urgent projects did not do well. The key issue, Miller and Lessard concluded, is the competence of the sponsor: 85% of the successes and failures in their projects' data set could be explained by the sponsor's abilities in (1) shaping strategy and coping with political, economic, and social turbulence and outside institutions[5], and (2) dealing with partnership and contractual turbulence[†]. Up to 35% of the total project expenditure (for a complex project) could usefully be spent on 'front-end' activities. Risk management, Miller and Lessard recommended, needs to cover more than just Monte Carlo analysis of cost and schedule (efficiency) but include, e.g. technical, market, finance, project management, social and institutional risks. Risk Management is "an important *ex ante* strategic concern for project sponsors requiring knowledge and reflection"[6]. Real-options approaches to risk analyses should be considered. In fact, the landscape was changing in precisely this manner.

Contracting and Procurement

To get real innovation, you need the specialist knowledge of your supply chain. Both the Harvard and MIT programs dealt extensively with supply chain issues. Which brings us to what has been a major omission in this account of the emergence of project management so far: the acquisition of resources, and that part of it in particular that deals with contracts and procurement.

Procurement of resources, and the contractual basis on which this is done, is and has always been a fundamental aspect of managing projects. We saw Plutarch referring to competitive tendering of work packages back in Chapter 2.

---

*Wouldn't one expect performance effectiveness logically to be better than performance efficiency? After all, effectiveness is the real driver. But this point is not discussed, either in principle or methodologically – for example, do the different measures share the same baseline? The question seems not to have been picked-up by the project management community.

†Grün on 'giant' infrastructure projects also focused on the sponsor – the 'owner' – and his ability to formulate and deal with change of goals, manage basic configuration, influence the socio-political environment, and provide an appropriate managerial capacity and management structure: Grün, O. (2004) *Taming giant projects*, Berlin: Springer.

It's important to recognise that this activity is not limited to the acquisition of external resources only. Projects at the initial concept stage typically only require a small amount of resources. The need then grows for more as work expands. These, however, could be brought in from elsewhere within the organisation, as well as being sourced from outside it. (Indeed outsourcing might be policy.) Nevertheless, even if acquired internally, they are typically supplied on some form of contractual understanding, if only on an implicit or informal basis (the matrix for example works best like this) and most projects do indeed procure most of their resources externally via an explicit contract.

The way resources are procured can massively affect the performance of the project. Historically, resources have been acquired on the basis of the lowest competitive bid, priced against a specification and set of drawings, usually evaluated in terms just of capital cost. (Plutarch again.) This becomes difficult and inherently riskier, though where the design is at a very preliminary state. Such was the case for much of the ICBM work e.g. which was performed on a 'cost reimbursable' basis. RAND (Marschak, Glennan, and Summers) pointed out in 1967 that there are essentially just two options available in the procurement of resources on projects having high technical (or other) uncertainty: either make a financial commitment early on, thereby allowing 'downstream' work to begin early but with the inevitability of uncertainty and risk of overruns; or make only a small commitment early-on, probably allocating a considerable contingency, while the project definition is worked-up[7].

The 'total procurement package' concept

The DoD at this time, the late 1960s, was finding that a number of contractors were bidding low in order to get a lock on the supply contract and then were ramping up their profits via change orders. Hence in 1966 DoD implemented the Total Procurement Package (TPP) practice, where suppliers were required to bid at the outset on the supply of the total 'system', including spare parts. This, unsurprisingly, proved to be a disaster, not just because of the technical game of poker that suppliers were being required to play but due to the cost inflation then becoming rampant. TPP was abandoned in 1972. The same story was played out in the UK nuclear industry at the same time: consortia were required to take fixed-price risks for the supply and installation of sophisticated, unproven technology with completely inadequate financial collateral for doing so. Incentive contracting was then implemented as the next best option despite a plethora of criticism from analysts.

The truth was, *the real problem was not the procurement strategy, it was the maturity of the project definition* and the risk which this added (which is why, over and over again, commentators kept recommending spending more time on front-end definition). This plus inappropriate risk management: risks not being allocated to those best able to bear them. To address these shortcomings, projects needed stronger overall management, from their earliest stages. Some industries were moving in this direction – defence/ aerospace for example. Others were miles behind – building say, at least where the project was 'led' by designers who often had neither interest nor competence in the management of projects*. The result was grievous project inefficiencies with claims, cost growth, and no downstream input into upstream activities (no buildability input – let alone operability). This said, building and civil engineering received two major aids to good project management with the introduction in the late 1980s to the early 1990s of 'construction management' and a new family of contracts for engaging construction services – the New Engineering Contract (the NEC) – introduced in the UK in 1993 based specifically on transparent and quick resolution of disputes, the positive role of the project manager and value of partnering.

'Construction Management' – the provision of downstream site construction advice in the design stages of the project – was developed in the building industry as an organisational answer to the extant lack of integration between downstream construction and front-end design in the late 60s, first in Toronto, then New York, coming over to the United Kingdom 10–15 years later (on London's Broadgate and Canary Wharf city centre office developments). In 1994, Sir Michael Latham published a major review of UK building practices, slamming the adversarial conflict and inefficiency inherent in the way the industry was run, endorsing the NEC and partnering as alternative good practices. Meanwhile, in the United States, the Construction Industry Institute, a 'club' of 28 leading owners and contractors in the engineering-construction industry, made similar recommendations about the same time, issuing ten reports on how best to implement partnering[8].

Partnering and the new Procurement Environment

Partnering has proved immensely popular and has truly revolutionised the basis on which many, though by no means all, projects are engaged.

Partnering meant several different things, though – see Chapter 13: pp. 179–80). To some, it was primarily a philosophy, spawned from Total Quality Management (TQM). To others it was a convenient way of making a guaranteed profit. A deeper rationale for it lay in Relationship Management, as espoused by Ian Macneil of North Western Law School and his theory of contracts[9]. Common to all was – is – an emphasis right along the supply chain on achieving mutually agreed objectives, developing trust, implementing joint approaches to work, empowering personnel and stimulating learning[10]. It smacked maybe of warmth and goodwill, above all of alignment, but it might not be without a real commercial edge. Take the case of BP and its Andrew Field.

---

*There are numerous examples. The British Library was one of the most notorious. The Scottish Parliament was another. See National Audit Office (2001), *Modernising Construction*, Report by the Comptroller and Auditor General HC 87 Session 2000–2001, London; and Murray, M. and Langford, D. (2003), *Construction reports*, Blackwell Science: Oxford.

The Andrew oil field

Andrew is an oil field in the North Sea. Discovered in 1974, given the then low price of oil, it initially looked too uneconomic to develop. However, in 1990, BP launched a series of initiatives to improve its general project financial performance. Andrew was put forward as a test 'guinea pig' and in May 1992 BP decided to apply a 'behavioural revolution' to the way it worked with its contractors on Andrew, appointing staff on a 'best man for the job' basis regardless of whether he were a BP employee or a contractor, and allowing key suppliers and contractors a share in the business 'gain' should the integrated team deliver cost savings, or shouldering 'the pain' should there be overruns. BP called this new way of working 'Alliancing'*.

As a result of applying these principles, savings of 20% were found, relatively easily; eventually, an expenditure target of £373 million was agreed. Intense coaching was provided to the team on developing and applying 'high-impact' behaviours, using the consultants JMW. These behaviours (commit, a-player-not-a-spectator-be, no-blame, continuously extend targets), coupled with extensive, far-reaching Value Management/Engineering (see pp. 83–4), enabled the project team to bring the target cost down first to £320 million and later to a further 'stretch target' of £290 million! The final cost was just below this; completion was six months ahead of schedule (Figure 5.1).

Undoubtedly, the 'behavioural revolution' had been a major force for change but, as the Brown & Root's, Andrew's EPC contractor, Project Director put it "We were not dealing with soft issues. We were chasing a serious financial goal with the opportunity to increase significantly our normal level of profits"[11]. Nothing wrong with that, only it does raise some interesting questions, made richer by subsequent events.

First, if £160 million saving (36%) could be effected, where should the estimate be set for the next project? In the automobile industry, targets were set to be progressively demanding, but here one has the benefit of a relatively stable, controllable production line. In projects, every project is different – unique. This points, in passing, to one of the other areas that are central to project management but which is often treated (at least organisationally) as separate from it: estimating (see Chapter 9). It also underscores the value of benchmarking suppliers' costs against industry norms so that there is a reality check on their prices. Second, how much of the capex (capital cost) saving was really at the cost of opex (operating costs)? Third, is the level

---

*Alliancing as illustrated here has characteristics which push it beyond normal partnering. Anvuur and Kumaraswamy propose (1) equal status or respect, (2) common goals, (3) cooperative work practices and (4) institutional support for the above as characteristics of Partnering (Anvuur, A. M. and Kumaraswamy, M. M. [2007], Conceptual model of partnering and alliancing, *Journal of Construction Engineering and Management*, 133, 3, pp. 225–234). [Project] Alliancing involves these plus, critically, "all (or almost all) of the contractors working on [the project] under a single compensation scheme" involving some gainshare or painshare (Merrow, E. [2011], *Industrial mega-projects*, Wiley: Hoboken, NJ, p. 257).

Figure 5.1   The Andrew project.
Source: © BP p.l.c.

of 'gainshare' awarded to the supply chain acceptable to shareholders – to governance? Fourth, how does this arrangement work when losses are incurred (as subsequently happened, on ETAP)?

Given the generous Gainshare arrangements, the project proved very profitable to the supply team, to such an extent that BP was criticised for being overgenerous. Interestingly, despite John Browne, BP's CEO, espousing partnering (and interestingly linking organisational learning to it[12]), with the opposition of Amoco personnel to the concept (BP bought Amoco in 2000), and problems on ETAP* and in Andrew's operations, BP did not pursue the practice and returned to traditional forms of contracting. In the process engineering industries as a whole, however, partnering continued to attract adherents, often and increasingly in the 2000s also merging it into framework contracts (where there is a long-term arrangement to use specified suppliers providing services at predetermined rates).

Despite such cautionary tales, partnering still seemed *prima facie* to make business, organisational and managerial sense, at least in those situations

---

*The original concept of BP ETAP was made up of several marginal fields which in isolation were not commercially viable. Improved ways of working and shared information facilitated a commercially viable solution sanctioned at £1.06 bn. The project was eventually delivered well within the original budget but some contracts experienced difficulties and the Alliancing team spirit was severely tested.

where the transaction was more value oriented rather than being primarily a commodity purchase, and it accordingly spread into most project-based industries. Thus for example, partnering became central to the SMART procurement initiative launched in the UK defence industry in 2002 (along with six other principles including a through-life approach to managing procurement projects and 'incremental acquisition'), leading directly to suppliers' involvement in new multidisciplinary 'integrated project teams' (IPTs) and greater attention to risk management. The defence procurement results were generally held to be good at least for a while until the cost of resourcing became questionable[13].

Risks and Opportunities

Risk was now in fact receiving much more attention generally. Probabilistic estimating had been present in PERT from the outset (and officially abandoned by DoD by 1963) but this is not the same as the formal project risk management process (of identification, assessment, mitigation and reporting) that was articulated in the 1986/87 edition of the PMBOK® Guide. By the mid-1980s formal procedures of risk management had become common practice with software packages available to model the cumulative effect of different probabilities. (Almost always this was assessed on predicted cost or schedule completion, rarely on value or business benefit – see next section and p. 191 and Chapter 20).

A House of Commons Defence Committee in the UK concluded in 1988 that "the application of a more disciplined approach by MoD to risk assessment is clearly required"[16], while the MoD's (the Ministry of Defence's) Jordan Lee Cawsey report, *Learning From Experience*[17], emphasised the need for improved risk management within the MoD in particular, and the defence industry in general. SMART Procurement had risk reduction and more effective risk management as one of its central pillars.

In the late 1990s/early 2000s in a conceptually important development, the project management community began to distinguish between uncertainty and risk – risk being the possibility of a negative event occurring ("hazard, chance of bad consequence, loss", etc. – OED), uncertainty being 'unknowns' (hence 'known-unknowns' and 'unknown-unknowns', etc.)[18]. This resulted in a number of reformulations of what many now termed 'risk and opportunity management', as for example in the fifth edition of the APM BOK (2004), resulting in new 'risk management' quasi-standards[19]. The idea was then promoted that we should be seeing the subject as 'opportunity management'[20].

Flyvbjerg *et al.*: Transportation Projects and Optimism Bias

Meanwhile, around the early 2000s, Bent Flyvbjerg and his colleagues at Aalborg University were introducing behavioural psychology to change the way we think about risk and estimating[21]. Drawing on data from over 300 road and rail projects, they showed that overruns of 50–100% in real terms was common and demand forecasts were often wrong by 20–70%. The key actor they found is – again – the sponsor and the key problem is accountability, not technical skills or data[22].

Cost overruns in their (public sector) data set can best be explained, Flyvbjerg and his colleagues suggested, by people getting trapped in an overly optimistic mindset ('optimism bias') and displaying a tendency to 'low ball' the estimate ('strategic misrepresentation'). (Similar data are found in Wachs, 1986)[23]. To counter this, they proposed a number of cost estimation techniques (e.g. 'reference class forecasting' based on the ideas of behavioural economist Don Lavallo and psychologist Daniel Kahneman of taking an outsider's view[24]) and, in the spirit of the times, they then advocated involving private sector funding as a means of improving estimating and budgeting realism*.

BOT/PFI

For another change was now occurring which was moving project management back towards a more holistic perspective: the method of procurement that involved the funding of public sector projects by the private sector. The so-called BOT/BOOT (Build–Own–(Operate)–Transfer) method of project procurement was originally developed in the Turkish power sector in 1984[14] following the development of limited recourse project financing (where funding is secured on the output of the project with limited or no recourse to any non-project entities) in the oil and minerals sectors in the 1970s. The intent was – and still is, though there are criticisms over the method's value-for-money – for private sector groupings to fund the project's construction on the basis that they would be given operational responsibility for the facility, generally only for a defined period (typically 25 or 30 years), for which they receive an income. The cost of building the facility is born by the private sector group on the basis of its future operating earnings. The Channel Tunnel was financed and built on this basis in 1987–1994[15] following experience with a number of UK trial projects

PFI – the Private Finance Initiative – was an institutional response to the BOT idea. Promoted on the basis of engaging private sector efficiencies, PFI offered private sector capital together with building and operating expertise for a range of public sector facilities – most notably schools, hospitals and prisons. The method was attractive to politicians since it relieved governments from the pressures of capital expenditure (but at the expense of an enlarged operating budget). The method is risky to the bidders, however,

---

*And Althuser and Luberhof look at government funding patterns – and hence power – in the USA since World War II, but with little ultimately to add to the practice of their funding or management: Althuser, A. and Luberhof, D. (2003) *Mega-projects*, Brookings Institute Press: Washington, DC.

and is expensive, not least in bidding. Performance requirements are defined in terms of output specifications. The legal arrangements require very careful drafting and bidding often proved chronically slow and very expensive. The cumulative impact of losing several bids in a row can be financially horrendous. Operating costs tend to be high too. Nevertheless, the prospect of getting benefits today and having someone else pay for them tomorrow proved irresistible to many governments. The idea soon morphed into PPP (Public Private Partnerships) as a more general way of getting private sector involvement in the delivery of public services. PPP is an ownership structure between government and the private structure while PFI is a procurement tool.

There were two putative conceptual benefits to project management from this set of developments. One, a greater emphasis on Whole Life Costs, on operating efficiency, and on benefits and effectiveness; two, the development of project companies as deliverers of services as opposed simply to products. Similarly in IT services and Facilities Management; and in aero-engines, the emphasis moved from capital cost to 'power-by-the-hour'.

Value and Benefits

This broadening of the treatment of risk, and forms of procurement, paralleled a growing interest in looking at how project upsides – the positives: value and benefits – could be better managed, specifically by Value Management/Engineering and Benefits Management.

Formally addressing value had in fact been a project management practice for some time. Value Analysis (VA) had been developed by General Electric in the late 1940s and was one of the techniques ushered in to the DoD by McNamara in 1960. It is a powerful concept and technique, frequently used in some sectors, hardly known about in others. There are several national and international standards on Value Management (VM)[25].

Simply put, value can be defined as the quotient of function/cost or quality/cost, performance/resources or similar. The aim is to analyse, in a structured manner using a wide selection of different stakeholders, the project's requirements and ways of addressing these more effectively – by getting more for less. Value Engineering (VE) focuses on the proposed engineering solution. The term 'Value Management' (VM) is used either as the generic term for VA and VE or as the activity of addressing the more strategic questions of whether the project should be being done at all, and whether a scheme or its development strategy could be improved. 'Optioneering' is a similar idea, although it lacks the workshop basis (a hangover from the US Government's original three- or five-day workshops). Analysing value improvement opportunities should naturally fit with opportunities to analyse risk. There has been little evidence of this yet, however[26].

In practice, VM was often confused with VE, and VE taken, as in the Andrew project, as a means of reducing capital cost. There was little evidence of the nominator in the quotient, i.e. benefit, being increased. True, the UK Government did require, as of 2008, that all capex projects should include estimates of whole life costs after 10 years' operation, but this is hardly a major endorsement of Value Management.

The trouble with value is that it is a notoriously vague term, not merely subjective in much of its assessment but even having several quite different meanings. To PMI members in 2008–2009, value may have meant 'the value of project management'. For PMI had commissioned a major study to determine this, the question being interpreted in terms largely of Return on Investment. Drawing on a worldwide team of 48 researchers who gathered data on 65 organisations and prepared 18 case studies on how companies value the benefits they derive from using project management, it concluded "that project management delivers value to organizations", but that quantification is very difficult (*sic*). The greater the maturity, the greater the likely return. And culture is very important[27]. The study curiously never seemed to address how project managers could work on their projects to enhance the project's (business) value, as meant by VM.

'Benefits Management' also became an area of strong interest in the early 2000s, at least in the IT/systems area. Arising out of the development of program(me) management in the mid-1990s (see p. 93), benefits are defined as "the measurable improvement resulting from an outcome"[28]. (Value is different: it is a comparative measure.)

Benefits Management involves identifying a 'Benefits Manager' who is responsible for measuring the benefits delivered (or 'harvested') and feeding this information back into the program for incorporation into future 'project tranches'[29]. The practice was largely ignored in construction but confined to 'systems' type program management while, on the other hand, Value Engineering, and VM, was ignored here and confined largely to the hardware-oriented construction and manufacturing industries.

In both cases, however, the focus is also shifting – from the preoccupation with efficiency (the iron triangle) to effectiveness (achieving business benefit), to use the distinction that Miller and Lessard (2000) made.

In an analogous way, Health, Safety and Environment became a central project concern too.

Health, Safety, and Environment

A series of high-profile accidents, not necessarily in projects, mostly in transport (shipping – for example Zeebrugge; rail – Clapham Junction), energy (oil and gas – Piper Alpha) and construction in the 1980s had propelled Health and Safety to be seen even more forcibly as central project criteria not just as important as the traditional iron triangle trio but much more so. 'Duty-of-care' legislation was enacted in several countries in the 1990s and early 2000s, with responsibility lying progressively on the project owner, which strengthened this further.

The ecological (as opposed to the organisational) 'environment' had become recognised in the late 1960s as an important dimension of project management responsibility. This was partly in response to the environmental opposition noted above regarding the nuclear, oil and transport sectors

– 'Whole Earth Day' in 1970 had seen significant opposition to the US SST*. But more formally it (the 'Environment') reflects the growing awareness that society cannot continue to consume resources in a way which in the long term is unsustainable. Although ecologists had been pointing this out for some time (the 1967 BBC Reith Lectures *A Runaway World?* were precisely to this point[30]), it was probably the UN report of the Bruntland Commission of 1987, *Our Common Future*, that really marked the establishment's acceptance of the overriding importance of sustainability – "sustainable development is development that meets the needs of the present without compromising the ability of future generations to meet their own needs"[31]. Hard on the heels of this major shift in the context in which projects needed to be managed came new legislation, e.g. in requiring more detailed Environmental Impact Assessments.

'Environment' became widely tagged in p.m. terms to Health and Safety as HSE (Health, Safety, and Environment). In some sectors HSE became HSSE – Security being added to cover risk of theft or misappropriation (e.g. of data or equipment) or threat to life or property. In any event, HSE/HSSE became inescapably supremely important across a large swathe of project-based industries from the 1980s. Not that there were no longer any HSE incidents. There were. BP's Deep Water disaster in 2010 in the Gulf of Mexico being one of the most notorious. (Its pathology reflecting precisely that of more general project failure: failure to maintain quality when under schedule pressure.)[32]

Figure 5.2   Apollo 11 view of Earth.
Note: This view of Earth was taken by the Apollo 11 crew shortly after lift-off on 16 July 1969 from a distance of 180,000 km. Most of Africa can be seen as well as parts of Europe and Asia.
Source: © Getty images.

---

*Many observers point to the view of Earth as seen from Apollo 11 (Figure 5.2) – the first time it had been photographed as an entire planet from space – had directly stimulated the rise in awareness of the uniqueness, the beauty, the awesomeness, and the fragility of our planet's environment. Rachel Carson's book *Silent Spring* (1962) is also widely acknowledged as an initiator of the environment alist movement.

Deep Water was a failure of technology caused by human error, and sloppy, inappropriate organisation, with disastrous safety and environmental consequences. We were seeing again how organisation and culture influences technology performance. (The US missile programs had learnt it 40 years before). In fact, considerable progress had been made in our understanding of how to manage technology projects, although in that most testing of domains, defence, options are often constrained by operational needs – urgency and new technology again, just as in the 1950s – to cause difficulties.

Defence Projects

Weapons systems are amongst the most demanding of all project types. Typically pushing the boundaries, they are usually wanted urgently and are often also subject to changing constraints from central funding.

This said, the SMART procurement reforms brought in to the UK by the MoD in 2002 (see above, p. 81) seemed largely to be successful. The National Audit Office's reports for the decade following their introduction reveal a 'not bad' verdict, with one or two whoppers (the aircraft carriers and the Typhoon), but otherwise a list of relatively minor issues straight out of these pages: need to beware 'optimism bias', and to improve partnering and risk management.

America's defence projects are an order of magnitude, or several, larger. But the issues are the same and cluster around a number of familiar, largely front-end factors. Thus Steve Meier, reporting on a CIA (Central Intelligence Agency) study published in 2008 based on analysis of 30 major CIA/DoD projects, concluded that *"most unsuccessful programs fail at the beginning. The principal causes of growth on these large-scale programs can be traced to several causes related to overzealous advocacy, immature technology, lack of corporate technology roadmaps, requirements instability, ineffective acquisition strategy, unrealistic program baselines, inadequate systems engineering, and workforce issues"*[33].

Meier makes the point that the findings also apply to IT projects[34]. So . . .

Software Projects and Standish

Not all IT projects are software development projects – many are a mixture of Off-The-Shelf systems, semi-bespoke, and fully bespoke – but they do represent a special class of difficulty.

Software projects had, until the 1990s, been an area where project management had yet to develop effective capability*. Frederick Brooks, formerly

---

*"Software is like entropy. It is difficult to grasp, weighs nothing, and obeys the Second Law of Thermodynamics; i.e. it always increases." Law XVII of *Augustine's Laws*: Norman Augustine (1986) New York: Penguin Books. Augustine was President and CEO of Martin Marietta, so he should know!

of IBM, had written on the challenges of software project management with his *The Mythical Man Month* (1975), first identifying the difficulty of estimating (including 'optimism bias') but also noting the ineluctable need for real (calendar) time: adding resource when things are going wrong can be counterproductive[35].

In 1984, the DoD, alarmed at the poor record of software engineering projects, set-up the Software Engineering Institute at Carnegie-Mellon University from which was to come in 1987 the hugely influential 'Capability Maturity Model' idea and tool. This, as we shall see shortly, was to result in a number of project management analogues (of variable effectiveness).

In the UK, a series of systems development methodologies were promoted in the 1980s (IPSE, PROMPT, SSADM[36]) from which was to come PRINCE – 'Projects in a Controlled Environment' (every project manager's wish) – as a project management methodology in 1990 and PRINCE2 in 1998*. Several 'systems houses', such as CSC, IBM and Logica, had their own in-house methodologies but there was scant sector-wide agreement on how best to approach software or systems projects.

All this should mean that Standish's findings, to which we now turn, should not come as a great surprise.

From 1994, the Standish Group, a private consulting company that benchmarked project management performance of US software projects, produced data which, at face value, were shocking. They showed that (US) software projects have only a 16–32% 'success' rate; that 50–60% were late, over budget and/or had less than the required features and functions (were what they called 'challenged'); while 20–40% 'failed', i.e. were cancelled prior to completion or delivered and never used[37] (Table 5.1).

Causes of these seemingly awful results are quoted by Standish as principally incomplete requirements and specifications, lack of user involvement, lack of resource, unrealistic expectations and lack of executive support. Though Standish is far from alone in reporting such a high rate of failure in software projects[38], the figures have nevertheless been challenged, partly on methodological grounds – the methodology and data are very

Table 5.1 Standish findings, 1994–2009

	1994	1996	1998	2000	2002	2004	2006	2009
Successful	16%	27%	26%	28%	34%	29%	35%	32%
Challenged	53%	33%	46%	49%	51%	53%	46%	44%
Failed	31%	40%	28%	23%	15%	18%	19%	24%

Source: Standish Group (1994–2009).

---

*PROMPT was originally developed in 1975. PROMPTII was adopted by the UK's CCTA (Central Computer and Telecommunications Agency) in 1979, as the default methodology for all UK government information systems projects. In 1989, PRINCE was created from PROMPTII and was made public domain as the UK's default methodology in the early 1990s. IPSE and SSADM are systems development rather than project management methodologies.

opaque and are not critically analysed (see Chapter 18, p. 242) – and partly as a failure of estimating rather than delivery[39]. Software estimating, as we've just seen with Brooks, has always been problematic[40], particularly because of the difficulty many software projects – and indeed systems projects in general – have in defining user and system requirements accurately at the estimating stage[41]. But now something different was to address this: the birth of Agile and its implications to project management. First, however, a word on technology and requirements management.

Technology and Requirements Management

Recklessly pursuing unproven technology and failure to elicit and manage requirements (so they are 'structured, traceable and testable'[42]) lies at the heart of many project difficulties. For 'if it isn't clear what is required don't be disappointed if you don't get it'. Eliciting requirements can be quite hard and may entail establishing what is appropriate. And by the time users can actually see – test-drive – the product, their needs may well have changed. So then either the requirements/scope have to change or the delivered product functionality has to be accepted as not adequate. In either case, innovation and development might be needed to generate the appropriate requirements.

Our knowledge about managing innovation has developed considerably since the 1990s, principally in the use of modelling (virtually and creating physical prototypes), data management (e.g. data mining in pharmaceuticals), and in getting closer to understanding customers' needs[43].

'Technology-Readiness' had become a mainstream system practice across the turn of the century. The US General Accounting Office concluded in 1999 that the DoD took greater risks transitioning emerging technologies at lesser degrees of maturity than private industry did and recommended therefore that the DoD adopt NASA's Technology-Readiness Levels (TRLs) tool as a means of assessing technology maturity prior to transition (see also p. 171)[44].

All this activity needs managing, either at the enterprise or the portfolio, program or project levels. But 'best practice' here is not well documented. Requirements management (engineering), on the other hand, is better understood, having been developed earlier, in the late 1980s to mid-1990s, and is simpler as a subject area[45]. Several systems development models were published – the Waterfall, Spiral, and Vee (see Figures 5.3, 5.4, 5.5)[46] – all emphasising a progression of moving from user, system and business requirements (requirements being solution-free), through specifications, systems design, and build, and then back through mirrored levels of testing (verification and, finally, validation against the original user requirements).

Should project management be responsible for ensuring that requirements are adequately defined? This is one of the defining issues of the project management discipline. In many organisations there is no question but that it is; in others, as in the PMBOK® *Guide, it either is not or, at best, its role is ambiguous.*

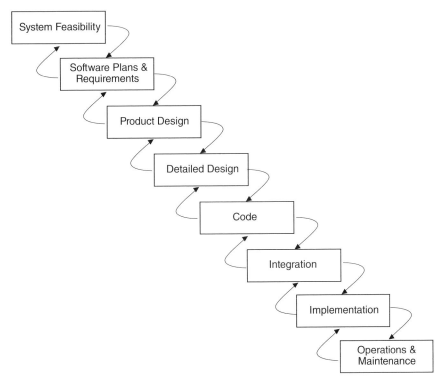

Figure 5.3   The Waterfall model.
Source: © Helms, J. W., Arthur, J. D., Hix, D. and Hartson, H. R. (2006), A field study of the Wheel – A usability engineering process model, *Journal of Systems and Software*, 79, 6, pp. 841–858.

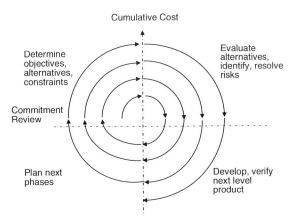

Figure 5.4   The Spiral model.
Source: © Boehm, B.W. (1988), A spiral model of software development and enhancement, *IEEE Computer*, 21, 3, pp. 61–72.

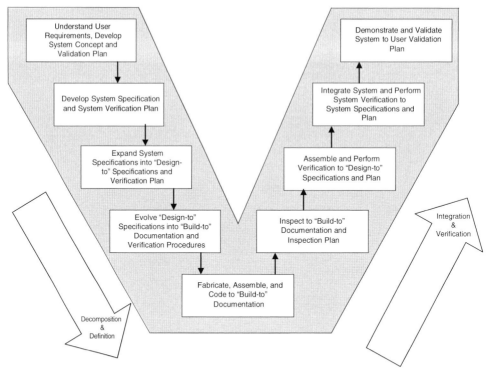

Figure 5.5   The Vee model.
Source: © Forsberg, K., Mooz, H. and Cotterman, H. (2000), *Visualising project management*, Wiley: Hoboken, NJ, p. 36.

Agile Project Management

The difficulty of accurately specifying software projects' requirements and of estimating the resultant time it would take to meet them led, around 2000–2001, to a group of software project managers essentially abandoning the traditional project management paradigm and declaring, in the Agile Manifesto (2001) – see box – the principles of Agile Project Management.

The Agile Manifesto includes:

- Highest priority is to satisfy the customer through early and continuous delivery of valuable software.
- Welcome changing requirements, even late in development.
- Deliver working software frequently, from a couple of weeks to a couple of months, with a preference for the shorter timescale.
- Business people and developers must work together daily throughout the project.
- Build projects around motivated individuals.
- Face-to-face conversations improve efficiency and effectiveness.
- Continuous attention to technical excellence and good design enhances agility.
- The best designs etc. come from self-organising teams.
- Teams should reflect regularly on how to become more efficient and change accordingly.

Agile is based on the premise that because it is so difficult to specify requirements using traditional methods, it would be better if they were specified through close interactions ('coupling') between the user and the software developer; that this user–developer 'team' works only on a selected number of requirements at any one time; and that solutions to these selected requirements should be delivered within a short time frame – a 90-day period is often given as a maximum. In this way, the problem of having to wait for, say, one or two years for solution validation, and then finding that the solutions are inadequate, is avoided. Instead, a very limited degree of functionality is selected and delivered in a very short time. *Delivering* this functionality takes priority, even if it means that more resource is needed (i.e. budget is exceeded), or, less commonly, that schedule, or even scope, is relaxed to maintain the budget (or schedule)[47]. The iron triangle is abandoned! Project management becomes in effect task or work-stream management.

Whether this is project management depends surely on what one means by the term; 'Project Management' is, as this book repeatedly reminds, after all a socially constructed concept. Evidence suggests, however, that by focussing exclusively on short-term tasks there is a danger of losing where one is in the bigger product development life-cycle[48] and that this may weaken value and benefits realisation*. It would seem to fly in the face of what so far has been the *de minimus* definition of project management: 'on time, in budget, to scope' delivery. Nevertheless, PMI in 2011 introduced Agile Project Management Certification†.

But if managing software projects may have raised particularly intractable difficulties for project management, in some respects it has helped it. Substantially.

Information and Communications Technology (ICT)

Microsoft made an enormous contribution to project management with its releases of *MS Project* during the 1990s. It was not alone: *PERT Master* and a number of similar personal computing systems brought project management planning and control tools directly to the project management professional in a user-friendly manner that had been unimaginable previously. Ten years before, project planners were only beginning to move away from punched cards and big mainframes. Whereas in the late 1960s and early 1970s, in the early days of APM and PMI, the majority of attendees

---

*This would not apply to RUP which operates within the product development life-cycle, unlike XP and Scrum which are organised around weekly and quarterly cycles: Leffingwell, D. (2007), *Scaling software agility*, Addison Wesley: Reading, MA.
†By earning the Agile certification, PMI said, practitioners can:

- Demonstrate their level of professionalism in Agile practices of project management.
- Increase their professional versatility in both project management tools and techniques.
- Show they have the capacity to lead basic Agile project teams.

at their conferences and seminars seemed interested primarily in planning and scheduling issues; now most of this could be done via shrink-wrapped software packages.

The early 2000s saw the major project management software suppliers begin providing enterprise-level platforms for use on multiple projects by multiple users; meanwhile, major ERP (Enterprise Resource Planning) suppliers such as HP, Oracle, and SAP began bundling-in specialist project management packages such as Microsoft, PlanView, Primavera, etc.

Simultaneously, another revolution was occurring with an unforeseen but dramatic impact: mobile telephony and the Internet arrived, thereby increasing project communications capabilities prodigiously, and with them project productivity. Modelling power also improved greatly, whether simply through Excel or via the broader efforts of CAD (CAM and CAE) and 4- or 5-D* simulation, asset configuration and concurrent engineering models, or performance management systems such as BIMS (Building Information Management Systems)[49]. Direct design-manufacturing computing technology provided the integration by 'hard-wiring' it between upstream and downstream parts of the development project structurally rather than necessarily via people (agents) such as project managers (see pp. 255–6).

Developments in the discipline of scheduling were not yet quite a thing of the past, however.

Critical Chain

One genuinely new and original development in scheduling that now appeared was Critical Chain, promoted by Eliyahu M. Goldratt from his 'Theory of Constraints' around 1996–1997[50]. Critical Chain Project Management emphasises the resources required to execute project tasks. A Critical Chain project network will tend to keep the resources loaded levelly, but will require them to be flexible in their start times and to be ready to switch quickly between tasks to keep the whole project on schedule. Key ideas include stripping contingencies from the activity level and managing them, as buffers, at the project level; and only working on one activity at a time, and doing so as fast as possible. But surely multitasking is sometimes unavoidable? And the practicality of scheduling first to resources and only second to task logic must often be questionable. Nevertheless, several commentators claim that CCPM (Critical Chain Project Management) can produce better results than traditional approaches to the management of projects[51].

One of the most interesting benefits of the approach seems to be the behavioural, motivational energy that implementing these ideas appears to develop, which of course is part of Goldratt's philosophy.

Developments, meanwhile, were happening at the other end of the spectrum: program management.

---

*CAD is Computer-Aided Design, CAM is Computer-Aided Manufacturing, and CAE is Computer-Aided Engineering.

Program Management

During the 1990s program(me) management* had begun receiving increased attention[52]. Where previously it had been treated (in USAF systems management terms†) more as heavyweight project management[53], and later by Wheelwright and Clark in terms of technology platforms[54], it now appeared more as about managing change, achieving outcomes and harvesting benefits: more obviously strategic, with a stronger change management orientation in distinction to project management's more 'nitty-gritty' world of execution[55]. This conception, captured by the phrase "programmes deal with outcomes, projects deal with outputs"[56], as promoted by PMI, APM, the UK Office of Government Commerce (OGC) and others (surprisingly, because the intellectual case for doing so was not a strong one), reconfirms the view of project management as execution management. But surely every project consumes resources and should be assessable in terms of its Return on Investment? Don't projects as well as programs produce outcomes and benefits? Is it not the job of project management to achieve benefits too?

This said, the rise of program management definitely reflected a growing awareness of the need for a different order of management when dealing with complex programs involving separate projects, whether enterprise specific (e.g. an organisational move, implementing change, etc.), developing major hardware (new weapons systems, urban development), scientific development (families of drugs), or addressing large-scale social change (meeting climate change requirements).

The New Accommodation Programme (NAP)

The re-morphing of the UK's Government Communications Headquarters (GCHQ), the government's intercept, code-breaking and communications centre, involving moving to a new set of facilities, embracing new ways of working, and the adoption of new systems, is an excellent example of

---

*The *Oxford English Dictionary* gives the first spelling of program with one 'm', noting that the French ending was introduced in the UK in the 19th century and stating that the single 'm' ending "is preferable as conforming with the usual representation of the Greek -γράμμα as in anagram, cryptogram, diagram, telegram". America, under *Webster*'s guidance, attempted to have spellings which were more phonetic with less redundant letters, as with program[me], but there has been much confusion and trading of what various dictionaries allow (see Mencken, H.L. [1937] *The American language*, Alfred A. Knopf: New York).

†Program Management, as a term, has been around since the beginning of the modern discipline of p.m. – we talk about the Apollo program for example – but not really as a distinct discipline. There is no evidence of hands-on program management in that other giant US public sector program of the time, the $27 billion Federal Highways Program of 1956–1991, other than in obtaining funding approval and supra 'system' level issues such as signage and overall design philosophy. See McNichol, D. (2006) *The roads that built America: The incredible story of the U.S. interstate system*, New York: Sterling.

the new systems approach to program management, illustrating many of the themes discussed in this chapter.

The NAP requirement was for over 4,000 staff to be moved along with hundreds of IT systems, all of which had to be kept fully operational all the time. Alongside the physical move, GCHQ management were determined to introduce new ways of working, including new processes and changes in behaviour.

The initial 1997 budget and plan were considered unrealistically generous and a tougher program management regime was injected and was in place by 2000 when the program was given ministerial approval. The program was to be managed as 'whole business change' program under the command of a 'single-point' Programme Director. 25% of NAP's budget was dedicated to program management activity, including procurement. A rolling five-year strategic 'blueprint' ensured alignment of goals was maintained.

The program was one of the earliest and largest PFI procurements of its time in Europe. The winning bid (of 18 tenders) was awarded on the basis of financial value and design innovation: funding needs were reduced by selling-off some GCHQ land for development; the architectural design was original, functional and attractive. Value Engineering was applied to enhance functionality while maintaining budget. The program management team managed the consortium contractor proactively on a partnering basis.

Over 100 user requirements were identified. These were linked to benefits, acceptance criteria and delivery milestones. A Business Change Manager oversaw this process and became the formal acceptance authority. Forty expected benefits were defined together with 20 dis-benefits. A benefits realisation plan was developed early in the program. The benefits gave rise to work packages.

Systems engineering was used to design, deliver and test the business systems and processes and was key to establishing their scope, based on the formally elicited requirements, benefits and schedule. A Design Assurance team verified that projects were delivered against an agreed design and in doing so identified and eliminated over 700 issues. A Systems Integration Authority combined responsibility for the building, the business processes and supporting technical services as one working entity.

Ultimately the £308 million program was completed in 2003 'on time, in budget, to specification'. "[For] many the move represented an opportunity to review and improve working methods. To maximise the benefits offered change had to be managed; but this was an area which methodologies such as PRINCE2 do not comprehensively address. GCHQ overcame this by forming a very strong change network."[57]

Developing Enterprise-Wide p.m. Capability: The US Department of Energy (DoE)/NRC Study

Things weren't going so well everywhere, however. The US Department of Energy's annual capital expenditure budget was, by the turn of the century, about $17bn a year. Yet its project management record was amazingly poor:

a 1996 study by the US General Accounting Office for example found that between 1980 and 1996, 31 of its major system acquisition projects had been terminated prior to completion, while 34 were continuing although over budget[58]. Cost overruns were running about 20–50%. Reasons were many: lack of a project management focus and culture, insufficient attention to front-end definition, no Value Engineering, poor scope management, inadequate estimating, weak contracting practices, and lack of training. By now, none of this may seem surprising. The special interest to this account of the growth of our knowledge of the discipline is the 1999 study by the National Research Council (NRC), a body within the United States National Academy of Sciences, of DoE project management and particularly Appendix C of the NRC report which attempted to summarise the "characteristics of successful megaprojects or systems acquisitions"[59]. Appendix 2 of this book gives a slightly edited version of these characteristics (success factors).

The NRC study published in the mid-1990s is worth dwelling on here not so much for its findings as for its methodology and attention to implementation in the enterprise – or rather, absence of consideration of either.

- The NRC Appendix C list is based on no scientifically credible methodology. It is in fact "based on the collective experience of more than a dozen highly knowledgeable professionals with experience in large-scale projects."[60] We shall be stressing in Chapter 18 the need for methodological rigour in the study of projects and their management. This NRC study is a prime example.
- The study also totally misses any discussion of implementation. It just assumes that its recommendations can be implemented. Having critiqued DoE and offered some characteristics associated with success and failure, what does the NRC recommend be done? Essentially the answer is, install a PMO – a Project Management Office – and create a project management Centre of Excellence. The challenges the PMO will have in trying to change DoE's vast, semi-contracted out p.m. culture are not mentioned at all. It's rather like showing a picture of a beautiful garden and saying 'create this'. In reality, 'doing this' is a whole skill area in itself – design the garden in its context (soil, light, vistas, mass, blossoming periods, colour, etc.), plant it, and nurture and maintain it. Creating and sustaining a project management culture in DoE would similarly require thought and care, but all this is simply ignored.

Implementing project and program management across the enterprise, not least via PMOs, was, as it happened, now beginning to be a major focus in the evolution of the discipline, as we shall now see.

References and Endnotes

[1] Miller, R. and Lessard, D. R. (2000), *The strategic management of large engineering projects*, MIT Press: Cambridge; Flyvbjerg, B., Bruzelius, N. and Rothengatter, W. (2003), *Megaprojects and risk: An anatomy of ambition*, Cambridge University Press: Cambridge; Meier, S. R. (2008), Best project management and systems

engineering practices in pre-acquisition practices in the federal intelligence and defense agencies, *Project Management Journal*, 39, 1, pp. 59–71; The Standish Group (1994), *The CHAOS Report*, http://www.standishgroup.com.

2 Womack, J. R., Jones, D. T. and Roos, D. (1990), *The machine that changed the world*, Macmillan International: New York.

3 Miller, R. and Lessard, D. R. (2000), *op. cit.*: 1.

4 *Ibid.*: 14–15.

5 *Ibid.*: 151–154

6 *Ibid.*: 31–34

7 Marschak, T., Glennan, T. K. and Summers, R. (1967), *Strategy for R&D: Studies in the microeconomics of development*, Springer-Verlag: New York.

8 See http://www.contruction-institute.org

9 Macneil, I.R. (1980), *The new social contract: An enquiry into modern contractual relations*, Yale University Press: New Haven, CT; Campbell, D. (2001), *The relational theory of contract: Selected works of Ian Macneil*, Sweet & Maxwell: London.

10 Hellard, R. B. (1993), *Project partnering*, Thomas Telford: London, p. 36.

11 Knott, T. (1996), *No business as usual*, BP: London, p. 143.

12 Prokesch S.E. (1997), Unleashing the power of learning: An interview with British Petroleum's John Browne, *Harvard Business Review*, 75,5, pp. 146–68

13 NAO (2001), *Ministry of Defence: Non-Competitive Procurement in the Ministry of Defence*, HC: 290 2001–2002, TSO: London.

14 Morris, P. W. G. (1994), *The management of projects*, Thomas Telford: London, p. 18.

15 Winch, G. M. (2010), *Managing construction projects*, Wiley-Blackwell: Chichester; Morris, P. W. G. (1994), *op. cit.*: 14.

16 Minutes of House of Commons Defence Committee, Session 1987/88.

17 Jordan, L. C. (1988), *Learning from Experience*, Report to the Minister of State for Defence Procurement, TSO: London.

18 Institution of Civil Engineers (1998), *RAMP: Risk analysis and management for projects*, Institution of Civil Engineers and Institute of Actuaries: London.

19 Simon, P., Hillson, D. and Newland, K. (1997), *Project risk analysis and management (PRAM)*, Association for Project Management, High Wycombe; Office of Government Commerce (2002), *Management of risk: Guidance for practitioners*, TSO: London.

20 Chapman, C. and Ward, S. (2011), *How to manage project opportunity and risk*, Wiley: Chichester.

21 Flyvbjerg, B. *et al.* (2003), *op. cit.*: 1.

22 Flyvbjerg, B., Garbuio, M., & Lovallo, D. (2009). Delusion and deception in large infrastructure projects: Two models for explaining and preventing executive disaster, *California Management Review*, 51, 2, pp. 170–193.

23 Wachs, M. (1986), Technique vs. advocacy in forecasting: A study of rail rapid transit, *Urban Resources*, 4, 1, pp. 23–30.

24 Lavallo, D. and Kahneman, D. (2003), Delusions of success: How optimism undermines executives' decisions, *Harvard Business Review*, July, pp. 56–63; Kahneman, D. and Lovallo, D. (1993), Timid choices and bold forecasts: A cognitive perspective on risk taking, *Management Science*, 39, pp. 17–31.

25 Thiry, M. (2004), Value management, In: Morris, P. W. G. and Pinto, J. K. (eds.) *The Wiley guide to managing projects*, Wiley: Hoboken, NJ.

26 Dallas, M. F. (2006), *Value and risk management: A guide to best practice*, Blackwell, Oxford.

27 Thomas, T. and Mullaly, M. (2008), *Researching the value of project management*, PMI: Newton Square, PA.
28 Office of Government Commerce (2003), *Managing successful programmes*, TSO: Norwich.
29 *Ibid.*
30 Leach, E. (1967), *A runaway world?* BBC Publications: London.
31 Brundtland, G.O. (1987), *Our Common Future (The Brundtland Report)*, The World Commission on Environment and Development.
32 BP (2010), *Deep water horizon accident investigation report*, BP: UK.
33 Meier, S.R. (2008), *op. cit.*: 1.
34 General Accountability Office (2006), *Improvements needed to more accurately identify and better oversee risky projects totaling billions of dollars*, GAO-06-1099; General Accountability Office (2006), *Assessment of major weapons systems*, GAO-06-391.
35 Brooks, F. (1975), *The mythical man-month*, Addison-Wesley: Reading, MA.
36 Office of Government Commerce (2000), *SSADM foundation: Business systems development with SSADM*, TSO, London.
37 Standish (1994): http://www.cs.nmt.edu/~cs328/reading/Standish.pdf
38 Brown, A.D. and Jones, M. R. (1998), *Doomed to failure: Narratives of inevitability and conspiracy in a failed IS project*, European Group for Organisational Studies; Fincham, R. (2002), Narratives of success and failure in systems development, *British Journal of Management*; Sauer, C. (1993), *Why information systems fail*, Alfred Toller: Henley-on-Thames; Waterdige, J. (1998), How can IS/IT projects be measured for success? *International Journal of Project Management*, 16, 1, pp. 59–63; Yardley, D. (2002), *Successful IT project delivery: Learning the lessons of project failure*, Addison Wesley: Reading Mass.
39 Jørgensen, M. and Moløkken-Østvold, K. J. (2006), How large are software cost overruns? Critical comments on the Standish Group's CHAOS reports, *Information and Software Technology*, 48, 4, pp. 297–301.
40 Boehm, B. W. (1981), *Software engineering economics*, Prentice Hall: Englewood Cliffs, NJ.
41 Davis, A.M., Hickey, A.M. and Zweig, A.S. (2004), Requirements management in a project management context, in: Morris, P. W. G. and Pinto, J. K. (eds.) *The Wiley guide to managing projects*, Wiley: Hoboken, NJ; Stevens, R., Brook, P., Jackson, K. and Arnold, S. (1998), *Systems engineering: Coping with complexity*, Prentice Hall: Hemel Hempstead.
42 Stevens *et al.* (1998), *op. cit.*: 41.
43 Dodgson, M., Gann, D and Salter, S. (2005), *Think, play, do*, Oxford University Press: Oxford; Thomke, S. (2003), *Experimentation matters*, Harvard Business School Press: Cambridge, MA.
44 General Accounting Office (1999), *Best Practices: Better Management of Technology Can Improve Weapon System Outcomes*, GAO/NSIAD-99-162.
45 Stevens *et al.* (1998), *op. cit.*: 41; Davis A.M. *et al.* (2004), *op. cit.*: 41;
46 Forsberg, K., Mooz, H. and Cotterman, H. (1996), *Visualizing project management*, John Wiley and Sons: New York.
47 Chan, F. K. Y., & Thong, J. Y. L. (2009), Acceptance of Agile methodologies: A critical review and conceptual framework. *Decision Support Systems*, 46, 4, pp. 803–814; Leffingwell, D. (2007), *Scaling software agility*, Addison Wesley: Upper Saddle River, NJ; Kruchten, P. and Kroll, P. (2003), *The rational unified process made easy: A practitioner's guide to the RUP*, Addison-Wesley: Reading. MA; Meso, P. &. Jain, R. (2006), Agile software development: Adaptive systems

principles and best practices, *Information Systems Management*, 23, 3, pp. 19–30.

[48] Dachler, D. (2008), Beyond Agile project management: The way forward, *Cutter IT Journal*, 21, 5, pp. 28–34.

[49] Eastman, C., Teicholz, P., Sacks, R., and Liston, K. (2011), *BIM handbook: A guide to building information modeling for owners, managers, designers, engineers, and contractors*, 2nd edition, Wiley: Hoboken, NJ.

[50] Goldratt, E. M. (1990), *Theory of constraints*, North River Press: Great Barrington, MA; Goldratt, E. M. (1997), *Critical chain*, North River Press: Great Barrington, MA; Leach, L. P. (2004), Critical chain project management, in: Morris, P. W. G. and Pinto, J. K. (eds.) *op. cit.*: 41; Raz, T., Barnes, R. and Dvir, D. (2003), A critical look at critical chain project management, *Project Management Journal*, 34, 4, pp. 24–32.

[51] Balderstone, S. J. and Mabin, V. J. (1998), A review of Goldratt's theory of constraints (TOC): Lessons from the international literature, In: 33rd Annual Conference of the Operational Research Society of New Zealand, Auckland, New Zealand.

[52] Ferns, D. C. (1991), Developments in programme management, *International Journal of Project Management*, 9, pp. 148–156; Artto, K., Martinsuo, M., Gemünden, H. G. and Murtoroa, J. (2009), Foundations of program management: A bibliometric view, *International Journal of Project Management*, 27, 1, pp. 1–18; Pellegrinelli, S. (2002), Shaping context: The role and challenge for programmes, *International Journal for Project Management*, 20, 3, pp. 229–233.

[53] Baumgartner, J. S. (1979), *Systems management*, Bureau of National Affairs: Washington, DC; Sapolsky, H. (1972), *The Polaris system development: Bureaucratic and programmatic success in government*, Harvard University Press: Cambridge, MA.

[54] Wheelwright, S. C. and Clark, K. B. (1992), *Revolutionizing product development*, Harvard Business School Press: Cambridge, MA.

[55] Lycett, M., Rassau, A., and Danson, J. (2004), Programme management: A critical review. *International Journal of Project Management*, 22, 4, pp. 289–299; Pellegrinelli, S., Partington, D. and Geraldi, J. (2010), Program management: An emerging opportunity for research and scholarship, in: Morris, P. W. G., Pinto, J. K. and Söderlund, J. (eds.) *The oxford handbook of project management*, Oxford University Press: Oxford; Pellegrinelli, S. (2011), What's in a name? *International Journal for Project Management*, 29, 2, pp. 232–240.

[56] Office of Government Commerce (2003), *Managing successful programmes*, TSO: Norwich.

[57] NAO (2003), *Government Communications Headquarters (GCHQ): New accommodation programme*, TSO: Norwich.

[58] GAO (1996), *Department of Energy: Opportunities to Improve Management of Major System Acquisitions. Report to the Chairman, Committee on Governmental Affairs, U.S. Senate. GAO/RCED-97-17. Washington, DC: Government Printing Office.

[59] National Research Council (1999), *Improving project management in the Department of Energy*, National Academy of Sciences: Washington, DC.

[60] *Ibid.*: 106.

Enterprise-Wide Project Management (EWPM)

The expansion of project management software from single-project to enterprise-level applications plus the impact of a number of sector-wide initiatives to improve project performance such as CRINE and ACTIVE, initiatives such as CII and ECI, and the work of the OGC, paralleled, and in some instances stimulated, an increased awareness that project management needed to be applied and managed at the enterprise level. The result, 'enterprise-wide project management', around the late 1990s and onwards, opened a large field of development ranging across functions such as staff planning, training and development, maturity assessment, and organisational learning. Competence and seeing project management as a career track receive more sustained attention. In addition, there was a progressive acknowledgement of the need to link the enterprise's, and the project's, governance and strategy. This sponsor-project strategic linkage is crucial and we'll begin the chapter with this.

Strategy and Governance

The early 2000s saw a growing interest in project strategy alongside the rise in interest in program management. Most project managers would encounter project strategy first in the 'Project Execution Plan', but logically this is the third of three or four stages of project strategy formulation. First is the sponsor's strategy; then the overall project development strategy, prepared early in the front-end; then the Project Execution Plan (the PEP); and finally, the Commissioning/Operations strategy. Researchers such as Karlos Artto, Christophe Loch and I began, independently, articulating the importance of voicing the project strategy around 2005, but the thinking was not mainstream. Strategy failed to appear in the standard *PMBOK® Guide* view of project management, instead appearing only in the new PMI standard on Program Management of 2006, the implication being that strategy is the prerogative of program management[1].

Reconstructing Project Management, First Edition. Peter W.G. Morris.
© 2013 John Wiley & Sons, Ltd. Published 2013 by John Wiley & Sons, Ltd.

Not only was the 'intra-project' responsibility and practice with regard to strategy fractured and often unclear, the responsibilities of those 'outside' the project but involved in its implementation also came to be seen as needing clarification. Particularly important is the sponsor. How, for example, does the project strategy abut and reflect the sponsor's or other stakeholders' strategy(ies)?

The role of the sponsor in establishing the project goals is key, as we've seen amply and extensively demonstrated in this account so far. Really, it's more than this: the sponsor establishes the tone, the culture, of the project, the way project management practices are respected and followed.

The early 2000s saw a growing recognition of the importance of project governance, not least as a result of instances of high-profile corporate malfeasance – the collapse of Enron and MCI WorldCom in 2001–2002 – and the legislation and corporate action which followed (Sarbanes–Oxley etc.). APM published recommendations on project governance in 2004, listing such principles as proper alignment between business strategy and the project plan (plan, note: no mention of project strategy); transparent reporting of status and risk; and periodic third-party 'assurance' reviews (and no mention of ethics)[2]. The late 1990s and early 2000s saw rising application of stage-gate reviews, peer reviews and peer assists as governance mechanisms.

PMOs

How is it that some projects or programs are done really well – the Hoover Dam, for example – but then the institutions behind them fail to perform consistently afterwards, as in the case of DOE? It can only be lack of focus and commitment, of capability and competence. The challenge for enterprises wishing to develop and deliver projects well, consistently, is just this. And the body that should take responsibility for doing this has increasingly been the PMO – the Project/Program Management Office.

PMOs began to be seen in the late 1990s/early 2000s as the enterprise group responsible for developing, maintaining and assuring p.m. capability and competence across the organisation.

Originally used as a reporting hub, the PMO moved first into providing 'p.m. first aid' where things were spotted as going wrong, then began to take responsibility for a number of EWPM functions. For example, for defining the enterprise's version(s) of best practice, organising training and support, initiating project review, and trying to ensure lessons-learned are fed into the enterprise[3]. Learning and development and tools such as methodologies and maturity models became increasingly popular now as means of improving p.m. performance across the enterprise.

Best Practice Guidelines and Maturity

Miller and Lessard had questioned whether it even made sense to try and define 'best practice' in project management; such is the effect of context and project type. Yet practitioners are bound to want to know how their practice compares with others' versions of good practice and to look at how improved practice could be implemented. Thus, in the United Kingdom, for example, the government put enormous effort, via its Office of Government Commerce, into defining Project, Programme and Risk Management recommended practices[4]. These are excellent documents but, as thousands became certificated, as, say, 'PRINCE2 Practitioners', the danger grew of people believing that passing a test after a four-day course meant being qualified as a competent project manager. The net results were indeed to 'spread the word'[5] but also perhaps to commodify the discipline.

Attempts to normalise standards of project management took a new and potentially exciting turn in the mid-2000s with the promotion of project management maturity assessment tools. The idea of assessing the level of an organisation's management maturity was first floated by Phil Crosby in his book *Quality is Free*[6]. DOD adopted the idea in a piece of work done jointly with the Software Engineering Institute (SEI) which, as we saw previously, it had established at Carnegie Mellon University in 1984, aimed at helping them assess the capability of potential contractors. The resulting Capability Maturity Model (CMM) for software engineering was released in 1991[7]. It proved popular, effective, and attractive, and stimulated the thought, if it works for software engineering, why not for project management? And so, a decade or more later, we saw first PMI and then OGC attempt to produce tools designed to do just this: to enable the levels of project management capability in an organisation to be measured[8]. (Several private sector maturity assessment tools were also released.) PMI issued its version of a maturity model, OPM3® – the Organizational Project Management Maturity Model – in 2003. It proved extremely complicated, however, and attracted considerable criticism. A second edition was released in 2008.

In 2006, OGC issued P3M3 – Portfolio, Programme and Project Management Maturity Model. This focuses on seven 'Process Perspectives': Management Control, Benefits Management, Financial Management, Stakeholder Engagement, Risk Management, Organizational Governance, and Resource Management.

The model is designed to work equally on program management and portfolio management – and the choice of Process Perspectives reflects the emphases of these (organisationally) higher p.m. areas. (A distinctly program management list!) Sadly, overall the sense is that P3M3 is unrealistically simple, missing completely many important p.m. functions, including several topics present in the APM BOK, not least nearly all the 'people' topics as well as Quality Management, Information Management and nearly everything to do with Procurement and Contracting[9] (Figure 6.1).

Academics still demurred and equivocated, however (and rightly so!). A research program, *Rethinking Project Management*, was conducted in 2004–2006, representing leading academics and practitioners in Europe and North America to reflect on the project management research agenda. Themes arising out of the exercise emphasised[10]:

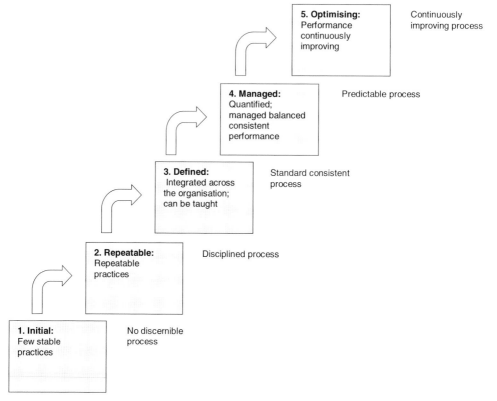

Figure 6.1   SEI Capability Maturity Model.
Source: Reproduced with permission from Paulk, M., Weber, C., Garcia, S., Chrissis, M. and Bush, M. (1993), Key Practices of the Capability Maturity Model, Version 1.1, *Technical Report, CMU/SEI-93-TR-025, ESC-TR-93-178*, Software Engineering Institute (SEI), Carnegie Mellon University, Pittsburgh, Pennsylvania.

- the complexity of projects and project management;
- the importance of social interaction among people;
- value creation (but noting that 'value' and 'benefit' have multiple meanings linked to different purposes);
- seeing projects as "multidisciplinary, having multiple purposes, not always pre-defined"; and
- the development of "reflective practitioners who can learn, operate and adapt effectively in complex project environments, through experience, intuition and the pragmatic application of theory in practice".

Some were even more dubious (say cautious) of its pretensions, as we'll now see.

Critical Management

Academic indignation at the seemingly mechanistic, functionalist, unreflective nature of much of the practical project management guidance now being pushed out, not least as epitomised in the proclamation of universal – 'global' in PMI's case – standards, began being voiced in a serious and sustained manner from around 2005, not least via a 'critical management'-driven discussion of projects and project management* led by Damian Hodgson of the Manchester Business School and Svetlana Cicmil of the University of the West of England[11].

Critical theorists, with their interests in how intellectual and personal drivers influence the way we shape practice and frame knowledge, questioned a number of features of the way project management as a discipline (if such, they wondered, it was) is articulated, promoted, and, dare one say, regulated. For example, the role of power and agenda in setting standards (Foucault). Or the conflict inherent in attempts to codify practice given the transitory character of much that is the essence of managing projects – projects as social constructs, 'invented not found', whose vision has to be created, communicated and then realised – building 'the future perfect' (Clegg); project and particularly program management as an activity more about 'becoming' than 'being' (Chia), creating than created[12]. (See Chapter 18 for more on these ontological and epistemological issues.) Attempts to reify knowledge as standards with little or no real adjustment according to context were obviously dangerous.

Yet the drive for help continued. The year 2011 saw the International Standards Organization (ISO) working hard on producing a revised international standard on project management, wrestling with the broader, essentially 'management of projects' scope taken by BS 6079[13] and the interests of the PMI delegates to ensure the new standard reflected the *PMBOK® Guide*. The resultant ISO standard 21500[14] covers projects, programs and portfolios and their management in a very summary manner and then treats benefits, governance, stakeholders, competencies, organisation, the life-cycle, and constraints. It then goes into PMBOK's p.m. processes (scope, time, resources, risk, etc.) and concludes with a short section on procurement (nothing on contracting strategy).

Meanwhile, an *ad hoc* grouping called the Global Alliance for Project Performance Standards (GAPPS) was working to develop globally applicable project management competency-based standards, frameworks and mappings independent of the big project management professional bodies[15].

The more sophisticated recognised the limits of such guidance. Project management knowledge is not context free. No management knowledge

---

*Critical Management is characterised by Hodgson and Cicmil (2006) in *Making projects critical*, Palgrave Mcmillan: London, p. 12, as being focussing on "who is included in and who is excluded from the decision-making process, analysing what determines the position, agendas and power of different participants, and how these different agendas are combined and resolved in the process by which decisions are arrived at". It is probably also fair to say that this is a description that would be shared by many others in project management who are sympathetic to Critical Management. However, it does not stand so well when compared with the broader definition given by Tsoukas and Knudsen in the *Oxford handbook of organization theory* (2003) and as quoted in the footnote on p. 238.

is[16]. Thus, respect for context and care in tailoring and application argued for giving more personal support to employees in learning and applying project management practice. Significant efforts in training and coaching to supplement more normative guidance were invested in by several large (and not so large) enterprises. Ramming knowledge in over a three- or five-day course was a start, maybe, but was far from adequate. Certification by PMI, OGC, IPMA and similar bodies became popular as a mark of competence (which, often, it was not*). In the UK, the Government established a 'Major Projects Authority' (MPA) in 2011, an institution focussed largely around project and program sponsors aimed at improving their surety of delivery.

Both the MPA and several companies (in oil and gas, aerospace, ICT, transport and construction) sought strategic relations with universities to create 'project academies' – educational programs involving research as well as executive education. Ensuring that the enterprise had the competencies required to manage its current and upcoming portfolio of projects and programs became increasingly recognised as one of *the* major challenges. Developing project leadership[17] and improving 'Emotional Intelligence'[18] took on a higher profile. And in particular, making better use of existing corporate knowledge and learning from project experiences received considerable attention.

Learning and Development

Learning lessons on 'how we did' on projects and programs has a long and noble track record. NASA put considerable effort into drawing lessons from its operations, and still does[19]. Sir Alistair Frame noted it as part of Rio Tinto's standard project management practice (see Chapter 3, p. 48). Wheelwright and Clark had it as an essential element in new product development. BP saw it as an integral part of its Alliancing philosophy. The 1990s, however, saw a particular rise in interest in knowledge management and organisational learning (see Chapter 16, pp. 220–3). Partly this was due to the rising capabilities of ICT tools in this area, notably the Web and personal computing, partly due to the growing sophistication of our understanding of organisational learning with key ideas from researchers such as Chris Argyris and Donald Schön with double-loop learning; Schön with 'the reflective practitioner'; Ikujiro Nonaka (following Michael Polyani), somewhat controversially, with tacit and explicit knowledge interaction; Karl Weick with sense making; and Etienne Wenger (and Jean Lave) with communities of practice (see Chapter 16, p. 222) and Gernot Grabher with project ecologies[20] (see Chapter 16, p. 247).

Projects were seen both as attractive vehicles for generating new knowledge[21] and, simultaneously, given their unique, temporary nature, as especially difficult challenges for organisational learning. While tools and techniques

---

*Because it ignored experience and/or was not focussed on the real requirements of individual project manager's roles (and/or was insufficiently testing).

might help, the consensus was that the real opportunity, and challenge, lies in leveraging tacit knowledge. Communities of Practice, peer assists and peer reviews, and project-based learning reviews became more common. The prize, as Chris Collison and Geoff Parcell of BP put it, is to find out how sharing behaviours, following smart processes, and leveraging technology can develop individual and organisational competency[22].

Project Management as a Career Track

Project management, meanwhile, became increasingly seen as a core competency, recognised within, and across, institutions as a career track in its own right. Demand outstripped supply. Recruitment and career development became increasingly important as well as competency uprating. National Vocational Qualification programs were introduced in Australia and the UK in the mid-1990s. University degrees in project management sprang up seemingly in their dozens.

PMI's membership continued to grow and grow (it had half a million members and credential holders in 185 countries by 2010), driven by excellent professional conferences, seminars and other events, very professional communications and marketing, and the success of its certification programs. In the UK, APM was the fastest growing of all the UK's professional institutions throughout the 1990s and 2000s, and in 2008–2009 applied for 'chartered professional association' status to put it alongside such established professions as engineering and medicine.

Which brings us approximately up to date (as of the time of writing, anyway). Where are we, then, in the construction of the thing we call project management?

References and Endnotes

[1] Artto, K., Martinsuo, M., Dietrich, P. and Kujala, J (2008), Project strategy: Strategy types and their contents in innovation projects, *International Journal of Managing Projects in Business*, 1, 1, pp. 49–70; Loch, C. and Kavadias, S. (2010), Implementing strategy through projects, in Morris, P. W. G., Pinto, J. K. and Söderlund, J. (eds.) *The Oxford handbook of project management*, Oxford University Press: Oxford; Morris, P. W. G. and Jamieson, H. A. (2004), *Translating corporate strategy into project strategy*, Project Management Institute: Newton Square, PA.

[2] Association for Project Management (2004), *Directing change. A guide to governance of project management*, APM: High Wycombe.

[3] Hobbs, B. and Aubry, M. (2008), An empirically grounded search for a typology of project management offices, *Project Management Journal*, 39, Supplement: S69–S82.

[4] Office of Government Commerce (2002), *Managing successful projects with PRINCE 2*, TSO: Norwich; Office of Government Commerce (2002), *Management of risk: Guidance for practitioners*, TSO, Norwich.

[5] Pellegrinelli *et al.* found that guidance on program management recommended by *Managing Successful Programmes* was often ignored and was more frequently shaped by context. (A result which is not particularly surprising, surely.) Pellegrinelli, S., Hemingway, C., Mohdzain, Z., Shah, H. and Stenning, V. (2007), The importance of context in program management: An empirical review of program practices, *International Journal of Project Management*, 25, pp. 41–45.

[6] Crosby, P. (1979), *Quality is free*, McGraw-Hill: New York.

[7] Humphrey, W. (1988), Characterizing the software process: A maturity frame-work, *IEEE Software*, 5, 2, pp. 73–79; Humphrey, W. (1989), *Managing the software process*, Addison Wesley: Boston; Paulk, M., Weber, C., Garcia, S., Chrissis, M. and Bush, M. (1993), Key Practices of the Capability Maturity Model, Version 1.1, *Technical Report, CMU/SEI-93-TR-025, ESC-TR-93-178*, Software Engineering Institute (SEI), Carnegie Mellon University, Pittsburgh, Pennsylvania; Paulk, M., Weber, C., Curtis, B. and Chrissis, M. (1995), *The Capability Maturity Model: Guidelines for improving the software process*, Addison Wesley: Boston.

[8] PMI (2003), *Organizational Project Management Maturity Model (OPM3)*, Project Management Institute: Newton Square, Pennsylvania; Office of Government Commerce (2006), *Portfolio, Programme and Project Management Maturity Model (P3M3®)*, Office of Government Commerce, London.

[9] Office of Government Commerce (2003), *op. cit.*: 5.

[10] Winter M., Smith, C, Morris, P. W. G. and Cicmil, S. (2006), Directions for future research in project management: The main findings of a UK government-funded research network, *International Journal of Project Management*, 24, 8, pp. 638–649.

[11] Hodgson, D. and Cicmil, S. (2006), *Making projects critical*, Palgrave Mcmillan: London.

[12] Chia, R. (1995), From modern to postmodern organizational analysis, *Organization Studies*, 16, 4, pp. 579–604; Tsoukas, H. and Chia, R. (2002), On organizational becoming: Rethinking organizational change, *Organization Science*, 13, 5, pp. 567–582; Thomas, J. (2006), Problematizing project management, in: Hodgson, D. and Cicmil, S. (eds.), *Making projects critical*, Palgrave Mcmillan: London; Clegg, S., Pitsis, T., Marosszeky, M. and Rura-Polley, T. (2006), Making the future perfect: Constructing the Olympic dream, In: Hodgson, D. and Cicmil, S. (eds.) *ibid.*; Pellegrinelli, S., Partington, D. and Geraldi, J. (2011), in Morris, P. W. G., Pinto, J. K. and Söderlund, J. (eds.) *op. cit.*: 1; Pellegrinelli, S. (2011), *op. cit.*: 5, p. 192.

[13] BS 6079 (1996), *Project management: Principles and guidelines for the management of projects*, BSI: London.

[14] BS ISO 21500 (2011), *Guidance on project management*, BSI: London.

[15] GAPPS: http://www.globalpmstandards.org

[16] Griseri, P. (2002), *Management knowledge: A critical view*, Palgrave Macmillan: Basingstoke.

[17] Gadeken, O. C. (1997), Project managers as leaders, *Army R&DA*, January–February; Keller, R. T. (2006), Transformational leadership, initiating structure, and substitutes for leadership: A longitudinal study of research and development project team performance, *Journal of Applied Psychology*, 91, 1, pp. 202–210; Lindgren, M. and Packendorff, J. (2009), Project leadership revisited: Towards distributed leadership perspectives in project research, *International Journal of Project Organisation and Management*, 1, 3, pp. 285–308; Müller, R., Geraldi, J. and Turner, R. (2012), Relationships between leadership and success in different

types of project complexities, *IEEE Transactions on Engineering Management*, 59, 1, pp. 77–90; Partington, D., Pellegrinelli, S. and Young, M. (2005), Attributes and levels of programme management competence: An interpretive study. *International Journal of Project Management*, 23, 2, pp. 87–95; Turner, R., Müller, R. and Dulewicz, V. (2009), Comparing the leadership styles of functional and project managers, *International Journal of Managing Projects in Business*, 2, 2, pp. 198–216.

[18] For example Hogan, R., Curphy, G. and Hogan, J. (1994), What we know about leadership: Effectiveness and personality, *American Psychologist*, 49, pp. 493–504; Goleman, D. (1998), *Working with emotional intelligence*, Bantam Books: New York, NY; Rutkowski, P. J. and Leban, W. V. (1999), Project managers and the wisdom of using emotional intelligence, in: *30th annual seminars and symposium*, Project Management Institute: Philadelphia, PA; Dulewicz, V. and Higgs, M. (1999), Can emotional intelligence be measured and developed? *Leadership and Organization Development*, 20, 5, pp. 242–252; Goleman, D., Boyatzis, R. and McKee, A. (2002), *Primal leadership: Realizing the power of emotional intelligence*, Harvard Business School Press: Cambridge; Leban, W. and Zulauf, C. (2004), Linking emotional intelligence abilities and transformational leadership styles, *Leadership & Organization Development Journal*, 25, 7, pp. 554–564.

[19] For example: Laufer, A., Post, T. and Hoffman, E. J. (2005), *Shared voyage: Learning and unlearning from remarkable projects*, National Aeronautics and Space Administration (NASA) Report, NASA SP-2005-4111, Washington, DC; Leonard, D. and Kiron, D. (2002), *Managing knowledge and learning at NASA and the Jet Propulsion Laboratory (JPL)*, Harvard Business School Case 9-603-062, Cambridge, MA; Lee, D., Simmons, J. and Drueen, J. (2005), Knowledge sharing in practice: Applied storytelling and knowledge communities at NASA, *International Journal of Knowledge and Learning*, 1, 1–2, pp. 171–180.

[20] Argyris, C. (1976), Single-loop and double-loop models in research on decision making, *Administrative Science Quarterly*, 21, pp. 363–375; Argyris, C. and Schön, D. (1978), *Organizational learning: A theory of action perspective*, Addison Wesley: Reading, MA; Schön, D. (1987), *Educating the reflective practitioner*, Jossey-Bass: San Francisco, CA; Nonaka, I. and Takeuchi, H. (1995), *The Knowledge-Creating Company*, Oxford University Press: New York; Weick, K. E. (1995), *Sense making in organizations*, Sage: London; Wenger, E. (1998), *Communities of practice: Learning, meaning, and Identity*, Cambridge University Press: Cambridge; Grabher, G. (2004), Temporary architectures of learning: Knowledge governance in project ecologies, *Organization Studies*, 25, 9, pp. 1491–1514.

[21] Nonaka, I. (1994), A dynamic theory of organizational knowledge creation, *Organization Science*, 5, 1, pp. 14–37.

[22] Collison, C. and Parcell, C. (2001), *Learning to fly*, Capstone Press: Oxford; Boyatzis, R. E. (1982), *The competent manager: A model for effective performance*, Wiley: Hoboken, NJ.

The Development of Project Management: Summary

Damian Hodgson and Svetlana Cicmil have a wonderful warning to would-be sense-makers of project management history.

> The consequences of throw-away statements regarding the existence of pyramids as evidence of the universal importance of projects are serious as they invoke an ahistorical concatenation of prehistorical work organization, Adam Smith's division of labour, Taylorism, Cold War project methodologies and the contemporary techniques and technologies associated with the discipline of project management. Such statements serve as a subtle legitimation of contemporary formulations of project management. (*Making Projects Critical* [2005], London: Palgrave, p. 45)

This is not what this account of the construction of what we call Project Management has been. Though the warning is salutary, this has been neither an ahistorical account nor a concatenation. Rather, it is historical and contextual, with examples chosen to illustrate the way different people have progressively built-up and shared their conceptions of what is required to manage projects and programs successfully. As such, it is an account of how both the emerging discipline of project management, and the general domain of the management of projects, have been constructed.

I would contend, however, that the historical account has been a legitimation of contemporary project management insofar as we have seen that *there are* concepts, practices, tools and techniques that are largely proven to help us manage projects better, and that these can be applied in ordered ways, as a discipline.

The art of applying them so that the rate of accomplishing projects successfully is not without its difficulties. Too many projects fail to meet their sponsor's objectives.

Reconstructing Project Management, First Edition. Peter W.G. Morris.
© 2013 John Wiley & Sons, Ltd. Published 2013 by John Wiley & Sons, Ltd.

Part of the problem is alignment: aligning the knowledge, skills, behaviours and experience of the p.m. staff together with the routines and infrastructure of the organisation on the one hand, with the characteristics of the project and the context it is operating in on the other.

Its not that we don't know what is required. There is plenty of knowledge. After reviewing all the studies and the lessons from the history we have just been looking at, it's clear that there is substantial agreement on many practices required to manage projects and programs effectively. What's missing – where we have difficulty and lack institutional agreement – are ways of combining and deploying the elements of this knowledge in ways that overall are optimal to the challenge presented by the project, or program, in its context.

Specifically, there is still a divergence of views regarding how the discipline should be conceived – how the knowledge should be framed and called in to play. For we have two fundamentally different accounts – paradigms; ontologies – of what we mean by the term project management:

- the execution-oriented model, à la the *PMBOK® Guide* and *ISO 21500*
- and the 'management of projects' one, as well as several other variants (as we'll see in just a moment).

Further, the execution-oriented model (the *PMBOK® Guide*) is, I believe, inadequate in that it does not present all the relevant knowledge needed to manage projects successfully. Indeed, it virtually ignores some of the most significant: the front-end and many relevant people-related and technology and commercial matters. The *PMBOK® Guide* is by far and away the most common view of project management, but it has a number of shortcomings, as discussed in Chapter 4.

Actually, the various models of project management, or what we take to be the discipline of managing projects, include, in addition to delivery–execution management and the 'management of projects', a number of other models: program management and portfolio management; projects as temporary organizations (the Scandinavian School); arguably Agile project management and even more arguably, Critical Chain Project Management; 'management by projects'; and Shenhar and Dvir's Adaptive Project Management.

And there are, quite reasonably, several schools of criticism of our knowledge about managing projects, particularly of the positivist, normative nature of much of the popular standards and guidance – the critical management school and the Scandinavian School, for example.

The various 'schools' have evolved, partly in response to external circumstances, partly in response to theoretical innovation. Table 7.1 shows how project management developed both as a formal discipline and as a domain of management at a fairly consistent pace since the early 1950s – through the efforts of many practitioners, whether working on projects and programs directly themselves, or as advisors or researchers. The

Table 7.1   The 10 stages of project management, or how the discipline developed: 1953–2013

1. PLANNING AND CONTROL: 1900–1970s
    - Early planning and control tools 1900+
    - (No evidence of p.m. before ~1953)

2. ENGINEERING COMPLEXITY AND URGENCY: 1953+
    - PM 'invented' by USAF for missile programs circa 1954 – systems thinking, planning and control; immediately followed by US Navy
    - Then DoD and NASA institutionalise: PERT, CM, WBS, PBS, EVA, C/SCSC. Stage-gate process. Leadership. All this, to address technical complexity + urgency. PM acted as a form of engineering management.
    - Environmentally sheltered.

3. THE ORGANIZATION THEORISTS: late 1960s++
    - Integration, contingency theory
    - Scandinavian School – temporary organizations (1990s)

4. ENVIRONMENTAL AWARENESS: 1970s
    - Environmental issues became intrusive and disruptive: TransAlaskan Pipeline, North Sea, Concorde. New environmental awareness began – stakeholder management, cost–benefit
    - PMBOK® Guide – process and execution oriented

5. FRONT-END DEFINITION: 1990s+
    - 'The Management of Projects' paradigm
    - New BOKs – APM, IPMA, ENAA/JPMS

6. LEAN MANAAGEMENT AND RELATIONSHIPS: 1990s
    - New Product Development/Toyota: Concurrent Engineering
    - Lean, TQM, partnering, relationships

7. ENTERPRISE-WIDE PROJECT MANAGEMENT: 1995+
    - ICT, PMOs, maturity, Knowledge Management, Project/Organizational Learning
    - OGC Program Management – Change Management, benefits, value

8. GOVERNANCE: 2000+
    - Sponsor, governance, strategy, reviews/audits
    - Financial stringency – BOT/PFI: WLC (Whole Life Costs); risk [behavioural economics]. Effectiveness [Miller & Lessard]
    - Japanese BOK

9. AGILITY: 2005+
    - Micro projects – Critical Chain, Agile

10. RELEVANCE: Today
    - Dramatic societal challenges – 2050 etc.
    - Projects and programs more interdependent, less 'mechanistic'
    - Funding a major issue
    - So too is competency, quantity of senior p.m. staff and integrated supply chains
    - Need 'dispersed intelligence and ownership' – Communities of Practice, ICT
    - More value-driven approach
    - Leadership crucial

Source: Author's own.

result is a field that is both extremely broad and relatively complex and deep*.

Yet despite the huge amount of work over the last 50 or 60 years, there is still some confusion in the domain, in the discipline.

- Does project management cover the management of the project front end, the definitional, development stages, or is it concerned essentially only with execution delivery? What should it be responsible for? How far does it stretch into concept definition at one end and operations at the other?
- If it does not cover the front-end, (a) is it fit-for-purpose; (b) do we need an enlarged discipline or body of knowledge to cover what we need to know about managing the overall project? (Since we have seen throughout this history that managing the front-end is key (i) in building-in value and (ii) in building-in (or out) future problems.)
- When and how can we get away from a PMBOK that is inadequately based around what was mistakenly chosen to be 'the knowledge that is unique to project management' rather than that which we need to know in order to develop and deliver projects successfully?
- Should there be, resources allowing (personnel, funds), one (senior) person responsible for, and in charge of, the project from beginning to end? (What does 'beginning to end' mean?)
- Does project management cover estimating and contracts and procurement? Are they part of the discipline? (Often they are treated organizationally as not.)
- Should project managers work to achieve effectiveness goals (value, functionality, business performance) or just stay with efficiency ones (on time, in budget, to scope, etc.)?
- Is program management just the management of projects having shared aims and possibly resources or does it have some special ownership of delivery of the business case benefits? If the latter, why? Shouldn't projects also have this concern?
- Is Agile a project management discipline or really just a form of task or work stream management?
- How can project management academics better provide timely, robust, theory-based normative guidance? When, and on what?
- What is the role of theory in explaining project management?

---

*This list is substantially different from the 'Schools' of project management proposed by Frank Anbari, Christophe Bredillet and Rodney Turner and separately by Jonas Söderlund as chronological accounts of the development of the discipline. I frankly believe that this account is more accurate. See Bredillet, C., Anbari, F. and Turner, J. R. (2007) Exploring research in project management: Nine schools of project management research, *Project Management Journal*, 27, 4; and Söderlund, J. (2002), On the development of project management research: Schools of thought and critique, *International Project Management Journal*, 8, 1, pp. 20–31.

There are other queries too. But, as I said at the beginning, maybe having such questions still open is a sign of the vibrancy of the subject. It offers plenty of research issues and, to practitioners, competitive edge opportunities.

These and other issues will be the focus of Parts 2 and 3, where we will first examine in detail individual practices and knowledge areas, in Part 2, before looking in Part 3 at the challenge which is overriding to the discipline now: how to combine these elements into a coherent, relevant whole.

Part 2

Deconstructing Project Management

Its elements deconstructed

Reconstructing Project Management, First Edition. Peter W.G. Morris.
© 2013 John Wiley & Sons, Ltd. Published 2013 by John Wiley & Sons, Ltd.

Introduction to Part 2

Our account of how project management grew so incredibly over the last 60–100 or so years, from a largely instinctive skill to a highly popular management discipline, offering benefit to practitioners and real interest to scholars of management and organization theory, has concluded that there are still substantial differences of view on what essentially the discipline is – less perhaps what constitutes good practice but more on how the discipline, or the domain, should be seen as a whole, and how its application might vary under different conditions and contexts.

This section of the book breaks out and analyses – deconstructs – the elements of what is involved in managing projects in order to understand them better before looking, in Part 3, at how they might be recombined and redeployed under today's conditions or to different ends.

The Domain

It's important to realise that there are two or three perspectives being employed here when we are talking about project and program management and the management of projects. Most project managers see themselves working in a world of execution implementation, a world where requirements have been stated and project targets set. This is where much of the professional project management institutes' mindset is, be it their members' events or research telling us that project managers are not seen to have a role in establishing project strategy, for example[1], or reporting that X% of those interviewed do not do something (like EVA).

It has become apparent from the historical account given in Part 1 that this view deserves being challenged. It is neither sufficient to today's realities, nor indeed entirely logical. And it doesn't chime with what you need to manage if you want your project to be a success (see Appendix 1). For we have seen that developing and defining these requirements and targets, in the project 'front-end', bears hugely on how efficiently and effectively the organization's goals for these projects and programs can be achieved. The front-end, therefore, can, and should, be managed. In fact, it is more than this: to understand how to create successful projects, we should be looking at the management of, well, not just the front-end but the whole project.

In parallel with this perception, many academics have become interested from an organisation theorist's perspective in projects as organisational forms. In doing so, however, they may not engage with the management responsibilities of people who are involved, or could be, in managing the 'whole project'. (Some would see the attempt as secondary to their interests, if that.) The continuing primacy given by many of them to projects being defined first and foremost as temporary undertakings is an example[2]. Indeed, I maintain that focusing on this as the principle criterion that characterises the domain of enquiry is wrong. An overnight scout camp could be a temporary organisation; it is not a project, except in a rather trivial sense. Instead, as I said in Chapter 2, the project development [life] cycle is, for me, the one thing that differentiates projects from non-projects.

Another danger is the scepticism which some researchers have towards the concept of project management as a discipline. The critical management school is the most obvious example, but even the 'projects-as-practice' school is too, in that by focussing on recording the minutiae of who really did what, there is little apparent reference to 'why'[3]. We have surely to begin addressing the 'why' and teasing out what are good practices relative to accomplishing projects and programs successfully.

What we shall be addressing, then, in Part 2 is a discussion of what is involved in managing whole projects. It is more than project management as execution management, and ultimately quite a different paradigm from, although informed by, the organizational interests of sociologists. It represents a corpus of knowledge analogous to that perhaps of an Army officer moving from front-line field command to the beginnings of staff command and generalship.

The intent is not to offer a 'how-to' text on all the elements of project management – many others have done this elsewhere – but to analyse the more salient features of the elements that make-up the domain.

I've called this 'Deconstructing project management'. The second and third of these words are confusing enough; what do I mean by Deconstructing?

Deconstructing Deconstruction

'Deconstruction' has acquired a specific meaning in literary, philosophical and sociological analysis following the work of Jacques Derrida from 1967 onwards[4]. Derrida used the term to refer to an activity that systematically breaks down a text, unpeeling and challenging one's overall understanding of it, often showing that elements in it have irreconcilable and contradictory meanings; that any text therefore has more than one interpretation; and thus that an interpretative reading cannot go beyond a certain point. The focus in Deconstruction is on the relationship – and potential conflict – between what is said about something and what actually happens in practice.

Here we are not being so dogmatically post-modern. I am using the activity of deconstruction to mean the breaking out of the elements that constitute the overall domain [of managing projects], and the analysis of them in terms of their 'actual practice' and relevance. (Relevance to project management practitioners and sponsors and to other project stakeholders.) We do not pre-suppose that there will be irreconcilable elements, but we may challenge established categories. Rather, in fact, we believe most elements of p.m. fit well together.

Approaching the Management of Projects

In order to dissect the management of projects into its elements, we need to have some logic of approaching it and hence accessing it. Not a fixed structure but some form of high-level narrative or schematisation of the way the parts relate. Without this, the analysis would be less informative, like discarded toys in a child's playpen. What then is our model of the domain; how do we conceptualise it?

In Part 1, we identified nine conceptual 'models' that have been proposed over the years as explicit approaches to managing projects, or as particular perspectives within the overall domain, namely: (1) project management as Delivery–Execution management, as typified by the *PMBOK® Guide*, (2) the 'Management of Projects' – Morris, (3) 'Management by Projects' – Gareis, (4) the Scandinavian School's projects as temporary organisations, (5) Program Management, (6) Critical Chain Project Management – Goldratt, (7) Agile Project Management, (8) Adaptive Project Management – Shenhar and Dvir, and (9) Critical Project Management – Hodgson and Cicmil, etc. (There are in addition dozens if not hundreds of books explaining the elements of project management all of which will use a model or a framework of some sort to organise their thinking. I do not think they have as strong a philosophical dimension to them as these nine however.) We also saw how the rise of the PMO mirrored a growing recognition of the role of building and supporting an enterprise-wide capability*.

Levels 1–3 in the Management of Projects

In 2011, Joanna Geraldi and I extended the enterprise focus by proposing (following Talcott Parsons) that the practice of managing projects could be thought of in terms of three levels: a technical core; a strategic wrap, shaping and shielding the technical core; and an institutional level of action aimed at creating, as far as one is able to do, an environment within which the project, and other projects and programs, will prosper and flourish[5].

The essence of the three levels is as follows:

---

*There are other models, if you can call a textbook a model, such as the books by Loch or Kerzner, but they lack the driving individualism, to say nothing of the absence of theory, of these nine or ten.)

- *Level 1*, the technical core, is pre-eminently delivery oriented. It is concerned with the management of the project's technical operations: writing code, testing, designing, building, fabricating, and, in project management terms, with scope, schedule, and cost planning and monitoring. The key concern is with how to deliver projects efficiently: *pace* Agile, 'on time, in budget, to scope'. This is the Delivery–Execution conception. It is the classic core of project management. (And analogous, to go back to our Army metaphor, to field command at the predominantly tactical level.) Agile and Critical Chain fit here. Knowledge, practice and research at this level tend to a largely normative or prescriptive and positivist position, as in the 1960s' systems project management, though it need not be exclusively so, as in Critical Chain. It is biased towards techniques and processes.
- *Level 2*, the project's strategic wrap, looks at managing projects as organisational 'whole' entities, (1) expanding the domain to include their front-end development and definition and (2) protecting the technical core from environmental turbulence. This is the 'Management of Projects' conceptualisation. Program management's concerns are very alive at this level. The Scandinavian School's interest in projects as temporary organisations is reflected here. Work at Level 2 recognises the relationship between the project and various stakeholders' strategies (not least the sponsor's). The importance of getting the front-end right is a major part of the strategic level. There is a strong concern for value and effectiveness. It is the world of the Project Director or Program Manager. The beginnings of generalship.
- *Level 3*, the 'institutional' Level 3, is about influencing and managing, as far as one is able, the context within which the project, and other projects and programs, occurs in order to enhance their effectiveness. Management of, and by, Projects, Program Management, and the Scandinavian School all resonate at this level. Management at Level 3 is primarily concerned with improving long-term project success through processes, standards, guides, agreements, behaviours, etc. outside of particular projects' or programs' individual management issues but predominantly in their institutional environment*. In the Army metaphor, there is a read-across with staff function work. This institutional context can be either the enterprise's own organisational environment, that is, projects in the parent organisation, or the wider environmental context within which the project is located; or both (Figure 8.1). Thus, in recognising the role and opportunities for management at Level 3, the focus switches "from organizations in their environment to the organization of the environment"[6].

---

*It was recognition of the significance of the institutional level that led the editors of *The Oxford Handbook of Project Management* to suggest that there might be a third wave of research in project and program management following a First Wave (Level 1: Execution) and a Second (Level 2: 'the management of projects').

Level 1: The technical core:
Like medics in a hospital, or workmen on a building site. Project management was initially very technically biased and in that it is still seen as heavily execution-oriented, it has, at this level, a predominantly technical character.

Level 2: The strategic envelope:
Parsons (1951, 1963) called this level 'management' – buffering the medics, organising the supply of materials to the building site, etc. In projects, Morris (1983) suggested calling this 'strategic' to capture the front-end project definition stages where the execution targets are set.

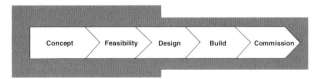

Level 3: The institutional context:
Management here is concerned with ensuring the long-term p.m. health of the organisation. Work will be in the 'parent' organisation and/or in the environment that the project is operating.

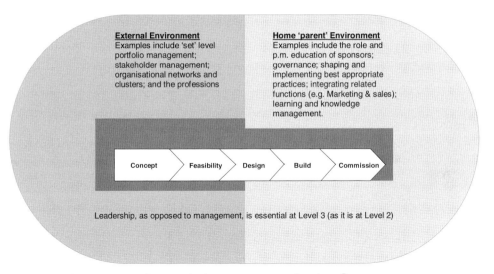

Figure 8.1   Levels 1 to 3 in the management of projects.[7]
Source: Morris P. and Geraldi J. (2011) Managing the institutional context for projects, *Project Management Journal*, 42, 6, Fig. 1, p. 23.

This schema reflects, incidentally, that proposed by the earlier systems theorists in the 1960s, particularly Katz and Kahn and by early writers on project management such as Cleland and King, and by Johnson, Kast and Rosenzweig, who proposed that we think of organizations at three levels: an internal core; an operating layer of stakeholders, suppliers, regulators and others; and a general environment[8]. It is different, however, from the more contemporary distinction of micro-, meso- and macro-organizational levels at which practice, and praxis, operates differently[9], being more precise as to the change in project management responsibility at these three levels.

Terry Cooke-Davies' three levels of success

The schema also corresponds with Terry Cooke-Davies' ideas of three levels of success in projects[10]:

1. Was the project 'done right'? – what Cooke-Davies called 'project management success'.
2. Was 'the right project done'? – what he called 'project success'.
3. And were 'the right projects done right, time after time'? – what he called 'consistent [or in our terms, 'institutional'] success'.

The first predominantly concerns the technical level, the second involves the strategic level, while the third calls for work at the institutional level. (It's important to recognise that these levels don't function in isolation of each other – just the opposite: actions in one level often affect actions at the other levels.)

Developing Projects

Another way to conceptualise – to reverse engineer (rather than 'deconstruct') – the relations between the major elements involved in managing projects is to think of the usual (Level 1 and Level 2) progression of work on projects. This I would propose can be conceived simplistically as follows (Figure 8.2) (more sophisticated sequencing will come in Part 3):

(a) Establish the business case for the project and the proposed strategy for its development and delivery to marry into this business case.

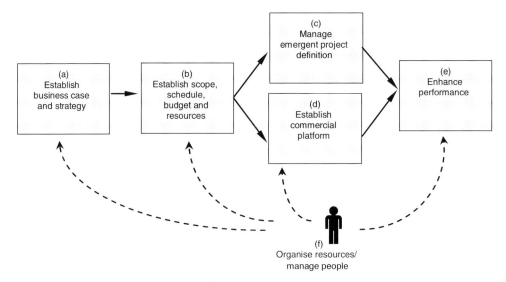

Figure 8.2   Sequence of project development.
Source: Author's own.

(b) Determine the scope of the project, plan the schedule (if not in detail then in outline or progressive detail); allocate resources; identify and allow for risks, allocate contingencies; and agree the budget.

(c) Manage the work on the emerging project definition (a major piece of which will be the pre-sanction product definition) and on preparing for downstream implementation.

(d) Establish the commercial platform through which the project work will be done.

(e) Enhance performance: build value, harvest benefits, manage risk, control performance and drive progress.

(f) Do all of this remembering that projects are done by, with and through (and for) people – that is, manage the people involved in the project.

Around these Level 1 and 2 activities, various Level 3 actions may occur:

(a) Assure the organization has the requisite capabilities – standards, guidance, methodologies, systems, procedures.

(b) Ensure there will be the required number of competent people available to meet the enterprise's overall portfolio needs.

(c) Assess performance and competence, gather 'lessons-[to be-] learned', organize up-skilling (training, education) and facilitate individual and organizational learning.

In a way, it would be obvious to start Part 2 of this book at Level 2 with strategy. I am nevertheless going to start with Level 1, the technical core of project management – predominantly scope, schedule and cost management, followed by organisation. This is where project management begins for many people and, as we saw in Part 1, it is in large part where it started as a discipline. The (a)–(f) sequence just outlined will kick in when we leave this technical core and go on to the Level 2 topics (Chapters 11–15).

References and Endnotes

1 Crawford, L. H. (2005), Senior management perceptions of project management competence, *International Journal of Project Management* 23, 1, pp. 7–16.

2 Blomquist, T. (1994), Tensions in three dimensions: A method for analyzing problems in renewal projects, In: Lundin, R. A. and Packendorff, J. (eds.) *Proceedings of the IRNOP Conference on Temporary Organizations and Project Management*, IRNOP: Umeå, pp. 71–78; Packendorff, J. (1994), Temporary organizing: Integrating organization theory and project management, in Lundin, R. A. and Packendorff, J. (eds.) *op. cit.*: 2, pp. 207–226.

3 'Projects as Practice' differentiates between Praxis ("the situated doings of an individual"), Practice ("norms, values, rules and policies"), and Practitioners. Its studies tend to focus on recording "the detail of work without making assumptions about how the work should be done" (Hällgren, M. and Söderholm, A. (2011), Projects-As-Practice: New approaches, new insights, in Morris, P. W. G., Pinto, J. K. and Söderlund, J. (eds.) *The Oxford handbook of project management*, Oxford University Press: Oxford, pp. 505–506); see also Hällgren, M. and

Söderholm, A. (2010), Orchestrating deviations in global projects: Projects-as-practice observations, *Scandinavian Journal of Management*, 26, 4, pp. 352–367; Blomquist, T., Hällgren, M., Nilsson, A. and Söderholm, A. (2010), Project as practice: Making project research matter, *Project Management Journal*, 41, 1, pp. 5–16; Hällgren, M. and Wilson, T. (2008), The nature and management of crises in construction projects: Projects-as-Practice observations, *International Journal of Project Management*, 26, 8, pp. 830–838.

[4] Derrida, J. (1967), *Of grammatology*, John Hopkins University Press: Baltimore, MD.

[5] Morris, P. W. G. and Geraldi, J. (2011), Managing the institutional context for projects, *Project Management Journal* 42, 6, pp. 20–32.

[6] Scott, W. R. (2008), Approaching adulthood: The maturing of institutional theory, *Theory and Society* 37, 4, pp. 427–442.

[7] Parsons, T. (1951) *The social system*, Free Press, Glencoe, IL; Parsons, T. (1953) Structure and Process in Modern Societies, Free Press, Glencoe, IL; Morris, P.W.G. (1983) Managing project interfaces – key points for project success, in Cleland, D.I. and King, W.R. *Project management handbook*, Van Nostrand Reinhold, New York.

[8] Katz, D. and Kahn, J. L. (1966), *The social psychology of organizations*, John Wiley and Sons: New York; Cleland, D. I. and King, W. R., (1968), *Systems analysis and project management*, McGraw-Hill: New York, pp. 22–23; Johnson R. A., Kast, F. E. and Rosenzweig, J. E. (1963), *The theory and management of systems*, McGraw-Hill: New York.

[9] Hällgren, M., Nilsson, A. and Söderholm, A. (2011), *op. cit.*: 3; Whittington, R. (2006), Completing the practice turn in strategy research, *Organization Studies* 27, 5, pp. 613–634; see also Giddens, A. (1984), *The constitution of society: Outline of the theory of structuration*, University of California Press: Berkeley and Los Angeles; Grabher, G. and Ibert, O. (2011), Project ecologies: A contextual view on temporary organisations, in Morris, P. W. G., Pinto, J. K. and Söderlund, J. (eds.) *op. cit.*: 3.

[10] Cooke-Davies, T. (2004), The "real" success factors on projects, *International Journal of Project Management* 20, 3, pp. 185–190; Cooke-Davies, T. (2004), Project success, in Morris, P. W. G. and Pinto, J. K. (eds.) *op. cit.*: 3, pp. 106–108.

Control

Control is more than merely monitoring. In the proper cybernetics sense of the term, control involves planning, monitoring (i.e., measuring and reporting), taking any necessary corrective action, and re-planning. Managing projects is fundamentally, rigorously and ruthlessly, about control – planning, monitoring, and correcting – particularly at Level 1. About being able to deliver efficiently as promised. If you can't do this, save with good cause, then you can't really claim to be a project management professional.

Most organisations wish to be in control. Projects, representing managed change, particularly want to be in control, but *the nature of that control is not the same throughout the project.*

- In the front-end, the definitional and development stages of the project's plans are being worked up and set*. Once the technical, financial and schedule parameters have been established, then the Design and Development people can be given substantial freedom to develop and innovate up to the point that the project is submitted for a gate review to ensure the emerging project is on track (and even then of course there will be innovation needs and opportunities on execution).
- Control in execution, however, is at the heart of the traditional view of project management: planning, monitoring, and correcting to make sure the project is completed on time, in budget, to scope.

Scope Management[1]

The basis of control is establishing the project's scope. What does scope mean? It means what is defined as being 'in' the project and what is not. A

---

*How much effort should be spent on the front-end? Millar and Lessard said anywhere between 3 and 35%, which perhaps doesn't help us very much. The truth is, projects are so varied it's hard to impossible to give any definitive figure.

Reconstructing Project Management, First Edition. Peter W.G. Morris.
© 2013 John Wiley & Sons, Ltd. Published 2013 by John Wiley & Sons, Ltd.

project's scope is what the project is supposed to comprise. This generally is defined by:

- first, the deliverables;
- second, the work associated with producing these deliverables.

The PBS and the WBS

The project deliverables are often represented via a *Product Breakdown Structure (PBS)* (see Figure 9.1), though other tools exist such as work flow diagrams. Describing the work associated with producing the deliverables is typically shown via a *Work Breakdown Structure (WBS)* (see Figure 9.2).

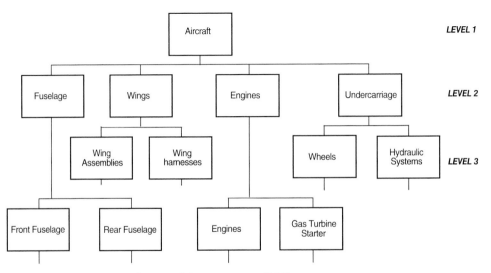

Figure 9.1   Product Breakdown Structure (PBS).
Source: Author's own.

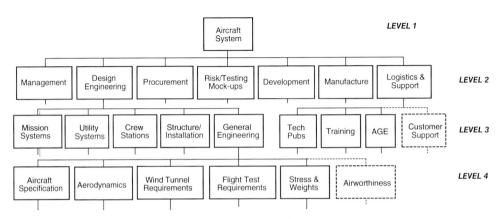

Figure 9.2   Work Breakdown Structure (WBS).
Source: Author's own.

The PBS contains only [product] deliverables. The PBS provides the product skeleton; the WBS shows how product elements will be organised. The WBS is the definitive expression of the overall scope of work. The WBS should be derived from consideration of the PBS, the project and product development cycles, make/buy and procurement strategies.

Defining the scope of the project and decomposing it into manageable pieces of work is a basic, core element of effective project planning, and therefore of effective project management. Construction project managers, for example, use the WBS to represent how the project will be 'bought out' by sub-contractors' work packages.

In general terms, the process begins with some form of 'visioning' – a vision statement and/or a set of objectives leading to project or program goals and strategy[2]. The project 'brief' or requirements (business, user and system) will have a direct impact on determining the required 'end items' and the emerging product definition. (Strategy, requirements, brief, are all discussed later.)

These may seem utterly simple techniques but they should not be thought the less of for that. A group of Facilities Management project managers I was training voted the WBS the most useful of all the new knowledge they had gained on the course. (OK: some course!)

Developing scope

Scope is developed as the project or program goals, strategy, requirements and the general definition are developed; as more detail unfolds the scope can be elaborated, breaking out into further functions or sub-functions (deliverable 'chunks'), tasks and work packages.

On many projects it is not possible to decompose the scope in a linear manner all at once: there will typically be development of requirements, or evolution of work packaging and product detail. Classically, project management has referred to this as the 'rolling-wave': progressive elaboration of detail (see Figure 9.3).

It is important to define what is not included within scope; that is, what is not agreed to be done.

Baselining scope

At some point, the scope needs baselining. This generally happens when there is formal acceptance of the scope, as in a proposal approved by 'governance' for budgetary and expenditure approval, regulatory approval, input into bid documentation, or contract award, and so on. From this point on the scope, including much of the detail regarding the general approach (strategy), schedule, and budget are 'frozen'. In the words of IPA, the process and mining project benchmarking company, "there needs to be at least one gate at which the business case can be assessed, [another] where the scope is closed, and finally, a gate at which the implications of the scope can be evaluated"[3].

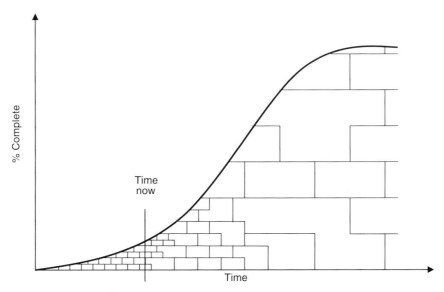

Figure 9.3   The Rolling Wave concept.
Source: Morris P. W. G. (1994) *The management of projects*, Thomas Telford: London, p. 267.

Scope change management

Having a baselined scope is fundamental to effective control of the project. Scheduling, cost control, procurement, performance management (Earned Value etc.), organisation, configuration management, and several other project management areas rely on a clear, baselined definition of scope.

As the project progresses and more detail builds up, there may well be a need to update the scope and re-baseline. Since the baseline represents the officially endorsed project 'target', sponsor approval of such changes is a prerequisite.

A fundamental rule of project and program management is that any changes to scope must be rigorously managed via change control. (Remember Configuration Management on Apollo: and see the next subsection). *Strict change control lies at the heart of effective project control. If changes are not properly evaluated in terms of their potential impact on the approved baseline targets but instead slip through un-analysed, the result will almost inevitably be missed targets and failed delivery.*

Note that although rigorous change control is fundamental to good project management, this rigour need not always imply rejection of changes, particularly in the early developmental stages of a project.

Configuration management[4]

Configuration Management is a special example of scope management being really a cross between change control and information management.

Configuration management defines the product, and then controls the changes to that definition: in the words of ISO 10007[5], "that applies technical and administrative direction to the development, production and support life cycle of a configuration item"*. It is concerned with managing the definition of hardware, software, processed materials, services, and related technical documentation – in other words, configured items – from design, through production and maintenance to disposal.

Configuration Management, as defined for example, in ANSI/EIA 649, comprises:

- *Identification* – item identifiers and attributes;
- *Change Management* – the change review process through which all changes must progress;
- *Status Accounting* – identification and customisation of information requirements, data retrievability, capturing and reporting of information;
- *Verification and Audit* – ensuring that the product design provides the defined performance; validating configuration information and data integrity; verifying the consistency between the product and its configuration information; and ensuring a known configuration provides a valid basis for operation and maintenance and life-cycle support.

Doing this effectively is important in a properly managed and delivered project or program. Configuration control is important during development and especially as the project moves into operations, since it will provide a record of the product build information which in turn will feed into the asset model of the product (facility etc.).

The project/program management team should ensure that proper processes, procedures and organisational structures are in place to identify, manage, account for and verify configured items. The actual configuration management work will, however, typically be carried out by specialists (e.g., the Configuration Manager, the Configuration Change Board, etc.).

The project or program manager is ultimately responsible to the project sponsor for scope management and should ensure the project team has a

---

*The term 'Configuration Management' seems hardly to be used as such in construction-related industries, nor in drug development, though it often exists as a function (asset modelling, the drug information model, etc.). The term – and in many ways the formal discipline within a project management context – appears largely to be confined to the wide spread of industry sectors employing a systems engineering framework, most obviously ICT and aerospace but also many manufacturing sectors such as automotives. This may be changing, however: Building Information Management Systems (BIMs) are becoming major pushes in this direction in construction. BIMs integrate architectural, structural, services and construction information models from supply chain members to create an integrated project/product model; this enables information exchange across the project which facilitates operations and maintenance.

clear understanding of their responsibility for scope planning, definition, documentation, development, verification and change control.

Several higher-level (maybe 'strategic' in the Level 2 sense) topics feed into the determination of scope – for example, project strategy, requirements management, innovation. We shall touch on strategy now in the early discussion of Scheduling but in general we'll leave discussion of these topics till we get to Level 2, Strategic section (Chapter 11).

Scheduling[6]

Scheduling refers to sequencing the activities in a project or program so that completion is accomplished in the optimal time. This can range from the detailed scheduling of activity durations and interrelationships to resourcing and the broader phasing and pacing of the project or program.

The effective planning and accomplishment of activities' timing and phasing is a central skill of project management; accomplishing projects on time is one of the most important expectations put on project and program management.

Scheduling is often, but far from always, done by specialist project planners.

Activity planning

Depending on the level at which the schedule is being developed, its preparation may be accomplished in a predominantly top-down manner, proceeding from milestone targets, or a predominantly bottom-up manner, working out the logic of activity durations, their likely duration, resource constraints, and so on. Construction for example, with its large number of discreet elements, will typically be worked out both bottom up and top down. Software development, on the other hand, has less predictable discrete activities and proceeds between short target review points (thus leading to 'agile' development: phases of a day or two to 90 days). Some projects may have relatively standard sequences of activity – drug development for example; others may be totally original, with everything having to be worked out from scratch. CPM and PERT reflect these orientations too: CPM (DuPont) being a bottom-up, construction linear build, PERT (Polaris) being top-down, R&D, probabilistically qualified, end-point oriented.

Phasing and fast-tracking

Program management, like high-level project planning, is typically more concerned with phasing, strategic pacing and the overlapping of different activities or blocks of activities. Program management is often significantly concerned with the interdependency between projects, 'tranches'* of projects, and activities.

---

*This is OGC's term, OGC being the [now defunct] UK Office of Government Commerce.

Strategic scheduling considerations will shape the setting of schedule targets, both as regards total and stage completion, and of intermediary milestones. These targets then need addressing in terms of the work required to be done, the resources required and available, and the length of time that accomplishing work activities will take. There may well be financial or operational reasons that drive the schedule targets or limit the resources available and these may result in very different durations from ones calculated simply from a bottom-up project perspective.

One very important project, and program, issue, however, is the overlapping between different blocks of activities, especially the way design and technical development is overlapped with full production. For example, the decision on whether or not to use:

- parallel development of design activities (Simultaneous Design/ Engineering);
- overlapping of stages or phases, such as design and production, as overall blocks – so-called Fast Track, common in construction (and often associated with 'construction management'*);
- Rapid Application Development prototyping – accelerated development and testing of (typically software) design without following the detailed formalisation of requirements and verification and validation normally recommended;
- Concurrent Engineering – most commonly found in New Product Development (automotives, defence/aerospace, and general manufacturing): a means of shortening the overall design-production development schedule by involving production in the design phase (avoiding 'over-the-wall' development[7]) using integrated teams (to bring in 'design-for-manufacturability'), integrating tools (such as Quality Function Deployment [QFD] and Failure Modes Effects and Criticality Analysis [FMECA]) and creating integrated information product/project models. (See Figure 4.8)

These are all well-documented phasing approaches. Other approaches, such as phased Hand-Over/Commissioning, or the chunking and phasing of projects within programs, are equally real while existing without specific terms having been developed for them.

Bottom-up scheduling

At a detailed level, the first task is to work out the listing, scope and sequencing of activities required to ensure timely completion of the project.

The most common and simplest method of illustrating and communicating a schedule is as a 'bar chart' (the Gantt chart). Bar charts, however, do

---

*Though in doing so, care should be taken in overlapping individual work packages which is dangerous – known as 'Concurrency' – which, as we saw in Part 1, caused many problems in the 1960s, 1970s, and 1980s in defence/aerospace and process engineering.

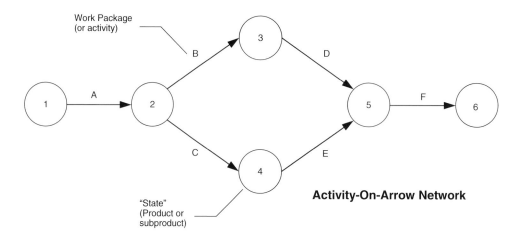

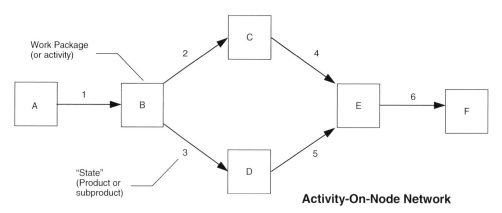

Figure 9.4   Activity-on-Arrow and Activity-on-Node networks.
Source: Morris P. W. G. and Pinto J. K. (2004) *The Wiley guide to managing projects*, John Wiley & Sons: Hoboken, NJ, p. 453.

not always show activity interdependencies: these are shown by activity networks. The method of notation used in these networks is either 'Activity-on-Arrow' or 'Activity-on-Node', as shown in Figure 9.4. Precedence, developed, as we saw, in the early 1960s, used Activity-on-Node. Precedence Diagram Method (PDM) is almost certainly the most popular method used today (though few would know it by that name). Other methods were developed later, such as GERT (Graphical Evaluation Review Technique[8]), to handle larger simulation.

Classically, project management identifies the 'critical path' in an activity network: the critical path has zero float (spare time) on it. Thus it is the longest sequence (path) of activities through the network and as such defines the minimum time within which the project as a whole can be accomplished.

(Hence 'Critical Path Analysis' [CPA] and 'Critical Path Method' [CPM].) CPM, developed as we saw in Part 1 by DuPont in 1957, is activity oriented, while PERT, developed at the same time by the US Navy, was initially event oriented, incorporating a threefold probability forecast for the accomplishment of an event. Nowadays such analysis – including resourcing – is nearly always done with the aid of scheduling software packages – many of which have muddied the terminology (e.g., the use of PERT) and indeed practice (e.g., regarding WBSs). (See Figure 9.5.)

Buy-in

There is often considerable iteration in developing an acceptable schedule, and a big behavioural input may be required in getting the schedule 'bought into', 'owned' and committed to by team members (and the functional groups they represent). The most immediate iterations will typically be with resourcing and budget. At a high level, the resourcing issue is more likely to be whether the required resources are available, and the cost implications, and if they are not available, what are the options and consequences? Then they need to be worked out in terms of the budget. Both resourcing and motivation are important features of Critical Chain Project Management which we shall discuss in a moment.

Resource scheduling

Resourcing can become a major part of scheduling (and, as we shall see, in Critical Chain Project Management, it is *the* driving part). Ed Merrow, calling on IPA data, finds schedules in process engineering projects too often to be unachievable, failing to be "fully integrated, networked, [and] resource-loaded"[9].

Project and program resources generally refer to manpower, machines (plant and equipment), money, and materials, though there may be other kinds as well, for example, legal approvals or languages[10]. Different resources need identifying for individual activities. Resources are then 'allocated' to activities and aggregated over time (day by day, week by week, etc.) to provide durations. Resources may then need to be 'smoothed' or 'balanced' to reduce peaks and troughs and reduce inefficiencies.

Resourcing is often a major feature of program and portfolio management. Programs may share resources and trade-off decisions regarding resource availability and the impact on the forecast business case are often an important part of the program manager's responsibilities. Portfolio management, while being less managerially directional on a day-to-day basis, similarly has to model the impact of future product capacity, capability and profitability given forecast resource constraints.

Many project management practitioners have found the bottom-up approaches previously described to be too chunky and heavy and insufficiently flexible for the more dynamic reality presented in much of the multi-project enterprise-wide world of many of today's projects and programs.

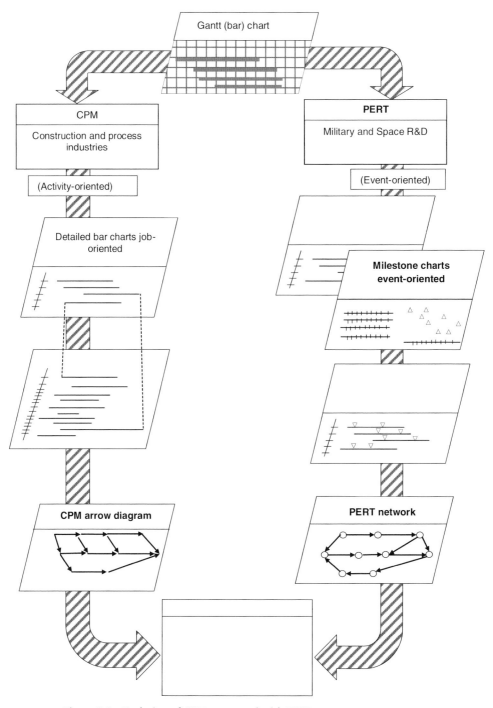

Figure 9.5   Evolution of CPM compared with PERT.
Source: © Archibald R. D. and Villoria R. L. (1967), *Network based management systems (PERT/CPM)*, John Wiley & Sons: New York.)

Software project management in particular has rebelled against the heavy formality of several of the techniques developed to manage large-scale, one-time, non-routine projects, and nowadays all kinds of management are expressed in terms of projects. The most obvious example is Agile Project Management, Chapter 5, pp. 90–1, but in the area of scheduling the important example is Critical Chain Project Management.

Critical Chain[11]

Though computer-based scheduling was propelled by the rising accessibility of desktop computing in the 1990s, there were no really new developments in scheduling until Critical Chain hit the deck in 1997 (see Figure 9.6).

Critical Chain was developed as an outgrowth of Eli Goldratt's 'Theory of Constraints'[12]. This is based on the premise that the rate of goal achievement in an undertaking is limited by at least one constraining process. Only by increasing flow through the constraint can overall throughput be increased. One can do this by:

a. identifying the constraint (the resource or policy that prevents the organisation from obtaining more of the goal);
b. deciding how to exploit the constraint;
c. subordinating all other processes to the above decision;
d. elevating the constraint (make other major changes needed to break the constraint).

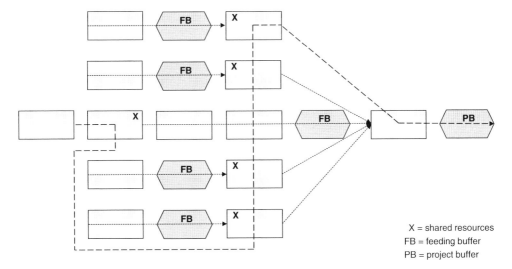

X = shared resources
FB = feeding buffer
PB = project buffer

Figure 9.6   Critical Chain.
Note: Resources and buffers are initially distributed across the project's activities. Both resources and buffers are amalgamated at the project level at some point.
Source: Author's own.

If, as a result of these steps, the constraint moves but is not released, then reiterate the process.

Essentially, Critical Chain Project Management comprises:

- A more integrated approach to resourcing (arguing that resource alloca-tion can change the critical path, which it obviously can – though this is not an insight unique to critical chain). While Critical Path and PERT emphasises task/activity sequencing, Critical Chain aims to keep the resources level but requires them to be flexible in their start times and able to switch between tasks and task chains quickly to keep the whole project on schedule.
- More centralised management of contingencies (not putting these to individual activities but collecting and managing them as one group – the buffer). Buffers are gathered at the end of each sequence of tasks that feed into the critical chain. When the schedule plan is complete and the project is ready to begin, the network is fixed and the buffers' sizes are locked, i.e., their planned duration may not be altered during the project. They are then used to monitor project and financial performance. If the rate of consumption is such that there is likely to be little or no buffer at the end of the project, then corrective actions or recovery plans must be developed either by the avoidance of multi-tasking (which is often of questionable practicability) or using behavioural factors in getting com-mitment to the schedule and motivating the team. Indeed there is anec-dotal evidence that much of the real value of Critical Chain is due to the motivational energy that its newness generates. (A 'Hawthorne' effect.)

Each element on the project is encouraged to move as quickly as they can: when they are running their "leg" of the project, they should be focused on completing the assigned task as quickly as possible, with no distractions or multi-tasking. Resources are encouraged to focus on the task at hand, com-plete it and then hand it off to the next person or group. The goal is to overcome the tendency to delay work or to do extra work when there seems to be time.

All this is clearly appropriate more, or even only, at a small-scale level of operations. For example, in practice, most people multi-task, and have to. Critical Chain would not seem very useful or appropriate in planning or running big projects and programs.

Last planner[13]

Although developed independently of Critical Chain in 1988, Last Planner fits well with it. Last Planner addresses the problem of tasks starting when, although it is their scheduled start date, they have not yet the required resources. Last Planner essentially prevents tasks starting if all resources are not yet available. (Called Last Planner because this is the last step in the scheduling process.)

The planning is very short-term: an horizon of about a week.

Estimating[14]

In the early 1970s, David Packard, the then US Deputy Secretary of Defense, identified program changes, poor risk identification and overoptimism in cost estimating as the three main causes of the alarming cost growth that DoD was experiencing. Packard ordered that methods to improve cost estimating be proposed[15]. If only life were that simple! There's certainly been progress since then but estimating remains a difficult area. (DoD costs continued to grow.)

Estimating is usually used to refer to cost estimating, though literally it could also apply to time or resource estimating (otherwise termed 'planning' which then gets confused between strategic planning and activity planning). We here use Estimating to refer primarily to cost.

Data may be drawn from parametrics, synthetics, or standards based on past records. Feedback and 'lessons learned' logically should be (but too often in practice are not) important inputs to the estimator's knowledge base.

Estimates can be, and generally are, developed both 'bottom-up' and 'top-down' – for different reasons: top-down to derive 'ball-park' order-of-magnitude estimates. For example, cost consultants keep data on completed projects such as construction costs per square metre of building space, or mile of pipe laid. Bottom-up estimating is used as a means of building-up estimates and as detailed checks to verify estimate realism. For example, contractors preparing a bid aggregate the costs of resources required, adding contingency (for unknowns and risks), overheads and profit. In construction this is pretty straightforward. In software projects it is not. Software estimating is especially problematic.

Software estimating

Software is harder to visualise and specify than most types of project, largely because of its intangibility and its concomitant challenges of description. Software estimates are often made on the basis of high-level user requirements. From these, the functional requirements can be set out and analysed and thus the resource and schedule estimated. But the picture of the software to be built is initially often quite fuzzy and thus software project estimation is generally a process of gradual refinement. Even after requirements have been identified it is often difficult to assess the amount of effort that will be required. Hence, as in the difference between the linearly based, deterministic CPM for construction versus the event-oriented, probabilistic PERT for R&D, software estimating is more fuzzy, more probabilistic than for construction*.

---

*Chris Chapman of Southampton University quotes an IBM 'old timer' in the 1960s as advising estimating by "working out the best estimates and then multiplying by 3". Nowadays, says Chris, its better to multiply by π since it looks more scientific! Chapman, C. and Ward, S. (1997), *Project risk management*, Chichester: Wiley, p. 187.

A variety of techniques exist for addressing software estimating, none wonderfully successful (but necessary for a contractor in preparing a bid). Two of the most commonly used are Functional Point Analysis and COCOMO.

- Functional Point Analysis counts the size of the project in 'source lines of code' (SLOC) – but this is quite hard to do at the outset of the project. An alternative method (known as Function Points), introduced in 1977 by IBM, measures the size of a project based on the features that are to be implemented; unlike SLOC, this is easier to define early in the project lifecycle, certainly before starting to code.
- COCOMO (II, successor to COCOMO 81) is based around Object Points (renamed 'Application Points') – a count of the screens, reports and language modules to be developed, each weighted by a complexity factor.

Of course, all these methods assume that there will not be major changes in requirements once the estimate has been given. This is often not the case however – which, as we've seen, became a direct cause of Agile project management[16].

Estimating and assessing performance

Which makes the point that the *estimating competency deployed in a project directly influences the measurement of project management success*. Or put another way, the effective management of projects entails effective estimating. Too often, however, the Estimating function is performed by specialists who are not part of the 'projects' function, and project management has little or no engagement in the formulation of the project estimate(s). It would be surprising if the project or program manager did not have an opinion on the validity of the estimates being developed, and it would be regrettable therefore if she, or he, did not contribute to the estimates' preparation.

There is generally an interaction between the way the work is to be performed (WBS, capability acquisition strategy, risk, etc.) and the aggregation of 'bottom-up' estimating 'norms'. Again, this highlights the need to involve the project or program management team in the estimating process.

It is important that the assumptions underlying the make-up of the estimate be documented and that an audit trail be available.

Uncertainty, contingencies and low-balling

Since estimating is about future events, there is bound to be uncertainty in the estimates. There are various methods of handling these uncertainties. Most estimates have, at a minimum, a contingency associated with them to cover the associated risk or uncertainties. Many give three-point estimates: either P50, P90, P10 probability confidence levels – 50% probability and so on; or minimum (optimistic), most likely (generally mode, not mean), maximum (pessimistic) estimates. Ideally the shape of the probability distribution should be considered. Some techniques require separate forecasts for risk (here meant as negative occurrences) and uncertainty (lack of valid information). Sensitivity analyses can be performed on the estimates to explore the impact of different assumptions.

The most obvious instance of decreasing estimate uncertainty is in the early definitional stages of the project. The well-known concept of the 'estimating funnel' represents the progressive reduction in uncertainty as additional definition information becomes available (see Figure 9.7).

It shouldn't need saying that getting estimates wrong is likely to have a major impact on project profitability/success.

But enthusiasm will often cloud judgement. Beware the optimist. Project people tend to see a third-full bottle as half-full; synergy as 1 + 1 = 4. Or as the slightly over-imbibed project manager put it: "a little too much would be just right"!

Bent Flyvbjerg and his colleagues at Aalborg University (he is now at Oxford University) have commented on this tendency for enthusiasm to lead to undue estimating optimism – 'optimism bias'. To counter this, Kahneman *et al.* have advocated 'reference class' forecasting – obtaining an outside view – as a way of getting a reality-check on proposed estimates[17].

Worse, Flyvbjerg has shown that estimates of some public sector projects – his data originally comes from the transport sector – are often deliberately pitched low as a means for getting the project launched ('strategic misrepresentation'), knowing that if the real estimate of the project cost were published, the project would never get sanctioned[18]. (This has to be as a

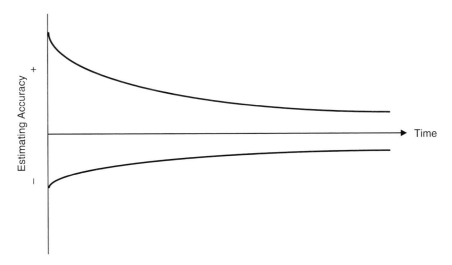

Figure 9.7   The estimating funnel.

Note: The uncertainty reduces – often quite quickly – as the project develops. The uncertainty is not equally balanced – typically there is greater range on the upside figures.

Source: Author's own.

result of poor and inadequate governance, that is, ineffective Assurance). Ed Merrow of IPA, however, believes that while this 'buy-in and hook' practice is "not unknown in private sector ventures, it is not very common, simply because there is usually no taxpayer available to foot the bill later"[19].) What this ignores, however, is the common practice of contractors low-balling their bid prices in the expectation that they will be able to get the contract (project) cost increased during the course of the project through changes and contract claims.

Budgeting

Budgeting is the activity of establishing from the cost estimate the financial budget that is to be authorised and against which expenditure may, within agreed procedures, be made. The completion of the project within its sanctioned budget is normally a central aim of project and program management.

Budgets are typically prepared either for the total project or program expenditure (predominantly capital costs, though possibly including some operating costs); for stages of a project or program; or for a specific time period. On many public sector projects (and some private sector ones, for example, on pharmaceutical drug development), the budget may only be available for the current financial year. This does not facilitate good project planning or control. Budgets, and plans generally, should pertain as far as possible to the full project life-cycle (see next section). Doing so facilitates realism in expenditure control.

Budgets should be authorised by the project or program 'sponsor'. It is, after all, their money that's being spent! Budgets, like scope and schedule, should be baselined; baselining and re-baselining should only occur with the approval of the sponsor, representing governance.

The budget will be broken down into budget items which will generally relate to the WBS, though there may be some differences. It should therefore be possible to develop the budget in greater detail (while still keeping the overall amount fixed) as the project develops and more information becomes available. In very 'foggy' projects* it may not be possible to budget far into the future with any certainty, making baselining very brittle. In these cases regular budget reviews or triggers need to be set to ensure projects do not go out of control from the budget perspective.

As commitments are made – contracts let, resources acquired, and so on – the overall level of project cost commitment should be scrutinised against the authorised budget.

Budgets have a direct input into Earned Value systems (see pp. 140–1) an inappropriately budgeted item will produce an inappropriate set of budgeted cost measures (BCWS and BCWP).

Generally, budgets should be prepared with the project or program manager's input, though this does not always happen. In any case, since the

---

*The term is Eddie Obeng's.

project or program manager are generally accountable for completing the project or program within budget, they must be totally familiar with the budget and should work with the cost management team (including estimating) to ensure the project or program completes within the authorised budget.

Cost Management

Cost management tracks the accrual of costs against the budget, often by line item or work package. It is concerned to know the true cost to the project of the work performed to date vis-à-vis the budget allowance for that work, and whether the forecast outturn for the project or program will be under, over, or on budget. This also involves understanding how and why actual costs have occurred, and advising management on the cost implications of responses taken or being considered to ensure that costs, or schedule, come within target. Cost management thus becomes a part of the overall project and program monitoring and performance measurement function.

Extending this point, cost control is more than just monitoring and measuring: it is about actively controlling costs so that they stay within budget. This may require proposals about technical or schedule options and often involves detailed consideration of what is allowed under the suppliers' contracts relevant to performance of the work. Thus cost management also often interacts with contract administration.

Cost management becomes much more of a significant task where contracts are reimbursable (even if only in re-measure form) than fixed price/ lump sum, for the obvious reasons of the greater checking and administration that the former require.

Cost management often has an important role to play in advising project and program management on 'triage options': that is, on looking at whether trade-offs in requirements/scope, or schedule, may be necessary in order for the overall project to stay within budget, scope or schedule.

Cash flow

Managing cash flow can be hugely important, in projects as in life generally. Cash forecasts are derived by allocating the budget across the schedule and taking account of leads and lags in payment (advances, delays in measuring, invoicing and paying, etc.).

Projects consume resources rapidly as they move through the development cycle (as represented by the typical 'S' curve – Figure 9.3) and are typically cash-negative until they begin generating revenue*. This can have

---

*Note that the effect of schedule delay, specifically completion delay, on profitability: interest is charged on the maximum amount of project indebtedness. This is what caused such trouble for Eurotunnel, for example with the delay on completion of the Channel Tunnel (Morris, P.W.G. [1994], *The management of projects*, Thomas Telford: London, p. 175).

disastrous consequences if not managed appropriately (e.g. by owners ensuring proper business plans and adequate lines of finance, and suppliers arranging lines of credit and stage payments).

Performance Management (Earned Value)[20]

Project management is about integrating all that needs to be done to achieve project success. Success has multiple measures but even at this relatively straightforward Level 1 level of planning and control, management will have to plan and control across several interdependent parameters. Performance Management is a broad term for doing this. Performance Management will inevitably involve topics above and beyond those of cost and schedule discussed so far (e.g., benefits management, technical performance most obviously but also, say, stakeholders' attitudes, environmental issues, and profitability) but for the moment we should at least look at these two. Not to do so would leave the discussion of planning and control incomplete.

Having established project targets, project control involves measuring performance against those targets and taking corrective action where necessary to ensure action stays in line with them.

The frequency of measurement and reporting will depend on the rate at which resources are being consumed and the need for control. Highly expensive or critical work may need recording or reporting on a daily basis, or for even shorter periods – as in plant shutdown or track renovation projects or systems testing, for example.

Two defining characteristics of project control are that it should:

- work off accurate measures of the 'real' position at a point in time,
- focus on the forecast position at project or program completion. ('Trend analysis' is particularly important in this.)

Thus in cost control, for example, cost measurements should be based on the cost of work actually done by a certain time (milestone for example) compared with the corresponding budget, having removed distortions due to payment advances, retentions, invoicing and payment delays; and the focus is then on forecasting 'outturn cost' against the project budget. Fundamental to doing this effectively is physically measuring and reporting the work actually done and still to be done. Knowing this, the costs (or more correctly, 'value') 'earned' and yet to be incurred, or the effect on the schedule of work still to be done, can be calculated and reported, their effect on the project meeting its targets assessed, and relevant action initiated.

'Earned Value' is a technique for integrating cost and schedule performance measurement. It does this by representing physical work accomplished in terms of the relative financial worth accrued*. Earned Value Manage-

---

*Construction 'Bills of Quantities' are a similar form of 'performance measurement' since physical progress is measured in value-earned terms, though they generally apply to the site works portion of the project rather than design or [overhead] management.

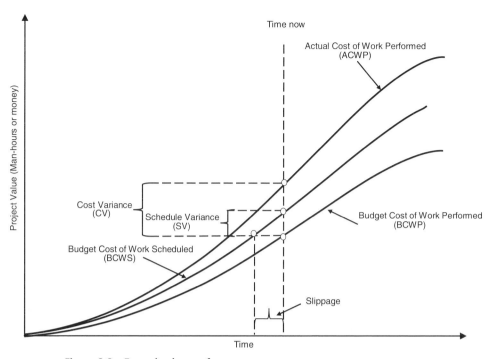

Figure 9.8   Earned value performance measures.
Source: © Morris P. W. G. (1994), *The management of projects*, Thomas Telford: London, p. 46.)

ment* is the process of representing physical progress achieved on the project in terms of a cost-based measure (i.e., money). Various rules and techniques are used to represent the value of work performed to date as a proportion of the total project value (e.g., how to 'book' the value of work done in the factory but not yet shipped, transported, installed or tested; or how to deal with the R&D tests done but whose results may be either totally negative or [quite] positive). Specifically, three basic measures are used: Budgeted Cost of Work Scheduled (BCWS), Budgeted Cost of Work Performed (BCWP), and Actual Cost of Work Performed (ACWP). (There are several other measures and ratios as can be seen in Figure 9.8. Comparing BCWP with BCWS gives the Cost Performance Index (CPI); comparing ACWP with BCWP gives the Schedule Performance Index (SPI). Earned Value Management only works effectively if the budgets have been estimated and allocated properly, if non-man-hour items (e.g., expensive equipment) do not distort control visibility, and if measurement is appropriate (e.g., handling time for testing) – and of course so long as 'project control' has full visibility of actual costs, which the owner doesn't have on Fixed Price contracts. There are other topics which could fairly be discussed at this technical core level (Level 1) such as Quality and Risk, and others which

---

*In many organisations, Cost/Schedule Control Systems Criteria (C/SCSC, also known as C-SPEC) are used as a specific version of Earned Value.

really are Level 2 performance measures, such as Benefits Management*. The Balanced Score Card (BSC) maybe should be mentioned here: it has proved itself to be immensely influential in corporate performance reporting, though not notably for projects or programs[21]. BSC measures performance against four clusters: financial, customer satisfaction, employee satisfaction, and the longer-term growth of the company (such as learning). Various attempts have been made to adapt this approach to project and program management but with varied and on the whole not very great success[22].

Basically, there is generally lots of opportunity for us to develop more comprehensive measures – particularly effectiveness measures: see Chapter 5, p. 76 – rather than for us to simply fall-back instinctively onto established formulae like EVA or BSC, and to have a much richer on-going analysis – for example a risk analysis of likely benefit realisation.

But we are clearly now moving away from Level 1 reporting. It is time therefore to move to the area which historically came immediately after the development of project planning and control, and which intellectually provides still one of the most fundamental bases for understanding the need for project management: organisation. We shall begin with discussion of the project (or product development) life-cycle.

References and Endnotes

[1] Archibald, R. D. (1976, 1997, 2003), *Managing high technology programs and projects*, Wiley: New York; Charvat, J. (2003), *Project management methodologies: selecting, implementing, and supporting methodologies and processes for projects*, Wiley: Chichester; Lewis, J. P. (2005), *Project planning, scheduling and control*, 4th edition, McGraw-Hill: New York; Kerzner, H. (2003), *Project management: a systems approach to planning, scheduling and controlling*, 8th edition, Wiley: Hoboken, NJ; Meredith, J. R. and Mantel, S. M. (2002), *Project management: a managerial approach*, 5th edition, Wiley: Hoboken, NJ; Shtub, A., Bard, J. F. and Globerson, S. (2004), *Project management: processes, methodologies and economics*, Prentice Hall: Upper Saddle River, NJ; Turner, J. R. (2000), Managing scope – configuration and work methods, in: Turner, J. R. and Simister, S. J. (eds.) *Gower handbook of project management*, 3rd edition, Gower: Aldershot; Williams, D. and Parr, T. (2006), *Enterprise programme management: delivering value*, revised edition, Palgrave Macmillan: Basingstoke.

*There is a view, generated by OGC and PMI, that projects do not generate benefits, only programs do. In reality, projects generate benefits as well as programs – though the opportunity, and need, to focus on benefits realisation is particularly important in programs. Upgrading an operating retail unit, for example, would, in most people's minds, be a project, yet benefits will need to be realised from the upgrade work as the project develops. Benefits, as described below on pages 189–190, should be identified, risk assessed, monitored, and reported on as the project or program evolves so that action can be taken to ensure they are realised ('harvested') to the optimum extent.

[2] Morris, P. W. G. and Jamieson, H. A. (2004), *Translating corporate strategy into project strategy*, Project Management Institute: Newton Square, PA.

[3] *Ibid.*: p.202.

[4] BS 10006:2003 *Quality management systems – Guidelines for quality management in projects*, BSI: London; Berlack H. R. (1991), *Software configuration management*, Wiley: Hoboken, NJ; BS 6079-1 (2002), *Guide to project management*, BSI: London; Kidd, C. and Burgess, T. F. (2004), Managing configurations and data for effective project management, in: Morris, P. W. G. and Pinto, J. K. (eds.) *The Wiley guide to managing projects*, Wiley: Hoboken, NJ.

[5] ISO 10007:2003, *Quality management systems – Guidelines for configuration management*, BSI: London.

[6] BS 6079-1 (2002) *op. cit.*: 4; Kerzner, H. (2003), *op. cit.*: 1.; Lester, A. (2004), *Project planning and control*, 4th edition, Butterworth-Heinemann: Oxford; Lock, D. (2000), Managing the schedule, In: Turner, J. R. and Simister, S. J. (eds.) *op. cit.*: 1; Lockyer, K. G. and Gordon, J. (1996), *Project management and project network techniques*, 6th edition, Financial Times/Pitman Publishing: London; Pinto, J. K. ed. (1998), *Project management handbook*, Project Management Institute: Newton Square, PA; Project Management Institute (2004), *A guide to the project management body of knowledge*, Project Management Institute: Newton Square, PA; Mahmoud-Jouini, S. B., Midler, C. and Garel, G. (2004), Time-to-market vs. time-to-delivery: managing speed in engineering, procurement and construction projects, *International Journal of Project Management*, 22, 5, pp. 359–367; Moder, J. J., Phillips, C. R. and Davis, E. W. (1995), *Project management with CPM, PERT and precedence diagramming*, 3rd edition, Blitz Publishing Company: Middleton, WI; Morris, P. W. G. (1994), *The management of projects*, Thomas Telford: London; Raz, T., Barnes, R. and Dvir, D. (2003), A critical look at critical chain project management, *Project Management Journal*, 34, 4, pp. 24–32; Shtub, A., Bard, J. F. and Globerson, S. (2004), *op. cit.*: 1.

[7] Anumba, C. J., Baugh, C. and Khalfan, M. M. A. (2002), Organisational structures to support concurrent engineering in construction, *Industrial Management & Data Systems*, 102, 5, pp. 260–270; Anumba, C., Kamara, J. M., and Cutting-Decelle, A. F. (2006), *Concurrent engineering in construction projects*, Routledge: London; Dowlatshashi S. (1994), A comparison of approaches to concurrent engineering, *The International Journal of Advanced Manufacturing Technology*, 9, 2, pp. 106–113; Prasad, B. (1996), *Concurrent engineering fundamentals*, Volumes I and II, Prentice Hall: New York; Gerwin, D. and Susman, G. (1996), Special issue on concurrent engineering, *IEEE Transactions on Engineering Management*, 32, 2, pp. 118–123.

[8] Williams, T. (2002), *Modeling complex projects*, John Wiley & Sons: Chichester.

[9] Merrow, E. (2011), *Industrial mega-projects*, Wiley: Hoboken, NJ, page 319.

[10] Gil, N. (2010), Language as a resource in project management: a case study and a conceptual framework, *IEEE Transactions on Engineering Management*, 57, 3, pp. 450–462.

[11] Goldratt, E. M. (1997), *Critical chain*, North River Press: Great Barrington, MA; Herroelen, W., & Leus, R. (2001). On the merits and pitfalls of critical chain scheduling, *Journal of Operations Management*, 19, 5, pp. 559–577.

[12] *Ibid.*

[13] Ballard, G. and Howell, G. (1998), Shielding production: essential step in production control, *Journal of Construction Engineering and Management*, 124, 1, pp: 11–17; Winch, G. M. (2010), *Managing construction projects*, Wiley Blackwell: Chichester, pp. 299–300.

[14] Clark, F. D. and Lorenzoni, A. B. (1996), *Applied cost engineering*, Marcel Dekker: New York; Humphreys, K. K. and Wellman, P. (1995), *Basic cost engineering*, 3rd edition, Marcel Dekker: New York; Smith, N. J. (ed.) (1995), *Project cost estimating*, Thomas Telford: London; Shtub, A., Bard, J. F. and Globerson, S. (2004), *op. cit.*: 1.

[15] Acker, D. D. (1980), The maturing of the DoD acquisition process, *Defense Systems Management Review*, 3, 3, pp. 7–77.

[16] Leffingwell, D. (2007), *Scaling software agility*, Addison Wesley: Upper Saddle River, NJ, Chapter 2.

[17] Khaneman, D. (2011), *Thinking fast and slow*, Penguin: London, page 251.

[18] Flyvbjerg, B., Holm, M. S. K. and Buhl, S. (2002), Underestimating costs in public works projects: Error or lie? *Journal of the American Planning Association*, 68, 3, pp. 279–295; Flyvbjerg, B., Bruzelius, N. and Rothengatter, W. (2003), *Megaprojects and risk: an anatomy of ambition*, Cambridge University Press: Cambridge; Flyvbjerg, B., Holm, M. S. K. and Buhl, S. (2005), How (in)accurate are demand forecasts in public works projects? *Journal of the American Planning Association*, 71, 2, pp. 131–146.

[19] Merrow, E. (2011), *op. cit.*: 9: p. 20.

[20] Bent, J. A. and Humphreys, K. R. (1996), *Effective project management through applied cost and schedule control*, Marcel Dekker: New York; Fleming, Q. W. (1993), *Cost/schedule control systems criteria*, McGraw-Hill: New York; Fleming, Q. W. and Koppelman, J. M. (1996), *Earned value project management*,: PMI: Newton Square, PA; Kerzner, H. (2003), *op. cit.*: 1; Kim, E., Wells, W. G. Jr. and Duffey, M. R. (2003), A model for effective implementation of Earned Value Management methodology, *International Journal of Project Management*, 22, 2, pp. 87–98; Shtub, A., Bard, J. F. and Globerson, S. (2004), *op. cit.*: 1; Webb, A. (2003), *Using earned value: a project manager's guide*, Gower: Aldershot.

[21] Norrie, J and Walker H. T. (2004), A balanced scorecard approach to project management leadership, *Project Management Journal*, 35, December, pp. 47–57.

[22] Kaplan, R. S. and Norton, D. P. (1991), The balanced scorecard: measures that drive performance, *Harvard Business Review*, 70, 1, pp. 71–79.

Organisation

Within the context of project and program management, organisation theory seeks to explain, amongst many other things concerning the nature of organisations, both how the integration required to achieve project goals varies – why it does and what form it takes – and how organisation structure varies depending upon the environment it is operating in and the technology it is using (so-called contingency theory). We shall begin, however, by looking at project management roles before turning to these issues.

Roles and Responsibilities

"This project's a mess," said the harassed executive, "no definition, no schedule, no idea how much money's going to be needed. Who's supposed to be in charge?" "You are," came the reply.

It helps to know 'who does what' on a project. Classically this was shown via the Task–Responsibility Matrix (which Cleland and King spent much time describing in their 1965 classic *Project Management and Systems Analysis*) which was developed from the Organisation Breakdown Structure (OBS) concept. Nowadays, however, accountabilities and responsibilities – there should only be one person accountable for a task or function but there may be several who are responsible for it – are generally shown on RACI (or RASI) charts. These show who is responsible, accountable, to be consulted on (or supported) or informed, in an organisation, or with regard to a key practice or process.

Sponsor-project manager roles

The two key roles on any project or program are those of sponsor and project/program manager. At the absolute heart of effective project management is the relationship between the sponsor and the project [delivery] team.

Reconstructing Project Management, First Edition. Peter W.G. Morris.
© 2013 John Wiley & Sons, Ltd. Published 2013 by John Wiley & Sons, Ltd.

The sponsor has a particular role in the client organisation. The **Client** is the term generally used for the person/organisation for whom the Supplier (Contractor) provides services. The **sponsor** is a member of the Client organisation: he, or she, is the owner of the project business case. He/she represents the funder's interests. (Though suppliers will have sponsors too, looking after the business success of their supply contract.) The sponsor has a key role in assuring proper governance of the project. Slowly, over the last 30 or more years, it has become progressively recognised that the client – and particularly the sponsor – can exert a dominant role in influencing the likelihood of project or program success[1]. (Interestingly, as I write this, the British Government is initiating a program with the University of Oxford to build sponsor competence in its top 200 or so civil servants.)

Significantly, the sponsor him- (or her-) self may have no particular competence in the management of projects. Nevertheless, his conduct, particularly in his demands and demeanour in chairing stage-gate reviews, can arguably make him the single most influential 'actor' on the project, with a disproportionately high impact on outcome success[2]. Yet few sponsors appear to get any project management training, and, worse, their incentive packages are often geared more to getting revenue-earning production capacity on-line quickly rather than adhering to project management process, for example, by requesting re-work to and re-submission of stage-gate documentation.

Since project management is not what the sponsor is there to provide, the owner may employ a project or program manager (or director) to manage the development of the project definition and, after its sanction by governance, its execution (see Figure 10.1). Note of course that this project/program manager is not filling the same role as the sponsor. The sponsor is the budget holder; the project management is the execution executive.

This said, principal amongst the sponsor's responsibilities will be[3]:

- defining strategic intent for the project/program,
- establishing values,
- ensuring appropriate assurance reviews and follow-up actions*,
- setting contingencies in the light of unknowns and risks prior to sanction,
- assessing progress and risks against the project/program business case.

---

*Resourcing and interruptions to project activity are two challenges commonly faced in such assurance reviews. Thus, the UK Government established a Major Projects Authority in 2012 to improve its assurance reports, the quality and visibility of data on its capital projects, and the quality of fits sponsorship in general. (As of May 2012, "there were 205 projects in the UK Government's Major Project Portfolio, with a combined whole-life cost of £376 billion, and annual cost of £14.6 billion. Thirty nine of those projects [had] a delivery confidence rating of 'red' or 'amber/red'". National Audit Office Press Release: *Assurance for major projects*, 2 May 2012.

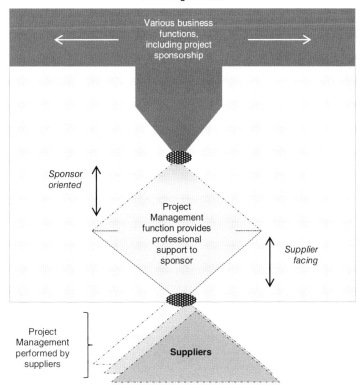

Owner organization

Various business functions, including project sponsorship

Project management can be performed both for the sponsor and for/by suppliers. On large projects several hundred people may be involved working on the project teams

Sponsor oriented

Project Management function provides professional support to sponsor

Supplier facing

Project Management performed by suppliers

Suppliers

Figure 10.1   Owner/supplier p.m. roles.
Source: Author's own.

Other p.m. roles

Project and Program Managers may work for the sponsor directly, or for one of the suppliers, whether prime or sub – or indeed, for someone else, for example, local government. There are several variants of this p.m. role.

- A **Project/Program Director** is generally used as a term for a director-level person responsible for managing a group of project or program managers. The Director may be expected to be especially concerned with establishing strategy, vision and direction, and providing leadership.
- A **Project Leader** is a term used in certain industries (such as drug development) for 'leading' the development of the project. The precise meaning of this may be a little elusive but often covers the role of developing the technical strategy and solution – rather like an architect in construction or a systems engineer (the Technical Design Authority) – with the project manager looking after the time, cost and scope issues; a distinction not unlike that found in movies between 'the Director' and 'the Producer'.

- A **Program Manager** is generally responsible for the overall development of a suite of interconnected projects (to achieve strategic change or a product family). The program manager will work closely with project teams, including project managers, but will generally tend to be involved in more strategic issues centred on inter-project prioritisation and resourcing while the project manager will be more involved in project-specific issues.
- The **Project Manager** is responsible for delivering the project to the agreed project targets (schedule, budget, scope/specification). The project manager is the 'single point of integrative responsibility'[4]: the person who pulls everything together so that the project achieves its Key Performance Indicators (KPIs). This means the project manager has a role in contributing to the design of strategies and solutions as well as controlling their evolution; leading people as well as managing for results.
- The term **Project Coordinator** is sometimes used to represent a function similar to project managers but with less authority: with a coordinating rather than managing role.
- **Project Controls** is concerned with planning, estimating and possibly measuring and reporting progress. (In construction, a Quantity Surveyor [QS] or Cost Engineer is more a Project Controls professional – though a QS also has a strong contracts management and financials dimension too – while an engineer would typically do the planning.) Many organisations have a **Cost Manager** as a separate role. A **Project Planner** is a specialist in preparing project schedules.

The **Project/Program Support Office** supports the enterprise through the examination of project or program status information and the provision of project help and support as may be required. (Various similar terms exist, clustered around the words Project, Program, Management, Support and Office.) The P(M)SO is often the holder of best project and program management practice in the organisation through the provision of guidance (methodologies, guides, analytics, competency grids, training, etc. and even project reviews).

There are many specialist roles that will be represented on projects. Among the more notable are the following.

- *User representative* – Users are very important and may need to be directly represented in the team. PRINCE2 has a Senior User representing "all those who will use the final product[s]" on the Project Board. A single representative may not always be practicable, however.
- *Technical design authority* – Such as The Architect in building, The Engineer in civil engineering, the Systems Engineer, and so on. These persons are responsible for the technical decisions on the project or program. The project or program manager should work with them on decisions and processes to meet the project or program requirements in the optimum manner.
- *Work package manager* – The person responsible for the performance of the work contained in a work package.

- *Resource manager* – In many matrix organisations there is a separate role for allocating resources between functional lines and projects and programs.
- *Contract managers/administrators* – Procurement of services or products for the project is often performed either by a central procurement function or by the project under the functional direction of a central procurement function. Contract administration, however, is invariably performed by members of the project team in some form, sometimes totally integrated into the team but also sometimes in separate departments.
- *Configuration manager* – Where configuration management is practised, the configuration management function will be managed by a configuration manager.

There are countless 'others' who may be involved with the project or program – engineers, scientists, architects, analysts, business analysts and business information managers, marketing managers, production managers, construction managers, expediters, regulators, inspectors, quality managers, and so on.

One of the most important terms in the management of projects and programs is that of *stakeholders*: those having an interest in, or the ability to influence, the outcome of the project or program: in some way they will all need identifying, then managing, and some may also need representing.

For some, the **Project Champion** is an important role[5]. The champion is someone who promotes the project, or program. This is more of an informal role than the formal ones presented above. It can be very important. Equally, it can be dangerous: for example, there may have been many occasions where project champions lost objectivity and pressed for the project when it should rather be being subject to more critical scrutiny.

The project management roles vis-à-vis technical and commercial roles have changed in recent years. The project management team is now more likely to be involved in a hands-on way with regard to ensuring requirements are properly elicited and their implications to the project properly evaluated than was the case, say, in the late 90s. Commercial management, on the other hand, has, if anything, become more separate. But as we shall see in Chapter 20, project management is now likely to be much more proactive in getting the best out of these parts of the project.

Roles are, of course, organisation- and context-specific. Hence they do need defining. There will need to be some tolerance within these definitions to allow for adaptability between different organisational contexts, projects and programs, and stage of the life-cycle.

Structure

The project's organisation structure defines, and is defined by, its reporting lines and relationships, processes, systems and procedures[6]. Structure helps us communicate, decide, direct and achieve this. Structure, particularly in projects, is not static – it responds to the environment and to the technology of production[7], as reflected by what stage in the life-cycle one is at, for example. And to the project's size, speed and complexity, and so on[8].

There are several basic ways of structuring projects. Across these structures, various actors strut. Most organisational writing separates the influence of people (actors) from structure. Anthony Giddens' Structuration Theory is one that does not. Giddens looks at how actors both shape and are shaped by structure (and how structure and actors are shaped by time and space). The theory, which is far from fully worked out, suggests that power enables leaders to shape process[9]. For example, in deciding the way the project life-cycle is laid out (for example, stage overlaps) and managed (for example, gate reviews). This in fact turns out to be one of the most important organisational features of the management of projects.

The project life-cycle

The life-cycle is fundamental to the management of projects. *The one thing that distinguishes projects from non-projects is their project life-cycle**. All projects essentially evolve through the same life-cycle sequence, albeit the terminology might vary a little (as we noted at the outset – Chapter 2, pp. 12–13); non-projects are not based on this development life-cycle. The sequence is something like: Concept, Feasibility, Design, Execution, Cut-over, or Commission; and in certain cases, Operations and Maintenance. Decommissioning or Disposing might also be included.

An alternative basis on which to organise the project is the process(es) by which project or program management should be undertaken, no matter what the type or stage of project or program. This is the basis used by PMI in its *PMBOK® Guide* and, though to a lesser extent, by ISO 21500[10]. The strength of this second approach is that it helps us to conceptualise the order in which project and program management activities are to be addressed. Its disadvantages are: firstly, that there is much that has to be done which is specific to each individual stage of the product development cycle, and secondly, that this sequence does not always hold.

The advantages of using the product development cycle, as in BS 6079, on the other hand, are principally: (a) it brings out more clearly the nature and characteristics of the work in the different stages – there can be a significant difference between the early stages and the later ones; and (b) it can show the management actions needed to control and direct the project as it evolves through its life-cycle.

For example, creativity of design and development are important in the front-end. Value-enhancing actions such as Value Management and Value

---

*Most people use the term 'the project life-cycle' but really it is the product development life-cycle, that is, the product development sequence, as described in the main body of the text. In fact, it could be argued that it is often not really a cycle at all since there is rarely any expectation that once completed, it or the project team will recycle to the front and repeat the process.

Engineering have a potentially very large role (remember the Andrew project – above, Chapter 5, p. 79). Management often needs a lighter, more facilitating touch in the more creative and option-exploring early stages of the development cycle. Leadership has a major role in the front-end in creating and communicating the project vision, motivating the project team, influencing stakeholders, negotiating high-level 'institutional' arrangements and shaping decisions. (See below, Chapter 15, pp. 199–204.) Later, as the project moves towards and into Execution, the emphasis shifts towards detailed planning and resource and capability acquisition (contracting and team formation, etc.). Control becomes much tighter – financially, in terms of detailed scheduling, change control, information management, testing (verification and validation) and so on. (The change from front-end developmental creativity to execution control is reminiscent of Burns and Stalker's shift from 'organic' to 'mechanistic' styles of management[11], see below, p. 156.)

Further, the product development approach shows how certain practices should be used to manage the progression as the project moves through its development cycle, particularly reviews – design and configuration reviews, risk reviews, value management reviews, bid reviews, quality assurance reviews, peer reviews, lessons-learned reviews and so on. (Lessons-learned reviews – learning from projects and feeding this learning to other projects and programs – is an important way of closing the loop – completing the cycle.)

Particularly important is the gate review process: 'stage-gates' are inserted between stages to provide formal review points for governance to check that the project is in good shape both with respect to the sponsor's business aims and with regard to the management of the project, and ultimately to sanction the capital expenditure represented by the project. (Agile, it will be recalled, rejects the stage-gate approach[12].)

Overlapping and sub-dividing stages

Although the gated development process strongly emphasises a staged progression ('heel-to-toe', in Packard's famous phrase[13]), sometimes overlapping of stages does occur. As discussed above in the section 'Scheduling', in Chapter 9, if poorly done, this overlapping can be bad for the project but there are ways of doing it usefully (to a limited extent), for example, by forming integrated teams and information models (as in Concurrent Engineering), or staggering work packages within stages so that while there may be stage overlap, there is not work package overlap. Similarly, it may be decided to initiate a follow-on stage before all the work in the current stage is totally completed, in which case stage maturity (design maturity etc.) must be assessed and all unfinished items cleared of potential disruptive effects on downstream performance.

It is common for some stages to be split into sub-stages – for example, the concept stage can be divided into two sub-stages, where the first would involve a wider range of options than the second, which would investigate the proposed products or services in more detail.

In ICT projects, systems development methodologies, such as the Water-fall, Spiral or Vee models (Figures 5.3, 5.4, and 5.5), can be integrated with the project development cycle[14].

Program management life-cycles[15]

Programs do not have their equivalent to the development life-cycle. They follow the characteristics of their projects, although projects may be under-taken in program 'tranches', as we have already noted.

Program life-cycles tend to be more integrative than project development cycles. Program management will often involve managing projects that are at different stages in their product development cycles. It would be a mistake, however, to ignore the product development nature of their project management activities, and the program management activities associated with them (which is what P³I – Pre-Planned Product Improvement – essen-tially is all about). For example, if considering a major version upgrade, all the strategy, technology, control, capability acquisition, organisation and people issues need to be considered both for the program as a whole and in relation to the information feeding back from the different phases of the projects within the program and from benefits realisation.

Structural Forms

As we've seen, the life-cycle provides the development path that the project will follow. It brings with it its inherent organisational characteristics; the organisation structure defines the roles and relationships between the project actors.

Functional, project and matrix forms

Traditionally, the most commonly discussed organisational forms with respect to projects have been:

- *Functional* – where resources are controlled totally from within their respective functional unit: all engineering personnel are in Engineering, testing is in Quality, procurement in Procurement, and so on;
- *Project* – where resources are allocated on a dedicated basis to a project (or program), from where they are controlled;
- *Matrix* – where resources are controlled functionally by their functional head but are allocated out to projects where they are controlled with regard to their project requirements by the project manager, as agreed between the project manager and the functional head.

These forms are not mutually exclusive: they may exist within or alongside each other – and the ensuing form termed 'hybrid'. (There are obviously other forms. Henry Mintzberg, for example, identified five fundamental 'structural configurations', one of which, the *Ad Hoc* form*, is essentially the project form[16].)

There has been much written about the benefits and challenges of each of these forms. The functional form gives greater control over functional resources but coordination and control between functions is often inefficient. The project form provides much greater cross-functional integration but is often expensive (resources are tied up on the project on a dedicated basis, and they may experience difficulties in being re-allocated to jobs when they come off the project).

Nowadays, the matrix, in one form or another, is probably the most commonly used form by which enterprises organise to undertake projects and programs[17].

There are different degrees of project or functional orientation in matrix organisations, hence leading to the terms strong, balanced or weak matrix, referring to the degree of project orientation.

In the early stages of a project, there may well be stronger functional, central management control over the project as key decisions are made which will establish the fundamental parameters defining the project. Later, as the project progresses towards and into execution, there will be a decentralising move giving more authority to project staff. Later still, as the project approaches the transition into operations, there will be a move back towards more functional control again. I called this move in relative project orientation [doing] the 'matrix swing'[18] (see Figure 10.2).

The matrix is classically held to suffer from the tensions of having two bosses (the functional head and the project manager)[†]. This is best dealt with on the basis of a 'contract' arrangement between the project and functions (which supply resources to perform to the time, budget and scope requirements defined by the project).

There are often challenges to the project or program manager's authority and power. Project managers may be more junior to functional heads and often have to exercise considerable influencing skills in pressing their needs.

The informal organisation (*de facto*) is generally particularly important in projects, and obviously especially so in the matrix organisation. It reflects the web of informal contacts between people working in the organisation. (The formal organisation [*de jure*] represents the official, formal authority and command lines.)

What the project and the matrix forms of organisation are doing essentially is providing integration. This is *the* fundamental organisational function of project management: to bring it all together; to provide, in the project or program manager, "the single point of integrative responsibility"[19] or accountability; and to furnish the control of, and supply the direction to, all that is necessary in order to achieve the project's goals.

---

*The other four are: simple, machine bureaucracy, professional bureaucracy, and divisionalised.
[†]Most of us have lived under two bosses: a mother and a father.

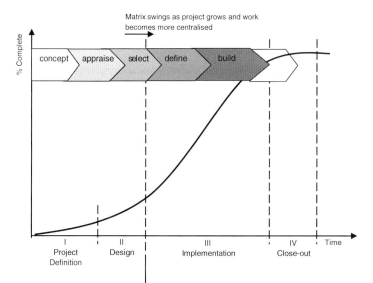

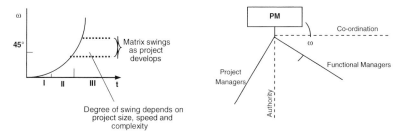

In the early, strategic stages of the project there is a need for strong centralised functional decision-making:

- Decide contracting strategy
- Basic engineering equipment supply
- Basic engineering elaboration
- Prequalification of bidders
- Development and installation of control systems
- Budget and schedule preparation
- Project organisation design and implementation

As contracts are negotiated and signed, work expands, the organisation grows and decision making becomes decentralised- the swing to a project orientation increases rapidly. Project managers are given clear scope, cost and schedule targets to achieve as the matrix moves from a primarily functional orientation to a more project one. Even so, a number of major functional decisions may still be pending – the swing does not happen overnight.

Project issues that may be still outstanding at the phase II/III interface:

- Top level negotiations
- Finance/governmental work
- Organisation expansion
- Administrative items
- Finalisation of project control procedures

Figure 10.2   The matrix swing.

Source: © Reis de Carvalho, E. and Morris, P. W. G. (1978), *Project matrix organizations – or how to do the matrix swing*. Project Management Institute Symposium, Los Angeles, Project Management Institute, Drexel Hill, PA.

Virtual organisations

Since the 1990s, more fluid organisational forms such as virtual organisations and networks have become more common. Virtual organisations are those whose members are geographically dispersed and who may meet physically only rarely, but which exist in a largely informal manner defined predominantly by their communications technology. Drug development teams in 'Big Pharma', for example, may meet regularly but only via video link or teleconferencing. The sense of identity and consequent commitment to the project is inevitably worse overall than were one, say, to go back to the Army metaphor for a moment, to be able to share battle conditions with one's colleagues in the front line.

Virtual organisations raise obvious issues of team identity and commitment and concomitant demands on team leadership[20]. They are important in much of contemporary team working.

Networks

Networks are alliances of discrete organisations that are in communication. Project networks collaborate to achieve the project's goals and objectives. Social Network Analysis uses concepts such as 'weak ties' and 'structural holes' to provide insight into topics such as the sourcing of innovation and the efficiency and effectiveness of communications[21]. Boundary positions are seen to be critical[22] (similar to the insight emphasising interface management in systems theory).

Research on project networks has proved particularly fruitful in such topics as the management of stakeholders, governance and the effect of parent company practices, especially procurement[23]. Procurement strategy and contract forms and administration may significantly pre-determine network patterns. Stephen Pryke of UCL, for example, has shown how construction management forms bring more efficient communications than traditional patterns of contracting[24].

Network organisations are particularly relevant in project supply chains. They can exert a dominating influence on the extended project structure, particularly with regard to determining who is needed, to do what; who has what decision-making authority, and what governance and control are needed.

Networks can be subject to economic distortion – "exploitation either by partners who have control of critical information or by certain suppliers who are able to create and take advantage of dependencies. . . . This is where networks built on friendship, reputation, or shared ideology may prove more effective"[25]. This, we'll see, applies to much contemporary project organisation.

Explaining the basic options for structuring project resources in organisations is just one contribution that organisation theory offers project and program management. Explaining how and why organisation structures vary is another, and, for the project management community, a particular strength is explaining how and why different forms of integration of others' work should be expected. This comes under the rubric of contingency theory.

Contingency Theory and Organisation Design

Back in the 1930s, writers like Mary Parker Follet and Chester Barnard proposed that there were 'universal' rules that 'determined' the design of organisations, rather as there are supposed to be universal traits of leadership (as we shall see below, Chapter 15, pp. 199–200). But we've learnt that organisations are forms of 'open systems' and that as such their structure is influenced by the technology they use (Level 1) and the environmental conditions they operate within (Level 3).

> ### Key Research on Organisational Integration
>
> * Trist and Bamford (1951): socio-technic system
> * Burns and Stalker (1961): organic/mechanistic
> * Woodward (1956): technology/environment
> * Emery and Trist (1960): systems turbulence
> * Miller and Rice (1967): socio-technic organisation design
> * Thompson (1965): integration and interdependency
> * Lawrence and Lorsch (1967): differentiation/integration
> * Galbraith (1973): integration as information processing
> * Davis and Lorsch (1984): matrix organization
> * Mintzberg (1979): adhocracy

Thus, Joan Woodward of Imperial College, showed in 1965 how technological complexity affected organisation structure and performance, while Tom Burns and Gareth Stalker of the Tavistock Institute looked at the impact of stability versus flexibility on structure, leading to the mechanistic versus organic forms referred to previously. Other Tavistock researchers such as Fred Emery and Eric Trist emphasised the pattern of relationships amongst people, coining the term 'socio-technic' systems to reflect the influence of both social and technological factors on organisation[26], later refining their analysis of the environmental context to emphasise environmental uncertainty and complexity. And as we have already seen, Paul Lawrence and Jay Lorsch of Harvard in 1967 dissected the coordination challenge in terms of the differentiation that characterises the organisations that need to be integrated, while James Thompson of Indiana University looked in 1965 at the impact of the kind of interdependence between organisational units – pooled, sequential and reciprocal. Jay Galbraith identified, as we saw, a range of integrating mechanisms, going from a liaison role, through expeditor to project coordinator, project manager and matrix[27]. (The project/ program management office is also an aid in achieving integration, but the PMO hadn't been invented then.)

Almost as a postscript: In 2007, Aaron Shenhar of Stevens Institute and Dov Dvir of Ben Gurion University used contingency theory to suggest not just that project structure is affected by technology and context (though this certainly), but that the whole project management approach will vary depending on the project's novelty, complexity, technological innovation and pace. (See Chapter 18, p. 241, for a commentary on this work.)

Project management – its processes, structures, practices, actors and actions – needs to be applied in ways that reflect the project's objectives, strategy, technical, commercial and other characteristics, and the parent institutions' characteristics, as well as those of the external (socio, economic, political, environmental, etc.) context. The mark of the skilled p.m. expert is in applying them in ways that fit – when; in what order; how rigorously? There may be standard approaches, but the project management executive has to tailor their application according to need and context.

Project Management Contingency: Getting the Fit

To get project management right, I suggest, *one needs to have people of the appropriate competence, supported by institutional systems and other infra-structural capabilities, which fit both the characteristics of the project and the environment in which the project exists.* (I know of no research which substantiates this proposition as such. It is, however, congruent with contingency theory and is based on what seems to me to be common sense.)

Having looked at practices and principles that operate at the Level 1 technical core, we should now turn to the strategic and institutional levels (Levels 2 and 3). First, Level 2.

References and Endnotes

[1] Müller, R. (2009), *Project governance*, Gower: Aldershot.
[2] Helm, J. and Remington, K. (2005), Effective project sponsorship: An evaluation of the role of the executive sponsor in complex infrastructure projects by senior project managers, *Project Management Journal*, 36, 3, pp. 51–61; Lampel, J., Miller, R. and Floricel, S. (1996), Impact of owner involvement on innovation in large projects: Lessons from power plants construction, *International Business Review*, 5, 6, pp. 561–578.
[3] See Crawford, L., Cooke-Davies, T., Hobbs, B., Labuschagne, L., Remington, K. and Chen, P. (2008), Governance and support in the sponsoring of projects and programs, *Project Management Journal*, 39, S43–S55 – the authors analyse *inter alia* the treatment of the sponsor role in the leading project management standards, finding it inconsistent between PMI's PMBOK (missing), OPM3 (missing), Portfolio Management standard (not discussed to speak of) and Program Management standard (confused); and PRINCE2 (spread over three roles) and MSP (good, albeit the SRO being ultimately accountable for the program). See also Bryde, D. (2007), Perceptions of the impact of project sponsorship practices on project success, *International Journal of Project Management*, 26, pp. 800–809.

[4] Archibald, R. D. (1976, 1997, 2003), *Managing high technology programs and projects*, Wiley: New York.

[5] Morton, G. H. (1983), Become a project champion, *International Journal of Project Management*, 1, 4, pp. 197–203.

[6] Davies, S. M. and Lawrence, P. R. (1977), *Matrix*, Addison-Wesley: Reading, MA; Gobeli, D. H. and Larson, E. (1987), The relative effectiveness of different project management structures, *Project Management Journal*, 18, 2, pp. 81–85; Larson, E. (2004), Project management structures, in Morris, P. W. G. and Pinto, J. K. (eds.) *The Wiley guide to managing projects*, Wiley: Hoboken, NJ; Might R. J. and Fisher W. A. (1985), Role of structural factors in determining project management success, *IEEE Transaction on Engineering Management*, EM -32, 2, pp. 71–77.

[7] Woodward, J. (1958), *Management and technology*, HSMO: London; Galbraith, J. R. (1968), *Achieving integration through information systems*, Working Paper No. 361–368, Alfred P. Sloan School of Management, Massachusetts Institute of Technology; Galbraith, J. (1973), *Designing complex organizations*, Addison-Wesley: Reading, MA; Burns, T. and Stalker, G. M. (1961), *The management of innovation*, Tavistock: London.

[8] Morris, P. W. G. (1973), An organizational analysis of project management in the building industry, *Build International*, 6, 6, pp. 595–616.

[9] Giddens, A. (1984), *The constitution of society: Outline of the theory of structuration*, University of California Press: Berkeley and Los Angeles, p. 283.

[10] BS ISO 21500: (2011), *Guidance on project management*, BSI: London.

[11] Burns, T. and Stalker, G. M. (1961), *op. cit.*: 7.

[12] Leffingwell, D. (2007), *Scaling software agility*, Pearsons Education: Boston; Augustine, S. (2005), *Managing Agile projects*, Prentice Hall PTR: London.

[13] Morris. P. W. G. (1994), *The management of projects*, Thomas Telford: London, pp. 130–131.

[14] Forsberg, K., Mooz, H. and Cotterman, H. (1996), *Visualizing project management*, John Wiley and Sons: New York.

[15] Bartlett, J. (2010), *Managing programmes of business change*, 5th edition, Project Manager Today Publications: Bramshill, Hants; Milosevic, D., Martinelli, R. and Waddell, J. (2007), *Program management for improved business results*, John Wiley & Sons: Hoboken, NJ; Thiry, M. (2010), *Program management*, Gower Fundamentals of Project Management Series*, Gower Publishing: Aldershot.

[16] Mintzberg, H. (1979), *The structuring of organizations*, Prentice Hall: Englewood Cliffs, NJ.

[17] Ford, R. C. and Randolph, W. A. (1992), Cross-functional structures: A review and integration of matrix organization and project management. *Journal of Management*, 18, 2, pp. 267–294.

[18] Reis de Carvalho, E. and Morris, P. W. G. (1978), Project matrix organizations-or how to do the matrix swing, *Proceedings of the Project Management Institute Symposium*, Los Angeles, Project Management Institute, Drexel Hill, PA.

[19] Archibald, R. D. (1976, 1997, 2003), *op. cit.*: 4.

[20] Muethel, M., Hoegl, M. and Gemuenden, H.-G. (2011), Leadership and teamwork in dispersed projects, in Morris, P. W. G., Pinto, J. K. and Söderlund, J. (eds.) *The Oxford handbook of project management*, Oxford University Press: Oxford, pp. 483–499.

[21] Burt, R. E. (1992), *Structural holes*, Harvard University Press: Cambridge, MA; Granovetter, M. (1973), The strength of weak ties, *American Journal of Sociology*, 78, pp. 1360–1380; Pryke, S., (2012), *Social network analysis in construction*, Wiley-Blackwell, Chichester.

[22] Granovetter, M. (1973), *ibid.*

[23] Winch, G. (2004), Managing project stakeholders, In: Morris, P. W. G. and Pinto J. K. (Eds.) *The Wiley guide to managing projects*, Wiley: Hoboken, NJ; Artto, K., Eloranta, K. and Kujala, J. (2008), Subcontractors' business relationships as risk sources in project networks, *International Journal of Managing Projects in Business*, 1, 1, pp. 88–105; Hellgren, B. and Stjernberg, T. (1995), Design and implementation in major investments – A project network approach, *Scandinavian Journal of Management*, 11, 4, pp. 377–394; Cova, B. and Salle, R. (2000), Rituals in managing extra-business relationships in international project marketing: A conceptual framework, *International Business Review*, 9, 6, pp. 669–685.

[24] Pryke, S. D. (2001), *UK construction in transition: developing a social network approach to the evaluation of new procurement and management strategies*, PhD thesis, Bartlett School of Graduate Studies, University College London, London; Pryke, S. D. (2001), Analysing construction project coalitions: Exploring the application of social network analysis, *Construction Management and Economics*, 22, pp. 787–797.

[25] Hatch, M. J. (1997), *Organization theory: Modern, symbolic and postmodern*, Oxford University Press: Oxford, p. 192.

[26] Trist, E. L. and Bamforth, K. W. (1951), Some social and psychological consequences of the longwall method of coal-getting, *Human Relations*, 4, pp. 3–38.

[27] Galbraith, J. (1973), *Designing complex organizations*, Addison-Wesley: Reading, MA; see also Davis, S. N. and Lawrence, P. R. (1977), *Matrix*, Addison-Wesley: Reading, MA; Youker, R. (1977), Organizational alternatives for project management, *Project Management Quarterly*, 8, 1, pp. 24–33.

Governance and Strategy

Chapters 9 and 10 have addressed the traditional core of project management: planning and control, and organisation. We now move to addressing Level 2 work around the front-end, specifically in this chapter to the proposed governance and strategy for the project's, or program's, development and delivery. Then, in Chapter 12, we shall look at managing the emerging project definition (a major piece of which will be the product definition) and in Chapter 13 at establishing the commercial platform upon which the project work will be done.

A logical place to start, since all decisions and all control flows from it, is how the project or program is going to be governed on behalf of its shareholders or stakeholders.

Governance[1]

Everyone, and every organisation, working on the project or program should have, and be subject to, its own form of governance. So, what is governance?

Governance sets the rules and makes the decisions that guide the way actions and relationships are to be conducted in an enterprise. As the OECD (Organization for Economic Cooperation and Development) says: governance "provides the structure through which the objectives of the company are set, and the means of attaining those objectives and monitoring performance"[2]. The Association for Project Management (APM) goes usefully further, proposing 11 principles of project governance (see Box)[3]. The first three are self-referential. Principles 4 and 5 stress the need for sponsor-project strategy alignment; 6 and 7 refer to the importance of people and information; 8 refers to assurance; 9 to risk; 10 to organisational improvement; and 11 to stakeholders.

Reconstructing Project Management, First Edition. Peter W.G. Morris.
© 2013 John Wiley & Sons, Ltd. Published 2013 by John Wiley & Sons, Ltd.

Not a bad list (we'll be looking at all of them later on), though curiously there is nothing on morality or ethics when of course much of the recent attention to governance has arisen as a result of the abuse of power, as in Enron and WorldCom in 2001–2002. Are projects morally agnostic (of course); are project managers? (They absolutely should not be.) The area is complex, however. Governance in many developing countries is an *ad hoc* affair. Corruption is still a major issue in many project markets. International standards of morality are not yet universally applied. Nevertheless, project management as a profession stresses the ethical responsibilities of the role[4].

APM Principles of Project Governance (2004)

1. The board has overall responsibility for governance of project management.
2. The roles, responsibilities and performance criteria for the governance of project management are clearly defined.
3. Disciplined governance arrangements, supported by appropriate methods and controls, are applied throughout the project life-cycle.
4. A coherent and supportive relationship is demonstrated between the overall business strategy and the project portfolio.
5. All projects have an approved plan containing authorisation points at which the business case is reviewed and approved. Decisions made at authorisation points are recorded and communicated.
6. Members of delegated authorisation bodies have sufficient representation, competence, authority and resources to enable them to make appropriate decisions.
7. The project business case is supported by relevant and realistic information that provides a reliable basis for making authorisation decisions.
8. The board or its delegated agents decide when independent scrutiny of projects and project management systems is required, and implement such scrutiny accordingly.
9. There are clearly defined criteria for reporting project status and for the escalation of risks and issues to the levels required by the organisation.
10. The organisation fosters a culture of improvement and of frank internal disclosure of project information.
11. Project stakeholders are engaged at a level that is commensurate with their importance to the organisation and in a manner that fosters trust.

Strategy[5]

The importance of project strategy has only recently (since the beginning of the 21st century) become acknowledged.

Projects and programs are undertaken for a purpose. This purpose – 'the business rationale' – should follow from the sponsor's objectives, goals and strategies. (Which is a principle of good governance.) In some sectors, most notably ICT (Information and Communications Technologies), these business goals and objectives are termed 'business requirements'. It is important therefore that those who shape the project or program do so in a way which best meets the strategic intent of the sponsoring bodies. In many ways, *the strategic intent lying behind the project or program is the single most important thing affecting its design and development.*

There may be several levels within the enterprise at which these objectives, goals and strategies are stated[6], and they may 'emerge' as well as be formulated in a 'deliberate' manner, to use the terminology coined by Henry Mintzberg[7]. Deliberate strategy, as in the annual capital budgeting process, is vital to projects since projects and programs are the mechanisms through which the capital plans get realised. But implementing strategy doesn't always go as planned. 'Events' arise* which change the strategic landscape. Some will be caused by the projects and programs themselves; some will 'come out of left field'. Project and program managers should have their antennae switched on, ready to identify these emergent changes and address their implications.

The project/program strategy needs to be developed from the earliest stages of the project or program and systematically updated and flowed down through the project team; and it should be reflected back to the stakeholders' strategies.

Plans are nothing, General Eisenhower is supposed to have opined; planning is everything. This said, the project/program strategy – sometimes called simply 'The Plan' – is the most important document in the overall planning, development and implementation of the project or program. It is the baseline document which should be used as the reference for managing the project development.

Using the term 'The Plan', however, misses the dynamic potential of strategy. (IT projects employ the "Project Initiation Document" [PID], a document that describes how the project is to be set up and managed – more of a statement of intent. Almost a strategic plan but not quite.)

What is the difference between a strategy and a plan? Actually this is a difficult one; there is not a ready consensus. The term strategy comes, after all, from the Greek *strategos*, meaning 'the General', or more literally 'the leader of the army'. Thus Mintzberg, for example, has for some time stressed the personal side to crafting strategy to fit the organisation's context and the conditions unfolding around its realisation[8]. 'Ploy', 'perspective' and 'pattern', all requiring insight and judgement, are as important, Mintzberg contends, as 'position' and 'plan'. Others approach issues in strategy implementation from a more process-oriented perspective[9], while yet others take a more synoptic view[10], stressing process but also bringing out the importance of the CEO's strategic judgement. Similarly, John Kotter emphasises the role of leadership while proffering an eight-step process to implementing strategy: a sense of urgency, organisation, vision and strategy, communicating, empowering, creating short-term wins, consolidating and institutionalising[11]. As Karlos Artto of Aalto University in Finland con-

---

*British Prime minister Harold Macmillan was asked in the 1960s, in his retirement, what represented the greatest challenge for a statesman. Macmillan replied: 'Events, dear boy, events'.

cluded, after an exhaustive study of project strategy, "Project Strategy is a direction in a project that contributes to success of the project in its environment". Strategy shapes, and gives momentum to, the project's course: the project and its strategy are dynamic[12].

Many view the Project Execution Plan (PEP) as *the* project strategy document, but really it only addresses the project implementation strategy – that is, the plan for the work to be performed after capital expenditure has been authorised. This and only this. Others, however, also have a separate development strategy plan which sets the strategy for the project or program pre-sanction. Others again roll the two together into one continuous strategy document.

Project and program strategies and plans should cover, at an appropriate level, all that is needed to be done to meet their objectives and goals: what the project has to achieve and how its success will be evaluated.

Generically, the strategic plan establishes project management's interpretation of the why, what, how, who, how much and when of the project development. The plan should include, at a minimum:

- a definition of overall objectives and goals
- statements on how these should be achieved (and verified)
- technical descriptions of the product (requirements, specifications, etc.) and the proposed development strategy
- project organisation (and the policy and strategy for the procurement of resources)
- key roles
- estimates of the time required, phasing and implementation strategy
- budget and related financial strategy issues (cash, insurance, bonds, penalties, etc.)
- change management policy
- quality policy and plans
- safety, health and environmental policies and plans
- risks and opportunities faced and strategies for managing these
- reporting requirements
- communications policy and document (information) management
- expected behaviours.

The program and project strategies and plans should be periodically and systematically updated, and reviewed and approved by the sponsors, not least at the stage-gates, where they might be re-baselined, but only with the sponsors' approval. (IPA found the variation in up-to-date-ness of the PEP to be one of the principle factors explaining differences in engineering construction productivity, the other being project controls[13].)

Key Performance Indicators (KPIs) and other metrics may be established to constitute the measures which will determine if and when this is done – for example, throughput or capacity measures to be achieved, timings, costs, safety measures, and so on. (KPIs will lead to various other performance indicators such as Functional Points, Design Quality Indicators, Environmental Quality Indicators, etc.)

Both the sponsor and the project/program manager and their team should formally accept and 'own' the strategy/plan and should flow it through their teams as the project or program develops.

The Front-End

While considerable confusion still surrounds the role of strategy in projects, few have a good understanding of the importance of the Front-End* and what needs managing there, even though they may have a blurred understanding that it is important.

But what do we mean by the front-end? Referring to Figure 11.1, there are two or three ways of defining it.

1. The period between the stage-gate that gives permission for resource to be spent in developing a scheme for sanction [SG1] and the sanction stage-gate itself [SG3].
2. Or, the period between the stage-gate that gives permission for resource to be spent in developing a scheme for sanction [SG1] and the gate confirming the proposed scope and giving authorisation to work-up the proposal for sanction [SG2].
3. Or from where requirements are elicited and then accepted by the subsequent sanction stage. The following chapter addresses the challenges of eliciting and managing requirements.

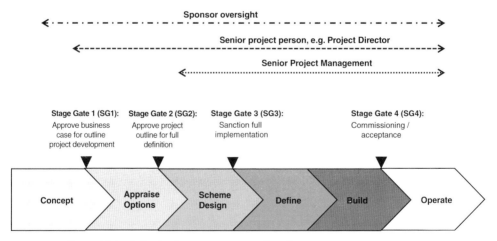

Figure 11.1   Roles in the management of projects.
Note: The Front-End is defined either as the stage SG1 to SG2, or SG1 to SG3; or to the stage where the project's requirements are defined.
Source: Author's own.

---

*Aristotle, as ever, saw it clearly, telling us that defining the question is half the answer (*Ethics*: Book 1.C.4): literally: "For the beginning is thought to be more than half the whole".

Given the confusion over what has to be done to manage the project front-end effectively, it should thus come as no surprise that there is little evidence of the project management community seeing an opportunity here for it to add value by bringing its distinctive knowledge and skill-sets to bear to improve the project's or program's emerging strategy. (By the term "project management community" I am meaning front-end managers, such as, say, Development Managers, not just downstream project managers.) An early alertness to identifying benefits (and unwanted dis-benefits) and of the concept of value and the ideas of benefits management (see GCHQ's NAP program, Chapter 5, pp. 93–4) are therefore important here. The subject of enhancing value and harvesting benefits is a bit downstream yet – we need to have something tangible to assess – so it is discussed below, after the technical development has got underway.

There is a natural mindset amongst project managers to see the project as something desirable, even if governance thinks the project should no longer be pursued. Hence the idea of the project champion discussed above (Chapter 10, p. 149). This is generally a mistake: projects are disruptive and expensive and it may well be best for them not to proceed, and for this decision, better it were made early, certainly well before 'sanction'. The default position should be that the project should not go ahead.

The project professional still has an enormously useful role however, even if the project is looking questionable: he/she should bring his/her project implementation knowledge to bear, advising the sponsor on the options open and the consequences of pursuing them. Avoid being a proponent; be a skilled development and implementation advisor.

References and Endnotes

[1] Crawford, L., Cooke-Davies, T., Hobbs, B., Labuschagne, L., Remington, K. and Chen, P. (2008), Governance and support in the sponsoring of projects and programs, *Project Management Journal*, 39 (Supplement), pp. S43–S55; Miller, R. and Hobbs, B. (2005), Governance regimes for large complex projects, *Project Management Journal*, 36, 3, pp. 42–50; Müller, R. Project governance, in: Morris, P. W. G., Pinto, J. K. and Söderlund, J. (eds.) *The Oxford handbook of project management*, Oxford University Press: Oxford, pp. 483–499; Turner, J. R. and Müller, R. (2004), Communication and co-operation on projects between the project owner as principal and the project manager as agent, *European Management Journal*, 22, 3, pp. 327–336; Gil, N., Tommelein, I. and Schruben, L. (2006), External change in large engineering design projects: The role of the client, *IEEE Transactions on Engineering Management*, 53, 3, pp. 426–439.

[2] OECD (2004), *OECD principles of corporate governance*, OECD Publications: France, http://www.oecd.org.

[3] Association for Project Management (2004), *Directing change: A guide to governance of project management*, Association for Project Management: High Wycombe.

[4] Brenkert, G. G. and Beauchamp, T. L. (2009), *The Oxford handbook of business ethics*, Oxford University Press: Oxford; Freidson, E. (2001), *Professionalism, the third logic: On the practice of knowledge*, University of Chicago Press: Chicago.

[5] Artto, K. A., Lehtonen, J. M. and Saranen, J. (2001), Managing projects front-end: Incorporating a strategic early view to project management with simulation, *International Journal of Project Management*, 5, pp. 255–264; Gardiner, P. D. (2005), *Project management: A strategic planning approach*, Palgrave Macmillan: Basingstoke; Mintzberg, H. (1983), *Structures in fives: Designing effective organizations*, Prentice Hall: Englewood Cliffs, NJ; Mintzberg, H. and Quinn, J. B. (1996), *The strategy process: Concepts, contexts, cases*, Prentice Hall: Upper Saddle River, NJ; Morris, P. W. G. (1994), *The management of projects*, Thomas Telford: London, Chapter 8; Office of Government Commerce (2002), *Managing successful projects with PRINCE 2*, TSO: Norwich, UK; Morris, P. W. G. and Jamieson, H. A. (2004), *Translating corporate strategy into project strategy*, Project Management Institute: Newton Square, PA.

[6] *Ibid.*: see the cases on BAA and Rolls-Royce, for example.

[7] Mintzberg, H. and Quinn, J. B. (1996), *op. cit.*: 5.

[8] Mintzberg, H. (1987), Crafting strategy, *Harvard Business Review*, July–August, pp. 66–75.

[9] Brown, S. L. and Eisenhardt, K. M. (1997), The art of continuous change: Linking complexity theory and time-paced evolution in relentlessly shifting organisations, *Administrative Science Quarterly*, 42, pp. 1–24; Bryson, J. M. and Delbecq, A. L. (1979), A contingent approach to strategy and tactics in project planning, *American Planning Association Journal*, 45, pp. 176–179; Eden, C. and Ackermann, F. (1998), *Making strategy: The journey of strategic management*, Sage: London; Loch, C. and Kavadias, S. (2011), Implementing Strategy thorough projects, in: Morris, P. W. G., Pinto, J. K. and Söderlund, J. (eds.), *op. cit.*, Chapter 9; Thompson, A. A. and Strickland, J. I. (1995), *Crafting and implementing strategy*, Irwin: Chicago.

[10] Artto, K., Kujala, J., Dietrich, P. and Martinsuo, M. (2008), What is project strategy? *International Journal of Project Management*, 26, 1, pp. 4–12; Burgelman, R. A. and Doz, Y. (2001), The power of strategic interaction, *Sloan Management Review*, 42, 3, pp. 28–38.

[11] Kotter, J. (1996), *Leading change*, Harvard Business School Press: Cambridge, MA.

[12] Artto, K., Dietrich, P. and Martinsuo, M. (2008), *op. cit.*: 10, pp. 4–12.

[13] Merrow, E. W., Sonnhalter, K. A., Somanchi, R. and Griffith, A. F. (2009), *Productivity in the UK engineering construction industry*, Department for Business, Innovation and Skills: London.

12

Managing the Emerging Project Definition[1]

Having planned the work, set-up the project controls, established the project (and program) organisation and governance, including formulating the strategy, we are now ready to start designing and developing! For some project and program managers, managing technical issues may not be something they feel very concerned with. For others, however, it is a major part of their work.

For years, there have been arguments about how involved project management needs to be in managing the technical development of a project. (Logic says the same argument should apply to other areas as well, such as Procurement or Finance – and it does!) Managers, it is claimed, needn't have substantive knowledge of the technology of the project or program. This is largely true: managers move between industries, often quite successfully. Yet there is also plenty of evidence that technical issues cause projects and programs to fail*. Why didn't management catch these? And, note, many of these problems are laid down in the early, Front-End stages of projects and programs. Managers, in fact, spend a considerable portion of their time on technology issues. Stephen Pryke of UCL, for example, found that "approximately 50% of all white collar management activity [during construction] involved design or specification issues"[2].

The answer surely is that project and program managers need to ensure the right processes and practices are being followed with respect to technical

---

*The US GAO found lack of alignment between corporate and project/program technology strategies to be a major source of cost growth and schedule slippage: *Assessment of selected major acquisition systems*. GAO-06-626 General Accountability Office 2006: Washington, D.C. In another study, the DoD and the CIA reported in 2008 that "programs with mature technology at key milestones had average cost overruns of 4.8% while programs with immature technology had average cost overruns of 34.9%" Meier, 2008, Best project management and systems engineering practices in pre-acquisition practices in the federal intelligence and defense agencies *Project Management Journal* 39, 1, 59–71.

definition and development so that obvious errors are avoided. (And the same goes for other aspects of the project or program that are also capable of causing severe damage.) Technical decision-making will remain with the 'technical authority' but the project or program manager must make sure that the right things are being done. More, the p.m. may want to press certain value-enhancing ideas which, unless really gone for, would be unlikely to materialise.

The beginning of technical definition and development is the elicitation of the projects 'requirements'. This is then followed by the management of the development work required to meet these requirements.

Requirements Management[3]

Requirements state project needs. Requirement definition is a hugely important step in the early stages of a project, and in reality, to the ultimate frustration of project people everywhere, it is a step that often fails to work properly. Either requirements change or emerge not as expected; the knock-on effect can be devastating, just as in Concurrency.

Requirements language stems from the discipline of systems engineering. Requirements precede solutions*. Requirements management begins with the process of defining the *users'* (the customer's and others') requirements for the project. These lead to *systems* requirements. These will be both functional and non-functional (e.g., safety, reliability) and will specify system functions, behaviour, constraints and interfaces and how to trace these back to user requirements[4]. Business requirements are defined in parallel. These requirements then lead to specifications.

The process essentially involves three stages: elicitation (or capture), prioritisation (or triage) and specification.

- **Elicitation** (capture) includes not just obtaining the needs, but also analysing, refining and recording them. Requirements should be solution-free and should be stated singularly in a form which is testable. There should be acceptance criteria defined for each requirement.
- **Prioritisation** (triage) is the process of assessing which requirements best satisfy the budget and schedule or other constraints, taking into account operating costs and taking a 'whole life' perspective, as, for example, the emphasis in defence procurement on Integrated Logistics Support (ILS).
- **Specification** is the process of refining and documenting the selected and agreed requirements. (The term 'Functional Requirement' is sometimes used here: the expressed need for a system to exhibit specific, often quantified, behaviour as a result of its interaction with its opera-

---

*For many suppliers, projects begin after the requirements are already defined, with specifications or designs already having been developed, depending on the stage of the product development cycle at which the particular project, or program, is initiated. This is particularly true for contractors and suppliers.

tional environment.) In some cases, specification is effectively performed through coding, designing, modelling or prototyping.

Requirements may be developed by, or elicited from, both the customer and/ or supplier. That is, part of the supply contract might involve eliciting the requirements – which of course would then need to be signed-off either by the supplier's sponsor (if, say, this is part of the bid) or the client (which puts the supplier at considerable risk.)

The specifications become the basis of design. (Requirements thus drive the work breakdown structure [WBS] – see Chapter 9.) The resulting realisation is then progressively tested (verified and validated) against the designs, specifications, and requirements as one works back up 'the Vee' (see Figure 5.5). Managing the verification and validation of the emerging solution against the requirements is considered to be part of requirements management, as is the whole process of structuring the requirements and managing proposed and approved changes to them during the system development. We shall discuss the management of the design, build and test activities in a moment under the section 'Solutions Development'.

Does requirements elicitation happen automatically or does it have to be managed; and if the latter, who manages it: the Systems Engineer; the Project Manager? Someone has to, and since the results of doing so may profoundly affect the fortune of the project or program, that someone had better have the interests of the overall project at heart. Or putting it another way, anyone responsible for the overall project success ought to ensure that requirements elicitation is happening properly. In other words, be a manager of the project.

Are unrestrained requirements a bad thing? We saw in Part 1 that technical issues (including unrealistic requirements) have been a major source of poor project performance. How ambitious should the user requirements be allowed to be? The answer must depend on how they relate to the project or program strategy, and the risk that is deemed acceptable and the risk management process employed. Unwarranted, undermanaged technical risk is foolish; technical risk that the project is prepared for might be a different thing. Innovation may be central to what the project or program is trying to achieve. (Again, discussed under 'Solutions Development').

Requirements should demonstrate integrity between each other, that is, should form an integral set without clash, misfit or dysfunctionality. Requirements should be comprehensive and clear, structured, traceable and testable. The requirements' definition should be progressively updated as the project develops. Proposed changes should be fully reviewed and, where modified, documented. Various database-type tools are available that help do this such as DOORS. 'Requirements creep' is a frequent problem and can cause major upsets if not controlled; requirements should be frozen as far as possible once development is initiated (and contracts are awarded).

There are subtle differences in requirements management practices between different industries and between programs and projects. Systems-based projects and programs differ in particular from construction.

In building and civil engineering, the equivalent of requirements is 'the brief'. 'Briefing' is widely recognised as often complex and difficult that may take time and be iterative[5]. Partly this reflects older practices, partly because it is often not possible to define user requirements readily and precisely (given the size of structure and potential variety of usage). Most process engineering projects (oil and gas, chemicals, paper and pulp, water, etc.) move straight from output specifications to flow diagrams and then to designs and specifications. In drug development and certain other R&D projects, the requirements are initially stated as general [therapeutic] targets; at the 'candidate nomination' stage-gate, the drug's technical specification is established (and is summarised in the insert 'label' that goes in the drug's packaging).

Requirements management is as much about interacting effectively with people as it is a technical subject. There is considerable evidence both in systems engineering and in briefing that difficulties in communicating what is really required is a major challenge[6].

Solutions Development

Requirements, by definition, are stated in terms which are solution-free. Once they start being translated into specifications, some degree of solution is being proposed. Developing the technical solution – Solutions Development – offers all kinds of scope for the project team to shape the project in ways which will optimise satisfaction of the requirements and avoid problems, help achieve project and program success and avoid mistakes.

Solution Development is a large subject area covering all aspects of the solution activities initiated in response to meeting the stated requirements.

Before finalising the specifications the project team may need to address the issue of innovation and the technical uncertainty inherent in the emerging solution. Innovation can be a major feature of the project or program and may require careful management.

Innovation

Innovation could be central to the project or program precisely because projects and programs can be powerful means of executing a company's innovation strategy. (And innovation needn't refer only to technology: market is another major dimension, or, indeed, method of production.) Andrew Davies and his colleagues at the Complex Product Systems research centre in the United Kingdom (Davies is now at UCL) have termed such projects 'strategic', 'vanguard' or 'base moving'[7]. The innovation may be exploitative, in which case it will be deterministic in the classic PMBOK, Wheelwright and Clark sense; or it may be exploratory, in which case the project becomes a "highly uncertain, reflexive, probe and learning process"[8] – but one which still needs managing!

Various strategies exist for handling innovation[9], for example:

- Prototyping – exploring the new innovation in an off-line capacity so the risk to the main project is constrained: 'Technology Demonstrator' projects and 'Rapid Applications Development' are two well-known examples of prototyping.

- The use of modelling, simulation and synthetic environments – Computer-Aided Design/Engineering/Manufacturing (CAD/CAE/CAM), in 3-, 4- and even 5-D and in virtual reality, for example.
- Phased implementation of new technology, as for example, in version releases of product upgrades (hence P^3I – Pre-Planned Product Improvement).

The incorporation of unproven technology into the mainstream project development is not recommended; technology demonstrators – to allow development and testing 'off-line' – are preferred where possible. As we saw in Part 1, the alternative, 'concurrency', is guaranteed to cause trouble.

The 'Technological-Readiness Level' idea TRL: (see Chapter 5, p. 88) applied in systems engineering by, for example, NASA, ESA and DoD, could be made more use of in other sectors (Figure 12.1). As in other maturity tools, assessment is made against pre-defined levels. (Most models have about nine levels.) Care needs to be taken in ensuring the TRL 'fits' its context, however – both with respect to the larger product system and its operational environment.

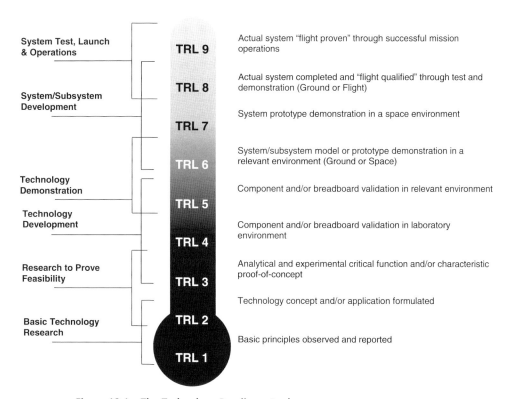

Figure 12.1   The Technology Readiness tool.
Source: © NASA.

The results of the 'technology proving' should feed into the project specifications. The specifications, like the user, systems and business requirements, should be 'owned' by the project/program manager and his or her team.

Specifications begin to dictate the emerging solution[s]. If a prototype becomes directly embodied in the emerging product solution, it should have been adequately tested, and there ought to be full documentation on it. (Lack of adequate documentation is a problem frequently associated with Rapid Application Development.)

The development from specifications into design and design into 'build' is represented in many different process forms. Systems development, for example, can, as we've seen, follow one of several available process models: Vee, waterfall, spiral, incremental or agile. (And it is worth noting that in many industries, design precedes specification: as in building, where the architect may move immediately from the initial briefing to pencil or computer-based sketching.)

Some claim that Design Management should be singled out as a stand-alone p.m. topic, but in what research I have seen it is really just the principles discussed here collected under the Design Management wrapper[10]. (This in no way is meant to suggest that it is not legitimate as a field of enquiry.) And also, if Design Management seems worth crystallising out, why not the next technical stage: manufacturing, build, etc.? Commissioning? (Admittedly, innovation is singled out but it is a broader topic than Design Management.) To an extent, they are addressed of course, but they are not quite as high profile in their claims as Design Management, mostly because of the 'front-end' power that Design Management has: the possibility of leveraging design's influence on the project outcome, which is so much greater than for the more downstream activities.

As we have seen, there is widespread agreement on the benefits of phased development, with 'stage-gate' reviews between phases and with periodic technical reviews*. This said, there may be subtleties that create difficulties in implementation. The most obvious is not involving teams from key successor, and predecessor, stages but instead just throwing the product 'over the wall' to the next stage at stage completion. This is generally inefficient, bringing a lack of integrative thinking, loss of key disciplines' inputs (manufacturing tooling, marketing and sales advice, etc.) and losing opportunities to shorten timescales (through lack of preparation, failure to order long lead items, etc.). Techniques such as concurrent engineering and fast-tracking have been developed to overcome this difficulty. Technical maturity at the end of a stage is potentially a problem, however, where innovation is significant.

---

*Other critical reviews that the project or program manager should ensure include formal Health, Safety and Environmental reviews, quality reviews, peer reviews, progress reviews, and lessons-learned reviews. See Chapter 16. Other reviews (peer, progress, lessons learned, etc.) are reviewed in Chapter 20, pp. 223–4).

Avoiding 'over the wall' development is an example of the integrative thinking which is so characteristic of good project and program management. Various techniques can help achieve this: for example, through-life management, whole life costing, integrated logistics support, and the "-ilities" (manufacturability, buildability, reliability, maintainability, operability, usability, etc.).

A core idea in systems engineering is that of the progressive testing of the emerging product against the expectations laid down in the product's or system's design, its specifications, and its requirements ('testing back up the Vee')*. Verification is the term used to ensure the product is being built right; validation ensures that the right product is being built. Validation is against user requirements; verification against system requirements, specifications, and designs.

Testing should be fully planned. It can be an absolutely critical schedule activity (e.g., obtaining product licence/certification) and can account for a lot of project time and effort. Too often inadequate testing takes place because of lack of time and, for the same reason, there may be an under-emphasis on validation.

In many industries – construction and process engineering, ICT and systems installation, for example – commissioning and handing-over the product/system is a major technical phase of the life-cycle and one that needs serious planning and management in its own right. Commissioning is often phased with operational data being collected and fed-back into some form of product/design lessons-learned system.

Similarly, in construction post-occupancy evaluation (POE) studies should be planned to feed appraisal data back into product/design.

For some program managers – those involved primarily in change programs – much of this may seem to be of limited applicability if the projects in their programs have little technical development content. For others, however, in drug development for example, where the same molecule (product) may be being 'developed' for different indications, delivery mechanisms, or dosages – all criteria for separate projects within the molecule program – most of these issues will be very real: specification, P³I, gate reviews, trial design, modelling, manufacturability, concurrent engineering, value optimisation, and, not least, testing.

All these topics are technical areas in their own right. To revisit the question posed at the beginning of this chapter, is it evident that the project's management should be involved in their management? The project team needs to ensure that, at a minimum, the right processes and practices are in place and are being followed, and that the broader implications of these technical issues on the project – on budget, schedule, scope, contract, documentation, and so on – are being effectively managed. Good leadership and teamwork, communications and decision-making can powerfully help achieve this.

---

*It's expensive though, and not just in monetary terms. Remember Mueller's decision on Apollo to delete a number of test launches in order to claw back schedule and go for 'all up' testing.

References and Endnotes

[1] Badiro, A. B. (1996), *Project management in manufacturing and high technology operations*, 2nd edition, John Wiley & Sons: Hoboken, NJ; Cadle, J. and Yeates, D. (2004), *Project management for information systems*, 4th edition, Prentice Hall: Upper Saddle River, NJ; Cusamano, M. A. and Nobeaka, K. (1998), *Thinking beyond lean: How multi-project management is transforming product development at Toyota and other companies*, Free Press: New York; Davies, A. and Hobday, M. (2005), *The business of projects*, Cambridge University Press: Cambridge; Eisner, H. (2002), *Essentials of project and systems engineering management*, 2nd edition, Wiley: New York; Forsberg, K., Mooz, H. and Cotterman, H. (2000), *Visualizing project management*, Wiley: Hoboken, NJ; Kirkpatrick, D., McInally, S. and Pridie-Sale, D. (2004), Integrated logistics support and all that, in Morris, P. W. G. and J. K. Pinto (eds.) *The Wiley guide to managing projects*, Wiley: Hoboken, NJ; McManus, J. and Wood-Harper, T. (2003), *Information systems project management: Methods, tools and techniques*, Prentice Hall: Upper Saddle River, NJ; Morris, P. W. G. (1994), *The management of projects*, Thomas Telford: London; Mooz, H. (2004), Verification, In: Morris, P. W. G. and Pinto, J. K. (eds.) *op. cit.*; Nobelius, D. (2004), Towards the sixth generation of R&D management, *International Journal of Project Management*, 22, 5, pp. 369–375; Prencipe, A., Davies, A. and Hobday, M. (2003), *The business of systems integration*, Oxford University Press: Oxford; Robertson, S. and Robertson, J. (2004), *Requirements-led project management: Discovering David's slingshot*, Addison Wesley: Reading, MA; Robertson, S. and Robertson, J. (2006), *Mastering the requirements process*, Addison Wesley: Reading, MA; Stevens, R., Brook P., Jackson, K. and Arnold, S. (1998), *Systems engineering: Coping with complexity*, Prentice Hall: Upper Saddle River, NJ; Tomola, F. and Sénéchal, O. (2004), Innovation management: A synthesis of academic and industrial points of view, *International Journal of Project Management*, 22, 4, pp. 281–287; Yardley, D. (2002), *Successful IT project delivery, learning the lessons of project failure*, Addison Wesley: Reading, MA.

[2] Pryke, S. D. (2001), *UK construction in transition: Developing a social network analysis approach to the evaluation of new procurement and management strategies*, PhD thesis, UCL.

[3] Archibald, R. D. (1976, 1997, 2003), *Managing high technology programs and projects*, Wiley: New York; Barrett, P. and Stanley, C. (1999), *Better construction briefing*, Blackwell Science: Oxford; Blyth, A. and Worthington, J. (2001), *Managing the brief for better design*, Spon: London; Davis, A. M., Hickey, A. M. and Zweig, A. S. (2004), Requirements management in a project management context, in Morris, P. W. G. and Pinto, J. K. (eds.) *op. cit.*: 1; Forsberg, K., Mooz, H. and Cotterman, H. (1996), *op. cit.*: 1; Office of Government Commerce (1999), *op. cit.*: 1; Office of Government Commerce (2003), *Managing successful programmes*, TSO: Norwich; Prencipe, A., Davies, A. and Hobday, M. (2003), *op. cit.*: 1; Stevens, R., Brook, P., Jackson, K. and Arnold, S. (1998), *Systems engineering: Coping with complexity*, Prentice Hall: Upper Saddle River, NJ.

[4] Stevens, R. *et al.* (1998), *ibid.*, Chapter 3.

[5] Barett, P. S. and Stanley, C. (1999), *op. cit.*: 3; Preiser, W. F. E. and Vischer, J. (2005), *Assessing building performance*, Elsevier Butterworth: Oxford.

[6] Johnson, S. (2003), Systems integration and the social solution of technical problems in complex systems, In: Principe, A., Davies, A. and Hobday, M. (2003), *op. cit.*: 1.

[7] Brady, T. and Davies, A. (2004), Building projects capabilities, from exploratory to exploitative learning, *Organizational Studies*, 25, 9, pp. 1601–1621; Davies, A. (2004), moving base into high value integrated solutions: A value stream approach, *Industrial and Corporate Change*, 13, 5, 727–756.

[8] Lenfle, S. (2008), Exploration and project management, *International Journal of Project Management*, 26, 5, pp. 469–478.

[9] Thomke, S. (2003), in relation to innovation in his book *Experimentation matter*, Harvard Business School Press: Boston, MA.

[10] Harpum, P. (2011), *The development of a model of design management for use in infrastructure and aero engineering industries*, PhD thesis, University of Manchester.

Procurement and the Project's Commercial Management

Virtually all projects and programs need to acquire and bring on board resources of the appropriate capability to help carry out the work. This applies for resources supplied internally as well as from outside.

Engaging resources is at least a major, and for some it is the dominant, part of managing projects and programs. It is like a game. There are rewards and penalties, rules and roles. Some cheat, or at least take advantage, where others wouldn't. Some play the game straight and true; others are always looking for an angle to make another dollar or two. Or three. Contracts describe what is to be provided and under what conditions. Some people put the contract in the desk drawer and forget about it, others use it as a means of extracting increased payments. The contract sets the rules but it is the individual who decides how play will be conducted. (The individual, the team, the organisation.)

Since the coming of Lean[1], with its emphasis on reducing waste, there has been a strong move to develop relationships between suppliers and buyers (owners); that is, to move from essentially transaction-type procurement to relationship-based contracting[2]. For the project owner, the sponsor, this generally makes a lot of sense, at least for goods and services which are not commodity items, because it helps the sponsor (owner) get the supplier to align his offer and his supply more closely to his needs. But it would be a mistake to say (however appealing intellectually or morally) 'relationship contracting good, transaction bad', because from a business point of view, hard-nosed contracting may sometimes be best. From the contractor's perspective, it will all depend on the contractor's strategy, on market conditions and on the resources that are available. 'Where is there more money to be made, and do we have the right kind of people available to work in this way?'

Transaction-based purchasing may also, in certain circumstances, be the best route for the owner (the buyer) too. Many buyers have found e-commerce to be enormously beneficial, for example. This is where the

Reconstructing Project Management, First Edition. Peter W.G. Morris.
© 2013 John Wiley & Sons, Ltd. Published 2013 by John Wiley & Sons, Ltd.

Internet is used to facilitate the buying and selling of products and services, either on a business-to-business basis (B2B) or business-to-consumer (B2C) basis. Applying the Internet to procurement services such as buying, auctioning, tendering, procurement (as well as project collaboration, document management) and payment, purchase prices have in some cases fallen substantially. This is obviously transaction-based procurement at its most raw, and the relative benefits expected compared with a more relational approach, which may still be being applied on other parts of the project or program, need thinking about carefully.

So. While many projects see resource acquisition merely as a handle-turning procurement or contracting activity, in reality it reflects fundamental strategic decisions about funding and contracting strategy.

Acquisition and Contracting Strategy[3]

First, we need our strategy. Acquisition strategy is a buyer activity to propose how best to acquire the resources – people, equipment, space, materials, money, and so on – of the appropriate capability needed to develop and deliver the project or program requirements effectively. As we've established, such resources may be provided internally from within the organisation; often, however, externally supplied resources will need procuring. The strategy for incorporating these resources will form a major element of the overall project or program strategy.

Corporate Social Responsibility has risen in importance in recent years. We now expect projects to be able to demonstrate that their procurement practices are totally ethical and transparent and that good governance and corporate accountability issues, regarding for example, sustainability and employment practices, are being observed.

Options for internally supplied resources form part of the capacity planning activity, related to the project or program, and will reflect:

- the capabilities and competencies needed for the p.m. function in the enterprise as a whole, as assessed through the resource planning process;
- where those capabilities and competencies reside;
- the amount of project work to be done, and the lumpiness in which it comes; and
- on what organisational basis these resources are to be supplied – transferred, seconded, matrixed, and so on – and on what basis they will be managed.

(Level 3) internal policies and procedures will govern how this is done. Governance and possibly the Resource Manager will be involved in approving and effecting resource allocations to the project or program.

The project or program manager should ascertain the levels and sources of resources needed and available for each stage of the life-cycle/schedule. Where there are not sufficient or appropriate resources available in-house, externally supplied resources may need to be acquired. This can have a major impact on the conduct and performance of the project or program, not least on the project or program structure. This could happen in the early, middle or even late stages of the project life-cycle and could result in major organisational change.

Funding availability can influence the choice of supplier, the amount, the timing, and basis on which one is prepared to pay for resources. Low cost providers do not necessarily mean lower quality resources: an innovative sourcing strategy (coupled with good bid management techniques) can secure lower cost resources of appropriate capability. Some form of supplier funding assistance may also be available: this may range from simple discounts to supplier credits to project finance, the ultimate being PFI. Supplier credits may cover relaxation of capital requirements or stretch into operational support arrangements and may involve specialist third-party or governmental assistance. In construction, systems, and defence/aerospace for example, this can be a major element of project initiation work often requiring significant estimating of work package elements, costs, and cash requirements by the project or program team to secure funding.

Work packaging

At an initial and high level, the decision needs to be taken as to whether there will be a single integrating supplier or whether work packages will be supplied by several separate organisational units. A prime contractor or turnkey contractor would be an example of the former; separate package contractors are examples of the latter. If separate work package suppliers are to be sought, there will need to be some form of project or program management function integrating them, probably organised as a separate work package in its own right. In the construction and systems engineering industries, it is common practice either to award a single turnkey contract for the supply and installation of a product, or to employ a construction or project manager or systems integration contractor to work as a project/program manager for the owner in letting and managing work package contractors. This work can be performed either against a firm price or against a range of variable cost bases.

The degree of active control that the owner wishes to exert over the evolving work package will affect the choice over which of these options is selected. If active control is sought, the turnkey or prime contractor option is less appropriate, particularly if the work is let on a fixed price basis. (The form of contract should align with the form of work package strategy to be adopted.)

Projects generally require the work of more than one supplier ('Teaming'). Generally this is arranged via subcontracts; sometimes either as a joint venture, a consortium or a partnership. A partnership is not the same as partnering, to which we shall turn next but in all cases, in any form of joint or inter-company work, great care needs to be taken to develop trust and collaboration. Objectives need to be clear and mutually workable as do risk and reward[4].

Risk allocation is a fundamental aspect of contracting strategy. Risk should be allocated to the parties best able to bear the risk. Avoid transferring risk to parties that cannot properly bear that risk: in the event of failure (maturing risk), collapse may ensue and the transferring party will end-up suffering anyway. (BAA – the British Airport Authority – took this to extremes in building Terminal 5 at London's Heathrow Airport in the early 2000s. Proclaiming Partnering as *the* way to contract, BAA, as owner, went further and said it would be counterproductive for the contractors to bear risk, since in the end it would be BAA that would have to bear the costs – not true, and what would seem on the face of it to be an extreme extension of responsibility. After T5 was completed, BAA then performed a *volte face* and put all its procurement on the basis of awarding contracts to the lowest capital bids.)

Work packages are usually considered on the basis of key differentiating features – nearly always technical expertise and commercial reputation but also sometimes geographic location. Potential suppliers will be identified against WBS elements, thereby giving rise to proposed work packages. Care will need to be given to ensuring that work package interface effectively, marry-up and are complementary without overlap, gaps or contradictions.

A time-based schedule of the actions needed to award the package contracts and achieve integrated development and completion will need developing. Similarly, budgetary and cost control schedules will need preparing.

Pricing options

Pricing can vary between fixed price ('lump sum' or based on a 'schedule of rates') to cost reimbursable (either totally [with possibly a fixed element such as fee or overhead and profit] or re-measure). The former transfers financial responsibility largely, though not necessarily wholly, onto the supplier; the latter more onto the owner/client, or the partnering organisations. Whoever takes the major responsibility will require more resources, more management and a larger organisation to discharge it[5]. Care is required, however. Agreeing a fixed price bid too early in the development cycle (too early in the estimating funnel before the design is sufficiently developed and change can be effectively controlled) will increase risk significantly, and the need for contingency allowances too, as Marschak, Glennan and Summers concluded in their review of DoD acquisition strategies in 1967 (see Chapter 4, pp. 59, 77).

Partnering and Alliancing

Since Plutarch's day or before, contractors had been selected typically on the basis of the cheapest bid, as we've seen. This leads readily to bidders under-quoting with the aim of making their money on variations. Which in turn leads often to adversarial relations. Partnering, which came in on the back of Total Quality/Lean Management in the 1990s, sought to change all that. And really, it has. (At least it has helped to bring about changes in behaviours and expectations.)

Terminology varies, as does practice between industries. "While there is broad consensus about the basic philosophy underpinning partnering – a commitment between firms to cooperate – there are contrasting views about a number of features . . . contracts, duration of agreements, need for formal team-building"[6]. The aim is essentially to get alignment along the supply chain so that everyone is working to the same ends (to reduce waste and increase performance).

The cost basis is often transparent (e.g., open book) with limited profit (and loss) opportunity, although sometimes, as in Andrew, potentially quite significant. The arrangements may be project-specific or multi-year. Performance targets will generally play an important role, not least in deciding upside and downside potential. Benchmarking, feedback/lessons learned, estimating, risk management, and Value Management and Value Engineering can play an important role in defining and achieving the improved performance.

Run well, its supporters claim that the results can be very satisfactory, especially form the buyer's side. Yet frustratingly there is no compelling evidence as yet that partnered projects are, on aggregate, more successful than non-partnered ones. Research suggests that actually there may well be better collaboration and cooperation on some contracts without partnering than is exhibited by some with[7].

Critics, such as Andrew Cox of the University of Birmingham, point out that partnering may by no means translate into a win-win situation for suppliers since contractors may not be able to make as much money as they did through more adversarial forms of contracting. And this is true. (Hence there is evidence of contractors staffing partnering projects with 'B team' personnel, reserving the 'A team' for the traditional forms of contract where the potential upside financial rewards are much greater.) Cox analyses this in terms of power relations[8] – for example, what power does the buyer have over potential suppliers to demand partnering over traditional forms?

In the end, partnering, like relationship management in general, only works because people want it to. As we said at the outset of this section: commitment. The important thing is how relationships are managed[9]. In this new world, trust becomes a dominant characteristic, as does shared learning, transparency and collaboration. A 'project charter' is often quoted as an important means of orientating the project team towards the common project goals. In some cases, project staffing is on the basis of 'best person for the job'. Team building and teamwork will generally receive heightened attention. (On the Andrew project, it will be recalled, the consultants JMW were brought in to facilitate 'high-performance' teamwork. Encouraged by the prospect of a sizable financial reward, good behaviours were promoted as an integral part of the Alliancing strategy – see Chapter 5, pp. 79–82)

But achieving real cultural change takes time, as we'll see in Chapter 19. Culture, trust, transparency and so forth are not things that can just be turned on as though by a tap. They require Board-level support and possibly years of proving in practice. Bovis, for example, a UK construction firm that had a strategic partnership with Marks & Spencer, the high-street retailer, since 1926, would not let its staff move easily between its 'M&S' division and its other contracting work, precisely because it took so long to develop the real trusting mindset. (Bovis was a committed relationship-oriented company. It always sought to build value and put the client's interests first. Better to keep the relationship alive for the long-term than score short-term transactional profit through claims.)

Procurement[10]

Having developed the desired strategic approach for acquiring resources, detailed work now needs to be initiated to get those resources in place. This applies for resources which are internally supplied as well as those that are external. In either case, the same principles apply, though the language is more explicitly procurement-orientated for external resources. (As noted for the matrix organisation, even where resources are being provided internally, some form of contract, if only informal, is the most effective way of establishing the basis on which they are to be supplied.)

Procurement is generally seen as the activity of acquiring new services or products. It covers:

- the financial appraisal of the procurement options available,
- the selection of the proposed contract form and the preparation of contract documentation,
- the selection and acquisition of tenders, contractors and suppliers,
- the administration of contracts.

Procurement involves selecting from a range of acquisition options. Buying is not necessarily the only – or even necessarily the preferred – one. Making, renting or leasing could be equally valid – and procurement may extend to storage, logistics, inspection, expediting, transportation, and handling of materials and supplies, and disposal at the end of the useful life of the product. It may cover all members of the supply chain. Operations and maintenance, for example, may need to be supported through a supply chain management process.

Procurement is often carried out by specialist departments having a matrix relationship to the project. Input from the project team is absolutely essential, however, in terms of advice on project performance criteria or timings (and budgets) for the project. Similarly, the project team will want to influence procurement activities to optimise their impact on the project outcome: bought-in components and systems are often a huge portion of the overall project budget and can significantly impact the project schedule, let alone such matters as technical performance, quality, risk, value, and even participant behaviour.

Tendering/bidding[11]

For many projects and programs, procured goods and services can represent the highest portion of the project cost. Tendering/bidding is where decisions are made which profoundly lock these costs into the projects as commitments. It is therefore essential that this activity be undertaken as competently and effectively as possible.

The approach to tendering should reflect the overall acquisition strategy: the business drivers governing the project or program should be properly reflected in the tendering documentation; and the contracting plan should identify what package contracts are to be awarded and when. Tender Event Schedules can show in detail the activities that are expected to lead to issuance of tender documentation, bid evaluation and contract award.

Individual work packages should be evaluated to see whether any scope exists, for example, via Value Management, for improving technical definition or commercial, schedule or other benefit.

The choosing of potential suppliers again reflects the 'relational versus transactional' debate. 'Widening the net' may introduce new competition and lead to improved offers; framework contracts and long-term alliance and partnering arrangements may, on the other hand, bring advantages, as we've seen.

Control of information is critical during the tender period, not least because bidders will be actively seeking to gain intelligence on their competitive situation. To ensure a level playing field and to maximise the competitiveness of bids, a single point of contact in both the owner and supplier organisation should be maintained. All other contacts should be actively discouraged. All questions arising should be formally documented and identical responses should be sent to each bidder. Information given during the tender period forms an integral part of the contract documentation and has to be very carefully considered and documented. Failure to manage the tender period in a fair and transparent manner can result in legal challenge. The tender documentation should reflect ethical procurement policies and practices.

Bids are typically prepared, and evaluated, in separate technical and commercial sections. Various evaluation schemes are used to assess both, taking care to separately identify and evaluate 'soft' areas such as contingent items, quality of proposed team, and so on. There may be some negotiation with the preferred bidder(s).

Contract Administration

A contract is an agreement between two parties under which one party promises to do something for the other in return for a consideration, usually a payment. A contract can be defined as "an agreement between two or more parties that is binding"*. A valid legal contract requires 'agreement'

---

*If it's a legal contract, then it's binding in law.

(offer and unqualified acceptance), an intention to be legally bound, 'consideration' (benefit and detriment), and competent parties, legality of purpose and certainty of terms.

Generally commercial contracts comprise: the contract agreement, which itemises the documents comprising the contract; a general specification and scope of work; the general conditions of contract – normally a recognised standard form of contract; special conditions of contract; and administrative and coordination procedures.

The provisions of a contract should:

- define the responsibilities of the parties
- allocate risk
- determine effective payment terms.

The project team should have a thorough understanding of the procurement process and post tender (bid) negotiation; assumptions made by the purchaser and the supplier; the buyer's expectations of the service relationship; and the contract terms and conditions, including specifically the legal implications of the contract for which they are responsible.

Key provisions under the contract that the project team will generally need to manage include time issues – principally commencement, schedule, suspension and completion; payment provisions; incorporating change; remedies for breach of contract; performance indicators; liquidated damages; termination; and bonds, guarantees and insurances including performance guarantees, retention guarantees, and surety bonds. Procedures should be in place which clearly define responsibilities for key activities required under the contract, for example, payment, budget review and control, change management, price adjustments, disputes resolution, compliance monitoring, and termination requirements.

The form of contract can have a huge impact on the roles played by project personnel and on the way they play them. Much of international construction, for example, is based on the FIDIC form of contract, which in turn is based on the British JCT forms in which the contractor bids usually to be, and is awarded the contract on the basis of being, the lowest price*. There is no overall project manager. 'The Engineer' administers the contract on the client's behalf. If things go wrong, disputes and claims are probable; things are as likely to get worse as better. The New Engineering Contract, introduced in 1993[12], however, is built around principles of project management (specifically the project manager) and partnering. Emphasis is on resolving rather than escalating disputes.

---

*FIDIC is the Fédération Internationale Des Ingénieurs-Conseils. JCT is the Joint Contracts Tribunal. For an excellent example of how poor dispute resolution can be so pernicious to project accomplishment, see Morris, P. W. G. and Hough, G. H. (1987), The Thames barrier, Chapter 5, in *The Anatomy of Major Projects*, John Wiley & Sons: Chichester.

Claims

A claim is an assertion under the contract of a right that may lead to a demand or request, usually for extra payment and/or time. A claim might not be a change – although sometimes a claim may be settled with the issue of a change order as it is the simplest way to vary the terms of the contract.

A *change* affects the contents of the contract – the scope or specification, the payment terms, the time and so on. A *claim* covers changes that have a material effect on the method of working.

There should be clear processes described in the contract for the management of claims, including their identification, notification, measurement, submission, and settlement.

Concluding the Contract

The mechanism for concluding and terminating a contract must be established at the very outset (during the Pre-Contract stage) and documented in the contract.

Usually the contract is progressively concluded, with responsibilities being handed from contractor to client through a series of approvals and acceptances. On small projects, the end may be marked by a single milestone and certificate issue; on medium and large projects, a progressive certification process is typical covering mechanical completion, Provisional Acceptance Certificate, and Final Completion Certificate.

References and Endnotes

[1] Womack, J. P and Jones, D. T. (1996), *Lean thinking*, Simon & Schuster: New York.

[2] Macneil, I. R. (1980), *The new social contract: An enquiry into modern contractual relations*, Yale University Press: New Haven, CT; Campbell, D. (2001), *The relational theory of contract: Selected works of Ian Macneil*, Sweet & Maxwell: London.

[3] von Branconi, C. and Loch, C. H. (2004), Contracting for major projects: Eight business levers for top management, *International Journal of Project Management*, 22, 2, pp. 119–130; Davies, A. and Hobday, M. (2005), *The business of projects*, Cambridge University Press: Cambridge; Loftus, J. (1999), *Project management of multiple projects and contracts*, Thomas Telford: London; Marsh, P. D. V. (2001), *Contracting for engineering and construction projects*, Gower Technical: Aldershot; Ministry of Defence (2004), *The smart acquisition handbook*, 5th edition, Ministry of Defence: London; Scott, B. (2001), *Partnering in Europe*, Thomas Telford: London; Venkataraman, R. (2004), Project supply chain management: Optimizing value: The way we manage the total supply chain, in Morris, P. W. G. and Pinto, J. K. (eds.) *The Wiley guide to managing projects*, Wiley: Hoboken, NJ.

[4] Smith, C., Topping, D. and Benjamin, C. (1995), Joint ventures, in Turner, J. R. (ed.), *The commercial project manager*, McGraw Hill: London.

5   Thompson, P. (1981), *Organization and economics of construction*, McGraw Hill: Marlow, UK.

6   Cox, A. and Ireland, P. (2006), Relationship management theories and tools in project procurement, in: Pryke, S. and Smyth, H. (eds.) *The management of complex projects: A relationship approach*, Blackwell Publishing: Oxford, pp. 251–281.

7   Gustafson, M., Smyth, H., Ganskau, E. and Arhippainen, T. (2010), Bridging strategic and operational issues for project business through managing trust, *International Journal of Managing Projects in Business*, 3, 3, pp. 422–42; Smyth, H. J. and Edkins, A. J. (2007), Relationship management in the management of PFI/PPP Projects in the UK, *International Journal of Project Management*, 25, 3, pp. 232–240.

8   Cox, A. and Ireland, P. (2006), *op. cit.*: 6.

9   Macneil, I. R. (1980), *op. cit.*: 2.

10 Baily, P. (2000), Procurement, In: Turner, J. R. and Simister, S. J. (eds.) *Gower handbook of project management*, Gower: Aldershot; Fleming, Q. W. (2003), *Project procurement management*, FMC Press: Tustin, CA; Ministry of Defence. (2004), *op. cit.*: 3; Venkataraman, R. (2004), *op. cit.*: 3.

11 Lowe, D. (2004), Contract management, in: Morris, P. W. G. and Pinto, J. K. (eds.) *The Wiley guide to managing projects*, Wiley: Hoboken, NJ; Marsh, P. (2000), Contracts and payment structures, in Turner, J. R. and Simister, S. J. (eds.) *Gower handbook of project management*, 3rd edition, Gower: Aldershot; Marsh P. (2004), *Contract negotiation handbook*, 3rd edition, Gower Publishing Limited: Aldershot; Association for Project Management (1998), *Standard terms for the appointment of a project manager*, APM Publishing: High Wycombe; Steel, G. (2004), Tender management, in Morris, P. W. G. and Pinto, J. K. (eds.) *The Wiley guide to managing projects*, Wiley: Hoboken, NJ; Simister, S. (2000), Bidding, In: Turner, J. R. and Simister, S. J. (eds.), *ibid*.

12 *NEC engineering and construction contract*, Thomas Telford Ltd: London.

Adding Value, Controlling Risk, Delivering Quality, Safely and Securely

Sir David Higgins, ex-Chief Executive of the Olympics Development Authority, the body that built the 2012 London Olympic facilities, and subsequently Chief Executive of Network Rail, the UK rail infrastructure provider, comments that value is created in the front-end: "that's when you think about what you want to do, you review the options, you work out how you'll procure it, you finalise the design. All the value is lost between [gates] 4 and 8"[1]. I'd put it more broadly. Management's task, especially but not solely at the Front-End, is essentially that of creating and building-in functional and economic value and controlling emerging risk.

Sounds obvious but how do we do this? To be sure, there are some simple actions that can be taken, for example, increasing financial value (NPV – Net Present Value) by shortening the project's duration; improving risk management by regularly challenging the list of major risks and the responses being proposed (or as we'll see in this chapter, by expanding risk management into uncertainty management and focussing on performance); just listening and challenging can be a very effective means of doing both. Over the last 20 or so years we have seen several initiatives aimed at formalising practices in this area. In doing so, some interesting questions have arisen about what we really mean and what we should be doing. Take value first.

Building Value, Achieving Benefits[2]

Over the last decade or two I have encountered several serious studies of value, each using a different meaning. The more usual one, as used typically in Value Management, is where functional performance is evaluated in terms of input cost – 'bang for the buck'. Another was the Thomas and

Reconstructing Project Management, First Edition. Peter W.G. Morris.
© 2013 John Wiley & Sons, Ltd. Published 2013 by John Wiley & Sons, Ltd.

Mullaly 2009 study for PMI on how companies perceive the value (worth) of project management (see Chapter 5, p. 84)[3]. A third was by a student of mine who assessed value in terms of Michael Porter's notion of the value chain[4], analysing where and how effectively value was added in the project production process (as one went from the raw materials input to the finished product output). There is a fourth interpretation too: the ethical, even meta-physical, one of the conception of value as a motivating desire, moral system or beliefs.

Value, in the sense we are using here, can be thought of, rather simplistically, as the ratio of benefits to investment, or quality over cost, function over cost, or performance over resources[5]. BSI 2000 defines value as "the relationship between satisfaction of need and the resources used in achieving that satisfaction": all still wide open when it comes to trying to develop objective measures.

Consider the well-known phrase "he knows the cost of everything but the value of nothing". The trouble with the phrase is that it assumes we can value things and he cannot. Our judgement is better than his. But value is subjective and relative. Van Gogh died in poverty: values change, and they may reflect views which are not only subjective but are transitory.

The difficulty in assessing value, in all bar the Porter instance, is that this subjectivity can lead to inconsistency and unreliability. The usual definition in Value Management (VM) is value as assessed "in terms of stakeholders' needs and objectives". The trouble is, stakeholders might have quite different needs and objectives, and hence criteria for assessing value. What might have seemed like good value to one person, or group, might not to another. This is illustrated in the classic conflict between assessing whole-life costs versus capital costs: how much do you spend in capital investment with the express intent of reducing operating costs? Dealing with this difference in role perspective is one of the reasons VM gives such prominence to facilitated workshops as vehicles for discussion (see Chapter 5, p. 83).

One way to make the assessment of value more objectively rigorous would be to tie it to the benefits that the project or program is meant to deliver. 'Benefits Management' is a comparatively new area, however, and, to the best of my knowledge, there is little published, at least as of the time of writing, on doing this (but see Ward and Daniel, 2006, amongst others in endnote 10 below). But in principle, the project's or program's desired benefits should surely play a major role in determining the proposed value of the project or program, or some aspects of them. Michel Thiry talks about tying value into the project's KPIs and CSFs[6]. KPIs (Key Performance Indicators, like operating performance) obviously relate. Critical Success Factors (CSFs) are less obvious, however, since some take CSFs to be input criteria (like having adequate resources) while others take them as measures of success.

Tying value to requirements would also seem sensible.

It would be logical, therefore, to begin with a discussion of Benefits Management but we'll start by trying to understand Value Management, which is still not well understood as a mainstream p.m. practice.

Value Management[7]

Value Management (VM) is the term used by the Society of American Value Engineers in the United States as an umbrella for Value Analysis (VA) and Value Engineering (VE)[8]. In other countries, it is seen as "the combined application of value methodologies and other methodologies at the organizational level . . . to improve organizational effectiveness"[9].

Value Management is based essentially on three key principles:

- a multidisciplinary approach to analysing value 'in terms of stakeholders' needs and objectives' – but see the comments above;
- a structured decision process to stimulate creative thinking;
- a focus on analysing functions rather than just accepting predefined solutions.

Value Planning (VP) comprises reviewing: (1) project objectives, scenarios and strategy(ies) to optimise project objectives and goals established by the owner (including making sure that the project requirements have been clearly articulated), alternative scenarios to attain those objectives and goals, and the resulting strategy(ies); and (2) outline designs, to provide an information base and understanding of the alternative solutions that will meet the objective(s).

Value Engineering (VE), on the other hand, comprises (1) Function Analysis – the identification of the functional attributes of the different solutions; (2) working on the detailed designs of a small number of alternative solutions for delivering the functional attributes; and (3) the evaluation of alternative solutions that arise during the implementation stage of the project, as additional new information becomes available during the development lifecycle.

Value Engineering can also include the evaluation of value management process outputs at handover – were the objective(s) defined by the value management process achieved, where was there the greatest improvement in functionality, etc.? These seem a different type of enquiry to the rest of VE, however, and it is surely inappropriate to mix this process-housekeeping with engineering-oriented analysis and redesign.

Typically VM, VP and VE reviews are carried out in a series of workshops, as we just noted. In many ways, however, value management can be considered as a state of mind – a disposition to seek out value. And many project people believe that the early Optioneering phase of a project's development (around Appraise in Figure 11.1) is essentially a form of value management.

Value Management is, at the time of writing, almost completely missing from the *PMBOK® Guide**. These omissions may seem, on first blush,

---

*In fact, Value Engineering is referred to twice as a technique, with no explanation and there is no discussion of the concept of value nor of the much bigger concept of Value Management.

surprising but, of course it reflects the PMBOK paradigm of project management as executing orders, not about shaping instructions! Value is hardly the concern of this paradigm. Reality, even amongst PMI's membership, is different. In research we conducted at UCL for PMI in 2005/2006, 55% of those surveyed* had a process for optimising the value of the project.

Chapter 20 takes an extended look at how project management would be reconstructed if its driving ethos were to be enhancing sponsor value.

Benefits Management[10]

Sir Ray McNulty, in his 2011 review of the UK rail industry, commented on the poor alignment being achieved in the industry between solutions and benefits[11]. Benefits are what the project or program is done for – what the benefits should be of all this capital investment and the work that's been put in. Benefits can be measured quantitatively, such as financially, market share, output capacity, and so on, and qualitatively, such as improving security or brand position. All projects and programs should achieve benefits that relate back to the sponsor's goals and strategies.

Managing benefits in project and program management involves clearly identifying what the project or program should be delivering, deriving measures for these benefits, and (very importantly) identifying owners for them. And then ensuring they are 'harvested' effectively and lessons learned from so doing are fed back so that future projects are changed and shaped (strategy, configuration, plans, etc.) such that future benefit realisation is improved (in a cost-effective – i.e., value-enhanced – way.)

Benefits Management thus comprises:

• having effective processes, organisation and techniques for benefits planning, management and harvesting;
• developing a benefits management plan/strategy;
• identifying and structuring benefits and relevant performance measures (business, technology, organisation, people and processes);
• implementing a system to track and act on benefits as they are realised, optimising the mix of benefits and identifying additional opportunities;
• aligning risk and changing management practices so that benefits (opportunities) are given proper review.

Value Management and Benefits Management should not be confused: VM is about optimising the designing-in of value, both to the project's development and the product's. Benefits Management is concerned with the realisation of benefits during project and program implementation, and in operations.

---

*We obtained data from 75 members of PMI Chapters in Europe – people at various levels of seniority, in small, medium and large enterprises in a diverse range of business sectors such as aerospace, automotive, IT, telecommunications, pharmaceuticals, retail, transportation and publishing; and academia and consultancy.

Benefits management applies to project management as well as to program management. (For PMI, program management is seen as being about the delivery of benefits; project management is not, though the terminology is confused*. This cannot be right: projects create benefits.)

Risk and Opportunity Management[12]

Risk is fundamental to projects. Projects are about creating the future and the future is full of unknowns, many bearing on the project. (Or program.) Yet amazingly it has only been in the last couple of decades that risk management has become a mainstream project management function, as we saw in Chapter 5 (p. 81).

The dictionary defines risk as the probability of a negative occurrence. Modern project or program risk management, however, as defined, for example, in the UK's Treasury documents, the ICE/CIA Ramp Guide, APM's PRAM Guide, and OGC's Management of Risk[13], considers the management of risk as including the management of upside opportunities as well as negative possibilities. (This twisting of English and normal usage of words is slightly bizarre!)

Negative outcomes (risks) and opportunities exist in project and program management because of uncertainty. Hence many people are now using the term 'uncertainty management', as we'll see in a moment.

There has been much heaving and huffing over the last few decades in defining a project risks (and opportunity) management process. This process is useful. It is a good place to start in the management of risk but it is not where the really difficult judgement calls are made. Chris Chapman and Stephen Ward of Southampton University, in their 2011 book *How to Manage Project Opportunity and Risk*, list the most well-known methods (Table 4.1) and, more interestingly, their shortcomings (Table 4.4), which include: lack of clarity on motivation (attitude to risk), level of detail (often too great), overdue emphasis on events, underemphasise on testing, lack of clarity on ownership, tendency to miss connections, and failure to capture iteration.

Essentially the Risk and Opportunity Management process covers:

* *identification* – through brainstorming, lessons-learned reports etc., leading to a 'risk register';

---

*According to PMI's standard on Program Management (PMI, 2006: p. 4), "some organisations refer to large projects as programs. The management of large individual projects or a large project that is broken into more easily managed subprojects remains with this discipline of project management . . . . If a large project is split into multiple related projects with explicit management of the benefits, then the effort becomes a program." What about projects that aren't split into multiple projects? And shouldn't all projects, and their subprojects, pay explicit attention to managing the benefits they are supposed to be delivering?

- *assessment* – using qualitative means (or quantitative tools, such as Monte Carlo simulation) to assess the probability of the risk occurring and its impact on the project should it occur. (This gives rise to the 'probability–impact' traffic light diagram). Other techniques include options management (to address possibilities at a point in time in the future) and risk efficiency (to look at the benefits of different risk management options);
- *allocate risk owners;*
- *manage project risks* either contracting them out, insuring against them, or retaining them, for example, changing the design, purchasing back-up, superior decision-making. Where risk is to be transferred, it should be transferred to the party best able to bear (and manage) it; the complete transfer of risk is rarely wholly effective or indeed possible. Risk control will involve not only risk monitoring (and taking action) and reporting to governance and other stakeholders but also contingency management (releasing monies set aside for use should a risk mature). The effectiveness of the risk management process should be periodically evaluated.

Chapman and Ward emphasise the way this process forces one to ask the right questions[14]. In my view, however, the real difficulties in risk, and opportunity, management are less in following this process as in:

1. Identifying the target zone – risk in relation to what? Typically risks, and opportunities, are identified in terms of their impact on the target completion schedule and budget. But there are other 'object functions'. For example, there is a raft of risks related to getting paid (client creditworthiness, political risk, currency risk, etc.); there are risks, and opportunities, that should be evaluated with respect to benefits. Or requirements or technology. Or Operations and Maintenance. Health and Safety risk is especially important, given the increasingly tough legislation in this area (see below). Managers of projects and programs need to think beyond the traditional 'iron triangle' measures.
2. Selecting the frequency of actioning the process. Should the risks be formally reassessed at stage-gates only or on a monthly cycle or some other periodicity? The answer depends to a great extent on the type of risk: credit risks should be appraised prior to signing contracts; O&M prior to Commissioning/Handover, etc.
3. Having an appropriate response to risks and opportunities. Should the response be risk averse or risk tolerant? And who decides? (Answer: the sponsor, or supplier management, advised by the project or program management team and functional specialists.) The role of governance is crucial here. And what about other stakeholders with their differing perspectives?
4. The actions taken to address the potential risks and opportunities.

In addition, there are fundamental questions as to how objective and rational decision makers are with respect to risk[15]. Behaviours and expectations (fuelled possibly by the prospect of bonuses or sanctions and penalties) affect judgement, for example, in prioritising risks and in shaping a willingness to accept risks for a perceived opportunity. 'Optimism bias' is a mooted example[16]. Chris Chapman stresses the importance of the subjective view: "visible, subjective uplifts are dispensed with only by the very brave, who can be made to look foolish as a consequence"[17].

We saw at the beginning of this section that 'uncertainty' is increasingly being used as a term to capture the management of both upsides and downsides[18]. Uncertainty really reflects unknowns. Thus, in addition to the category of identified risks, there are the categories of known–unknowns and unknown–unknowns, as we saw in Chapter 8. Chapman and Ward claim a paradigm shift in their 2011 book by arguing that focussing on risk management is too limiting. Instead they propose a seven-stage 'performance uncertainty management process' (Table 4.2)[19] – PUMPs. Intellectually sensible, in practice I can't see managers lowering their attention to risk (as properly defined as the probability of a negative occurrence), nor being willing to substitute PUMPs for the more general progress meetings supported by value management and risk management.

Quality Management[20]

Arguably, nothing has had such an effect on project management as Quality (courtesy of Toyota, and NASA). Partnering, Lean Management, Stage-Gate reviews, audits – all stem, via Wheelwright & Clark (1992) and Cooper *et al.* (1986), from the work of Edward Deming with Toyota in the 1950s[21]. And of course Mueller's 'all up' testing "with massive quality control and staffing implications" (Chapter 3, p. 38). Certainly Total Quality Management made a massive impact on the general business scene in the 1980s and 1990s, and though its freshness, and influence, may have now diminished, its potency lives on – through for example, Six Sigma and relationship management as we shall see shortly.

Modern Quality Management (QM) is the coordinated set of activities required to direct and control an organisation with regard to quality. In its broadest sense, Quality relates to how a set of characteristics fulfils its requirements. It relates potentially not just to output quality but input and to process. Usually we distinguish under Quality Management between Quality Control (QC) and Quality Assurance (QA). Both are mature functions in most organisations.

Quality Management is defined by the ISO (9000, -1; 10005, -6, -7) as comprising the following four activities.

1. *Quality planning* – sets quality objectives and specifies necessary operational processes and resources to fulfil these objectives. In a project- or program-driven enterprise, management should set performance objectives for all parts of the organisation, including the project or programs it undertakes. These should be defined in the Quality Plan which states

what the intentions are for achieving quality and what the control and assurance processes are. ISO 10006 *Guidelines for Quality in Project Management* outlines quality management practice vis-à-vis project management (not very persuasively).

2. *Quality control* – actions to ensure quality requirements are being met. Progress in achieving organisational objectives may be measured through regular management reports while (product) output quality may be measured through direct comparison with specified requirements. Quality control operates in most areas of project management, for example, in 'Solutions Development' through Quality Function Deployment (QFD); in 'capability acquisition' through supplier validation; and in 'people' through accreditation, performance reviews and competency development.

3. *Quality assurance* – providing confidence that quality requirements are being fulfilled. This is normally done by reviewing or auditing actual progress against detailed quality objectives and the processes and resources needed. Peer Reviews, Peer Assists and Design Reviews are examples. ('Peer Assists' help build the project definition; 'Peer Reviews' follow after Sanction and are intended to ensure what is being delivered conforms with what was sanctioned.)

4. *Quality improvement* – is about increasing the ability of the organisation to fulfil quality requirements. As such it is clearly a Level 3 activity – that is, one led at the enterprise level rather than from the project level. It is a key aspect of quality management and is a mandatory requirement of certification under ISO 9001.

Six Sigma is another Level 3 Quality activity. It builds directly off TQM and, using a variety of analytical tools to reduce variation, is a powerful force in many project-based enterprises, directly affecting the way projects and programs are undertaken[22]. Typical Six Sigma concerns include project selection, evaluation and deployment; project control (information utility); strengthening the customer focus; and measurement of financial results.

Lean and Six Sigma are often combined rather vaguely, giving us terms such as 'the lean project manager' who is then tagged as a p.m. principally concerned with adding value. And in truth, Total Quality does indeed cover many different ways of developing an improved offering. The ethos is an important one and is the subject of Chapter 20.

Quality is often quoted as the third criterion of the 'iron triangle', but it is probably too broad an attribute to be totally happy filling this spot*. Its very breadth can be troublesome too: tight time and cost targets can too easily compromise quality. Sometimes, as in safety critical software, quality can be of such an extremely high standard that it dominates all other considerations.

---

*Functional performance seems nearer the mark and is better surely than scope, which many people have difficulty in comprehending.

Health, Safety, Security, and Environment (HSSE)[23]

Increasingly, the broader perspective of Corporate Social Responsibility, covering health and safety, environment management and pollution control, community development, social accountability, ethics and integrity, and stakeholder engagement, is being pushed up the political agenda in all sectors of business. It should be seen as a major responsibility of all project and program managers. Health, Safety and Environment (HSE), however, has quite specific and detailed connotations in project and program management.

HSE management involves determining the standards and methods required to minimise, to a level considered acceptable by legislation and the public, users and operators, and others, the likelihood of accident or damage to people, equipment, property, or the environment. HSE management applies across just about all industries: aerospace, autos, construction, energy, transport, foods, pharmaceuticals, waste and many other sectors.

Employers have a duty of care to those employed to work on their behalf and for the environment within which they operate. Employers and their staff can end-up in prison if they are judged to have been derelict in their Health and Safety duties. There are processes and practices that legislation requires to be implemented to ensure accidents do not happen and the environment is not inappropriately damaged. Construction Design and Management (CDM) regulations are now embodied in European law[24] and impose legal requirements regarding safety and health on designers, owners, project managers and others. In the process engineering industries, HAZCON and HAZOP processes are followed for formally identifying major hazards in construction and operation.

Environment has, in many companies, and in project management in general, been added to Health and Safety to form a kind of catch-all first aid function. Achieving sustainability and carbon emission targets is now exerting a substantial impact on the conduct of many projects. Environmental requirements such as those for noise, dust, protection of flora and fauna, waste and sustainability *must* now be actively incorporated within the project's planning where these regulations apply.

Some projects go further and add Security to this topic. This covers everything from site (and personnel) guards to intellectual property*. (There should be an information management strategy in place from the outset of the project or program covering *inter alia* communications, document and information management, and knowledge management.)

Much of the standard of application of HSSE work is strongly influenced by the culture and proactive nature of the project's home organisation –

---

*Care needs to be taken to ensure due process is paid to all legislative and regulatory conditions pertaining to the project or program information, for example, the requirements of data protection and Freedom of Information (FoI), and expectations raised through Intellectual Property Rights (IPR) and commercial-in-confidence information. Among other implications, these raise obvious issues of access rights to data (Data Protection and IPR) and communications requirements (FoI).

though it is absolutely critical that project and program management set a clear and uncompromising stance with respect to implementing requisite standards. We thus see a strong interaction with Level 3.

References and Endnotes

1 House of Commons Transport Committee (1 March 2011), http://www. parliamentlive.tv.

2 BS EN 12973:2000 *Value management*, European Committee for Standardization (CEN) Technical Committee CEN/TC 279, BSI: London; BS EN 1325-1:1997 *Value management, value analysis, functional analysis vocabulary*, BSI: London; Connaughton, J. N. and Green, S. D. (1996), *Value management in construction*, CIRIA: London; Hamilton, A. (2002), Considering value during early project development: A product case study, *International Journal of Project Management*, 20, 2, pp. 131–136; Highways Agency (1999), *VFM easy guide*, Highways Agency: London; HM Treasury (2000), *Central unit on procurement: No. 54 value management*, HM Treasury: London; Male, S., Kelly, J., Grongvist, M., Fernie, S. and Bowles G. (1998), *Value management benchmark: Framework document*, Thomas Telford: London; Office of Government Commerce (2002), *Value for money evaluation in complex procurements*, Office of Government Commerce: Norwich, UK; Society of American Value Engineers (SAVE) (1997), *Value methodology standard*, SAVE International: Northbrook, IL; Thiry, M. (2004), Value management, in Morris, P. W. G. and Pinto, J. K. (eds.) *The Wiley guide to managing projects*, Wiley: Hoboken, NJ; Thiry, M. (1997), *Value management practice*, Project Management Institute: Sylva, NC; Woodhead R. and Downs, C. G. (2001), *Value management: Improving capabilities*, Thomas Telford Publishing: London; Institute of Value Management http://www.ivm.org.uk; Bennington, P. and Baccarini, D. (2004), Project benefits management in IT projects – An Australian perspective, *Project Management Journal*, 35, 2, pp.20–30; Office of Government Commerce (2003), *Managing successful programmes*, TSO: Norwich; Ward, J. and Peppard, J. (1999), *Strategic planning for information systems*, John Wiley & Sons: London.

3 Thomas, J. and Mullaly, M. (2008), *Researching the value of project management*, PMI: Newton Square, PA.

4 Porter, M. E. (1985), *Competitive advantage*, The Free Press: New York, pp. 11–15.

5 Thiry, M. (2004), Value management, in Morris, P. W. G. and Pinto, J. K. (eds.) *op. cit.*: 2, Chapter 36.

6 *Ibid.*

7 BS EN 12973:2000 *Value Management*, European Committee for Standardization (CEN) Technical Committee CEN/TC 279, BSI: London; BS EN 1325-1:1997 *Value management, value analysis, functional analysis vocabulary*, BSI: London.

8 Society of American Value Engineers, http://www.value-eng.org.

9 BSI, (2000), *op. cit.*: 7.

10 Bradley, G. (2010), *Benefit realisation management*, 2nd edition, Gower: Aldershot; Bradley, G. (2010), *Fundamentals of benefit realization*, TSO: Norwich; Jenner, S. (2009), *Realising benefits from government ICT investment: A fool's errand?* Academic Publishing International Ltd: Reading, UK; Jenner, S. (2010), *Transforming government and public services – realising benefits with project portfolio management*, Gower: London; Sward, D. (2006), *Measuring the business value of information technology*, Richard Bowles, Intel Press: Santa

Clara, CA; Thorp, J. (1999), *The information paradox: Realizing the business benefits of information technology*, McGraw-Hill: Toronto, Canada; Ward, J. and Daniel, E. (2006), *Benefits management: Delivering value from IS & IT investments*, John Wiley & Sons: Chichester.

[11] Mcnulty, R. (2011), *Realising the potential of GB rail*, Department for Transport: London.

[12] Bartlett, J. (1998), *Managing programmes of business change*, Project Manager Today: Hook, Hampshire; BS 6079 (2000), *Guide to project management – part 3 risk management*, BSI: London; Chapman, C. B. and Ward, S. (2003), *Project risk management: Processes, techniques, and insights*, John Wiley & Sons: Chichester; Chapman C. and Ward S. (2004), Why risk efficiency is a key aspect of best practice projects, *International Journal of Project Management*, 22, 8, pp. 619–623; DOD (2002), *Risk management guide for DOD acquisition*, 5th edition, Department of Defense, Defense Acquisition University: Fort Belvoir, VA; Hillson, D. A. (2003), *Effective opportunity management for projects: Exploiting positive risk*, Marcel Dekker: New York; Institution of Civil Engineers (1998), *RAMP: Risk analysis and management for projects*, Institution of Civil Engineers and Institute of Actuaries: London; ISO (2002), *ISO/IEC Guide 73, Risk management – Vocabulary – Guidelines for use in standards*, BSI: London; Office of Government Commerce (2002), *Management of risk: Guidance for practitioners*, TSO: Norwich; Simister, S. J. (2004), Qualitative and quantitative risk management, In: Morris, P. W. G. and Pinto, J. K. (eds.) *op. cit.*: 2; Williams, T. (2002), *Modelling complex projects*, John Wiley & Sons: Chichester, Chapter 5.

[13] Faculty of Actuaries, Institute of Actuaries, the Institution of Civil Engineers (1998), *Risk Analysis and Management for Projects*, Thomas Telford: London; OGC (2002), *Management of risk: Guidance for practitioners*, TSO: Norwich.

[14] Chapman, C. and Ward, S. (1997), *Project risk management: Processes, techniques and insights*, John Wiley and Sons Ltd: Chichester.

[15] Winch, G. and Maytorena, E. (2011), Managing risk and uncertainty on projects: A cognitive approach, in Morris, P. W. G., Pinto, J. K. and Söderlund, J. (eds.) *The Oxford handbook of project management*, Oxford University Press: Oxford.

[16] Flyvbjerg, B., Bruzelius, N. and Rothengatter, W. (2003), *Megaprojects and risk: An anatomy of ambition*, Cambridge University Press: Cambridge.

[17] Chapman, C. and Ward, S. (1997), *op. cit.*: 14.

[18] Ward, S. and Chapman, C. (2003), Transforming project risk management into project uncertainty management, *International Journal of Project Management*, 21, 2, pp. 97–106; Chapman, C. and Ward, S. (2011), *How to manage project opportunity and risk: Why uncertainty management can be a much better approach than risk management*, John Wiley & Sons: Chichester.

[19] Chapman, C. and Ward, S. (2011), *op. cit.*: 18.

[20] Anbari, F. and Hoon-Kwak, Y. (2004), Success factors in managing six sigma projects, In: Slevin, D. P., Cleland, D. I. and Pinto, J. K. (eds.) *Innovations: Project management research*, Project Management Institute: Newton Square, PA; Association for Project Management (2004), *Directing Change: A guide to governance of project management*, Association for Project Management: High Wycombe; BS ISO 10005:1995 *Quality management – Guidelines for quality plans*, BSI: London; BS ISO 10006:2003 *Quality management systems – Guidelines for quality management in projects*, BSI: London; BS ISO 10007:2003 *Quality management systems – Guidelines for configuration management*, BSI: London; BS ISO 9000:2000 *Quality management systems – Fundamentals and vocabulary*, BSI: London; BS ISO 9004:2000 *Quality management systems – Guidelines for performance improvements*, BSI: London; BS EN ISO 19011:2002 *Guidelines for*

quality and/or environmental management systems auditing, BSI: London; Dale, B. G. (1995), *Managing quality*, 4th edition, Blackwell: Oxford; Heumann, M. (2004), Improving quality in projects and programs, In: Morris, P. W. G. and Pinto, J. K. (eds.) (2004), *op. cit.*: 2.

21 Deming, W. E. (1986), *Out of the crisis: Quality, productivity and competitive position*, Cambridge University Press: Cambridge; Cooper, R. G. (1986), *Winning at new products*, Addison-Wesley: Reading, MA; Wheelwright, S. C. and Clark, K. B. (1992), *Revolutionizing product development*, Harvard Business School Press: Cambridge, MA.

22 Anbari, F. and Hoon-Kwak, Y. (2004), *op. cit.*: 20.

23 BS ISO 14001:1996 *Guide to Environmental Management Systems – Specification with guidance for use*, BSI: London; BS ISO 14001:2004 *Guide to Environmental Management Systems – Requirements with guidance for use*, BSI: London; BS OHSAS 18001:1999 *Occupational health and safety management systems specification*, BSI: London; Croner (2004), *Health and safety manager*, Croner Publications: Surrey; European Construction Institute (1999), *The ECI guide to managing health in construction*, Thomas Telford: London; Gibb, A. (2004), Safety, health and environment, in Morris, P. W. G. and Pinto, J. K. (eds.) *op. cit.*: 2; Health and Safety Executive (1999), *Health and safety law*, 2nd edition, HSE Books: Suffolk; Health and Safety Executive (2003), *Health and safety regulations. A short guide, rev 1*, HSE Books: Suffolk.

24 Directive 89/391/EEC.

People

Now we come to the fun part: people! Remember: 'projects are built by people, for people, through people'. Sir Alistair Frame of Rio Tinto Zinc had it right in 1988 (Chapter 3, p. 48): "the most important part of project management is the people".

One should never compromise on people. But one always does. (Why? Because one rarely gets the perfect fit; it's difficult to assess how people will really perform in a new situation; people have careers and many resent being pinned down; people, unlike all the other factors of production, are animate – they have egos, are passionate, wilful, emotional, make mistakes. And many people don't actually have their competency assessed in terms of the specific job they are being asked to perform.)

We have seen how even in quite technical areas such as requirements management, procurement, HSE, risk, scheduling and critical chain, interpersonal skills and behaviours can be of overriding importance. Project personnel often talk of managing people as the soft skills; in fact, they are the hardest. Doing it well requires self-knowledge, knowledge of the sponsor's objectives, strategies and goals; knowledge of the project and knowledge of the subject. Intuition, patience, and communication.

At Level 2, the strategic level, project managers should have a major input into selecting people *you* can trust to perform well on the project. Demonstrate that you can be trusted too. Be emotionally intelligent. Meld your people as a team. Shape and share the project vision. Inject drive and dynamism. Facilitate decision-making (in a timely fashion, utilising the appropriate knowledge). Communicate clearly. Be sensitive to changing external and internal conditions. Leverage informal forces. Manage conflict creatively.

There is a vast amount of academic literature available, but not much based in or around projects or programs.

We begin with leadership and then turn to team-working. We shall touch on stakeholder management, individuals, communications, emotional

Reconstructing Project Management, First Edition. Peter W.G. Morris.
© 2013 John Wiley & Sons, Ltd. Published 2013 by John Wiley & Sons, Ltd.

intelligence, influencing and negotiating, conflict management, decision-making, delegation and empowerment, motivating, and culture.

All these are important at Levels 1 and 2, and, in a slightly modified way, at Level 3 too. Chapter 16 shifts the 'OB' (Organisation Behaviour) focus to Level 3: building enterprise-wide project and program management capabilities. Competencies are examined as is the concept of project (and program) maturity. Knowledge management and project/organisational learning are reviewed. Culture is discussed.

Leadership[1]

The essence of project and program management is change: shaping the proposed way of achieving the desired change, and delivering that change. Leadership is central to achieving this.

Leadership is the activity of forming, shaping and giving voice to goals – establishing and 'selling' a vision; and motivating and influencing others to follow in the realisation of that vision – doing what needs to be done to fill out that vision and deliver it.

Management, on the other hand, is the art of getting others to do what one cannot necessarily do oneself, by organising, controlling and directing resources.

It would be wrong to separate the two too strongly – indeed, theorists talk of 'managerial leadership'[2]. For what is clear is that program and project management cannot succeed without effective leadership – at all levels of the project team! ('Distributed Leadership'.)

Many theories of leadership exist; indeed, it is one of the most studied subjects in management, though there is much scope for debate and disagreement*. Broadly, there are two schools of leadership thinking: universal and contingency.

'Universal' theories of leadership suggest there are enduring leadership traits and behaviours which apply in all situations: leaders are born, not made. Charismatic and transformational theories of leadership are examples[3]. How often does charisma and the power of personality lead to ill-founded or ill-executed projects? Countless times†!

---

*A review of *The Handbook of Leadership* concluded that "after 5,000 studies . . . the confused state of the field can be attributed in part to the sheer volume of publications, the disparity of approaches, the proliferation of confusing terms, the narrow focus of most researchers, the high percentage of irrelevant or trivial studies, and the absence of an integrating conceptual framework". Yukl, G.A. (1989), *Leadership in organizations*, Prentice Hall: Englewood Cliffs, NJ.

†To quote page 329 of *The Management of Projects* (Morris, 1994): "Dixon conducted a detailed review of military leadership. Militarism, he concluded, tends to appeal to persons neurotically absorbed with sex, dirt, aggression, self-esteem and death. Unfortunately these traits tend to work against effective leadership!" (See Dixon, N. F. [1976], *On the psychology of military incompetence*, Jonathan Cape: London.)

Theories of Leadership

- Getting others to follow you (Thompson and Tuden)
  - Shaping vision
- Attributes of leadership – internal to the leader
  - Trait theory
  - Behaviours:
    - motives, need for power & affiliation; self-control
    - styles & performance: authoritarian, democratic, laissez-faire (Lewin)
    - Managerial Grid: people v. production (Blake & Mouton)
  - Emotions: mood; Emotional Intelligence
- Attributes of leadership – internal to the leader
  - Relationship vs. task oriented (Fiedler)
  - Path-goal: achievement-oriented, directive, achievement, participative, supportive
  - Situational (Hershey Blanchard): leadership style and development level; fit to followers
  - Functional: environmental scanning, organising, coaching, motivating, intervening
- Transactional v. Transformational

'Contingency' theories, on the other hand, suggest that leadership styles are, or should be, contingent upon the task, the 'business' need, the environment, and the people needing leading (the team, stakeholders, etc.); that is, leadership styles and behaviour change depending upon the differing needs of the situation[4]. Thus Situational Leadership, an example of contingency-based leadership, proposes that managers need to vary their leadership style depending *inter alia* upon the maturity of their subordinates[5]. We should therefore expect to see a variety of leadership styles being variously applied on projects. Jeff Pinto of Penn State University has shown clearly that effective project management leadership needs to vary depending upon the stages of the project, the task, and/or the level of organisational support[6].

John Kotter of Harvard University[7] positions leadership as involving: establishing direction, aligning people, motivating and directing, and producing change. These are very much the characteristics of front-end development management.

Management he characterises as planning and budgeting, organising and staffing, controlling and problem-solving, and delivering consistent results. These fit with what many see as the characteristics typical of project managers. (Others might say the leadership skills are more resonant of program management and the latter again as project management. It would be a mistake, however, to believe that projects can be developed and delivered without effective leadership, at all levels.)

This suggests something that is really important to project management as a discipline: project managers – project leaders etc. – will often have to adapt their style to fit a whole raft of factors – not just the nature of the project but the form of contract, the characteristics of the people being managed, and above all, the nature of the tasks being undertaken. This will mean for example that *the style of management and leadership required for the front-end will almost certainly be substantially different from that required for downstream execution.* Yet in practice the discipline is too often judged as relevant or not by the appropriateness of the style of project management being displayed (and project management usually behaves as execution management à la Kotter). This is the tail wagging the dog. In fact *the discipline requires that the style of management should adapt to the needs of the project task being managed.* Few project managers, in my experience, even acknowledge this, let alone do it. And in truth, for many people it is a very difficult thing to do. (A leopard doesn't easily change its spots!)

Teams

Many of the theories of leadership involve the way that leadership interacts with team behaviour. Teams and team members also create visions, form judgments, make decisions and motivate people. It would be a mistake to position leadership as something that is neither done by, nor is the responsibility of, the project management team, in its widest definition, including all stakeholders as well as the 'core' project team. Everyone can act like a leader. Often in projects they positively need to.

Given the spread of skills needed on most projects and programs, teams are inevitable. There can be very few projects of any significance that are not, or were not, undertaken by teams. Teams are made up of different individuals working together in an integrated manner so that, ideally, the overall team performance is greater than the sum of the parts[8]. (Together Everyone Achieves More – aagh!!)

The basis of effective teamwork is interdependency – everyone playing their part professionally and expertly; trusting and respecting each other's contribution; communicating, acting and deciding as a unit. All lead in their areas of expertise. There may be a chairman, a captain, or a team leader, but in effect leadership is distributed. Leadership in projects and programs is multi-headed, not least because so much of project work is team based. The project or program manager, as integrator (the 'single point of integrative responsibility' [accountability], Chapter 4, p. 58), has lead responsibility for shaping his or her team and for getting it to perform as effectively as possible.

Research suggests that effective teams are empowered and cohesive, and their members exhibit trust towards one another, work interdependently, communicate openly and strongly, are results (goal) oriented, are competent, have high energy, and celebrate! Colocation has been shown to be associated with improved team performance[9].

Team selection should, as far as possible, be based on individuals' competencies; that is, based on the knowledge, skills, behaviours, and experience required of a person to perform a role required on the team. In reality, one's options may be limited, however: alternatives may be scarce and in large matrix organisations for example staffing may be strongly determined by the functions/lines or Resource Manager. (Belbin's famous classification of team roles – company worker, chairman, shaper, plant, resource investigator, monitor-evaluator, team worker, completer-finisher – may seem slightly orthogonal to this competency view; it really reflects a behavioural perspective[10].)

Teams take time to form. The "forming, storming, norming, performing" sequence (or cycle) is well known but the linearity – 'inevitability' might be a better word – implied by the sequence doesn't always work, particularly in the later phases. Actually teams seem to form and norm remarkably quickly and effectively on projects – its part of the project management culture[11].

Whatever model of team one takes, 'initiation' is important and one of the very first roles in initiation is clarifying purpose and team member selection. If the team's purpose is not clear it is unlikely to be successful.

Some early form of team-building activity may well be beneficial. Groups differentiate on the basis of liking, status, power, roles and leadership[12], and these issues need addressing in the early team formation stages.

'Project chartering' is a popular form of helping the team buy-in to the project purpose. The 'Project Start-Up' concept has promoted the idea of combining behavioural team-building activities with project planning. Some companies have extended this idea to produce intense value management-driven, motivational team-building workshops, the goal being to work out how to achieve exceptional performance[13]: the method is frequently used in partnering-based project organisations, as we saw on the Andrew project. ('Integrated Project Teams' – IPTs – have become a feature of some partnering-type arrangements[14].)

Start-up and chartering initiatives should not only make clear what the purpose of the project is but also how the team's performance will be measured and how it expects its members to behave: how conflict will be handled, how decisions will be made, how communications will be handled, diversity respected, trust honoured, and so on. The key to so much of project management performance is how we behave and time spent on making clear the expectations of team behaviour will be time well spent.

A commonly voiced difficulty is that in large or complex projects (and programs) team members enter and leave throughout the life of the project, and so team-building is a challenge. While this is generally true it may help to distinguish between the core team and the non-core; and periodically to refresh the team with behaviourally oriented activities such as milestone events, socials and so on.

Many projects – perhaps most – involve some degree of virtual working. Virtual teams are often of a more temporary and dynamic nature; physical bonding is obviously less and team bonding correspondingly harder. Every effort should be made to create team-building opportunities in such projects.

Whether virtual or real, team working can, and will, experience challenges. Many will revolve around problems of prioritisation, lack of team cohesiveness, competence, personality, or inappropriate behaviours. For example:

- failure to keep up and participate (lack of mental toughness, conflicting priorities, lack of shared purpose, vocabulary);
- lack of motivation (often caused by conflicting priorities, may be caused by an inability to articulate the importance of objectives and/or criteria for judging success, and difficulty in assessing the individual's contribution to success);
- misaligned behaviours (possibly to an extent unavoidable but probably due to insufficient focus on expected behaviours);
- contractual or other organisational influences (whether actual or just perceived) may impact behaviour and relationships within teams.

These can all be addressed, and should be. If there is a blockage, unblock it. Effective teams can create huge performance potential benefits.

Stakeholder Management[15]

Not everyone 'on' the team will or need be a 'core' member. Non-core team members may extend well beyond the immediate project or program. Stakeholders will be both core and non-core.

Stakeholder Management has been one of the major areas of growing awareness in project management practice since the 1990s. A stakeholder is an individual or group of individuals within or external to the project or program who have an interest in its outcome, for whatever reason*. Stakeholder management refers to the activity of identifying, analysing and influencing, and as far as possible meeting, the expectations of stakeholders and their particular interests and needs (through such techniques as influence diagrams) and using 'influencing' behavioural practices[16]. The cost of managing all potential stakeholders can be high, however, and some 'cutting-to-fit' might be necessary. (See Chapter 16, p. 224 for a note on stakeholder management from an institutional theory perspective.)

Culture[17]

Culture can often be recognised but be difficult to define. Individuals have their own distinctive culture; so do teams; so do lines, business units, and organisations. Culture is usually evident on most projects and programs. In some instances it can be crucial – the official report into the BP spillage on

---

*BS6079 defines stakeholders as those with a vested interest in the success of the project. ISO 21500 describes stakeholders as "the individuals, groups or organizations impacted by or impacting the project" (p. 23); these may be *pro* or *contra* (p. 24). Regulators and planning authorities are examples of important stakeholders who are not necessarily interested in the project's success as such but rather on its compliance and impact.

Deep Water in 2009 talks overwhelmingly of the cultural problems behind the terrible problems suffered, and inflicted, by this project[18].

Several frameworks have been proposed for explaining the layers of factors that shape culture: for example the iceberg model where readily observable behaviours sit above less visible influences such as strategies, styles, systems and organisation, while these sit above deeper beliefs, values, and social, economic, and political and religious influences[19].

Cultural factors are obviously evident in international projects and programs – McKinsey's data on 1600 developing-world macro-projects showed that "a striking aspect of these macro-projects is the human side – the linkage between managers from different cultures"[20] – but technology, professional formation, business ethic (public, private, not-for-profit), size, gender, age, and many other factors exist in a cultural hierarchy and are to be found on most projects.

Geert Hofstede analysed data on organisational culture in IBM, where he was an HR manager, in the 1970s. He showed that differences in attitudes could be explained along four dimensions: power distance, uncertainty avoidance, individualism, and masculinity, implying that culture could be managed to some extent by addressing these factors. Several researchers have subsequently looked at how national cultural attributes may be managed and leveraged using Hofstede's model[21].

Post-modernists are not sympathetic to the idea of managing culture (or trying to at least), arguing that culture is either too deeply ingrained to be susceptible to being managed, or that to do so is ethically questionable. But as I have argued, in projects one may well be dealing with issues having millions of dollars' worth of ramifications, or potentially bearing directly onto people's lives (as in HSE allocation.) At UCL we have used Edgar Schein's three-level model of organisational culture (one, artefacts – physical, verbal, ritualistic, etc. – resting on, two, values and behavioural norms, which in turn rest on, three, beliefs and assumptions) to show how a Health and Safety culture can be strengthened so that performance is improved[22].

If there is a question over whether culture can be managed there cannot be any surely that it can be – that it is – shaped. Education, parents, mass media, church, mosque and temple are all powerful shapers. Enterprises shape culture. Projects are often too short-term and transitional to shape culture much, however. (Look at how long it took Sam Phillips and George Mueller to get NASA's culture to change from one of scientific investigation to one of project accomplishment. Lord King, Chairman of British Airways in the 1980s, famously said it took a least a generation to change a company's culture.) But project personnel can:

- recognise the diversity of culture present in the project and its stakeholders;
- respect it for what it is and learning to understand it;
- use this knowledge to respect the various cultures present and to behave appropriately in integrating the diverse cultures back onto achieving the project goals.

Individuals' Skills and Behaviours

Amid all this discussion of teams, it is important to remember that projects are worked on by individuals, often working alone for lengthy periods.

It ought to help if the individual understands his or her own character. Many project personnel have at some point taken a Myers-Briggs personality test or similar. The results can help teams and groups too. McKinsey, the consultancy company, regularly gets staff to share their Myers-Briggs profile when forming to start a new assignment.

Management should help individuals to meet their personal goals and maintain their well-being, making sure that these goals are aligned as fully as possible with the project's, the program's and the enterprise's, and helping the individual stay healthy and motivated in spirit, mind and body.

Individuals need to deploy a host of what might loosely be termed skills and behaviours to which we turn to now and which are absolutely crucial to the effective definition, development and delivery of the project or program*. The first one we'll look at is knowledge regarding Communications, which actually is only partially about behaviours. It tends more to be about policies and procedures.

Communications

Truly, few things have changed as much in the last century as communications. In the two World Wars communication was slow and unreliable. Now C^{3}I provides real-time, global 'command, control, communications, and intelligence' capability – at horrendous cost. Meanwhile social networks bring down governments and the mobile phone has revolutionised the way we communicate. The iPhone produces a universe of information at our command.

Effective communication with all stakeholders is fundamental to project and program success.

Both the need for and the difficulty of communication increases as the organisation gets more complex[23]. (Though this is a watered-down version of contingency theory: that the need for, and difficulty of, managing projects increases with increased novelty, complexity, technological innovation and pace, or similar.)

---

*I am rather crudely here lumping together cognitively derived behaviours (i.e. those based on rational reasoning) with those that are behaviourally derived, i.e. derived from stimulus–response ('associationism' learning). I am trying to address skills that individuals employ rather than capabilities reflecting organizational structure and form.

What to communicate, what media to use, what orientation to give, what not to disclose, and so on will vary depending on the topic and the intended recipient. Different communications strategies might be appropriate for different types of stakeholders. Openness is to be preferred for the project team, where trust is so important, but recognise the potential danger of knowledge being communicated which can do damage – commercial, motivational, and so on. Openness does not necessarily mean full disclosure about everything. Be open and frank on a need-to-know basis, wrapped around with a good dollop of diplomacy!

The project's communications strategy should be a part of the overall project or program strategy and expectations should be clearly established early in the project. Communications efficiency is about ensuring the recipient understands clearly what you wanted him, or her, to understand.

Formal meetings are one important aspect of communication but can, if not correctly managed, result in the waste of time, money and energy. Certain meetings play a structural or process role in projects, for example, the inaugural ('start-up') meeting at project launch.

Informal communication networks can be an extremely potent form of communication. Social Network Analysis maps the pattern of actual communications, indicating thereby part of the communication efficiency[24]. Effective face-to-face and small group communication is vital. Listening skills are as important as verbalising, writing and presenting. Beware message distortion.

Trust[25]

Despite the remarks above on guarding what you communicate, trust is fundamental to truly effective project performance[26]. Being able to trust someone to perform has always been important in projects, regardless of the form of contract being employed but with the growth of Relationship Management (from Lean, TQM, etc.) bringing a dramatic growth of partnering, alliancing, SMART procurement, IPTs, and so on, trust has risen to be a core feature of project behaviour.

But beware! Trust is intangible. The verb 'trust' implies a willingness to be vulnerable with the expectation of positive outcomes. The noun 'trust' is about attitude; a belief-informing action. Trust is observed indirectly in behaviour but directly in actions. There is often a lag in observation, however: Othello trusted Iago. Trust has to be earned.

Emotional Intelligence (EI)[27]

EI is a salad of many other behavioural skills. (Of course, most overlap.) Peter Salovey and John Mayer define it as "the ability to perceive and express emotions, to understand and use them, and to manage emotions so as to foster growth"[28]. In their model, Salovey and Mayer emphasise perceiving, facilitating, understanding and managing emotion. In contrast, the (more popularised) Daniel Coleman/Richard Boyatzis model is competency-oriented, focusing on self-awareness, self-management, social awareness, and relationship management[29] – Anthony Mersino adds Team Leadership in his study of EI in project management[30].

Superficially, the ideas seem very attractive but most of the terms are difficult to define, ground operationally or measure, and in effect are promoted above their agreed understanding. Nevertheless, the work seems helpful – a useful lens through which to look at behaviours. Merino for example emphasises stakeholder relationships, reducing emotional breakdowns, working with team members, decision-making, communication, motivation, and benefits management as areas where improved EI will benefit project management.

Influencing (power) and negotiating[31]

Influencing is the ability of one person to get another to do something they want them to do (particularly when there are no significant power differences between the two). Negotiating is getting two or more people to come to an agreement by at least one of them revising their initial position, recognising that the objective is to get to a win-win situation for all parties.

Negotiating is more evident in formal transactions and influencing in informal ones. Both overlap. Both are important in projects.

Projects and programs abound with influencing and negotiating situations: they are inherent both in their change dynamics and in the informal nature of power and authority that is so typical of the project team.

"There is no more elemental concept than that of power"[32]. Power and authority may be positional (legitimised [*de jure*], reward based, or coercive based) or personal (referent, expert, informational) (*de facto*). Other sources of power include resource power, dependency power, centrality power, and non-substitutability power[33]. Influencing leverages informal power (personal, centrality, non-substitutability), generally on situation-specific and short-term transactions. Persuasion, ingratiation, pressure and guilt are typical means of exerting influence.

Negotiating should be prepared and managed carefully. There are well-developed techniques for effective negotiation. For example, preparing for negotiation involves:

- *Identifying project stakeholders' interests* – what are the basic needs that an agreement should satisfy?
- *Identifying issues* – what are the areas available for negotiation on both sides, and what might be traded?
- *Prioritising issues* – deciding how much to concede on one issue in exchange for concessions on another; make larger and more frequent concessions on less important issues than on more important ones.
- *Developing proposals* – having several prior to starting negotiation: an initial offer presented at the start of negotiation; a target proposal that represents your preferred outcome; and the least favourable proposal.

- *Deploying negotiating strategies* – possible concessions, what should be asked of the other party, what middle ground is there; when is withdrawal necessary; what timetable should be followed; what are the resources and costs to be expended in relation to what benefit?

The art of negotiation is in achieving to the greatest extent possible what you want from a transaction while leaving all parties sufficiently content that the relationship subsequently works well.

Conflict Management[34]

Conflict can – and probably will – occur at all levels in projects and programs, largely because there may be many different parties working together with their own separate aims, which at some point collide, or diverge. Projects, programs, and contracts can easily engender conflict. But this can be good. The art is to leverage the energy to the project's goals. (Never waste a good crisis!) Indeed it may be useful to inject some tension into the project to improve decision-making.

Conflict Management is thus the art of managing conflict creatively and productively. The art of conflict management is to channel conflicts so that the result is positive, preferably synergistically so, rather than destructive.

Analyse where conflict might arise and what might be the cause. Some will be obvious: certain stakeholders with conflicting goals; cost pressures arguing with expensive technical solutions; and time pressures. Look at the types of people involved, the way that they interact, and the types of conflict-handling skills which are most likely and appropriate.

When seeking to resolve conflict seek to use one, or a combination, of the following actions.

- *Competing (Forcing)* – used when quick decisive action is required, for example in emergencies.
- *Collaborating* – used when it's too important to compromise or to gain commitment or drive towards understanding.
- *Compromising* – used to avoid disruption, achieve temporary settlements or as a back-up to collaboration.
- *Avoiding (Withdrawing)* – used for trivial issues or as a medium to reduce tensions. Alternatively, it allows others to resolve the situation.
- *Accommodating (Smoothing)* – used when things are more important to the other party or to preserve harmony.

Different cultures often have different tendencies to use one or more of these modes; and some may be confused with others. Avoidance, popular in many Asian cultures, is sometimes mistaken for accommodation, for example, and only later does it become evident that decisions that project team members thought had been reached in fact had not!

Problem-solving/decision-making

A lot of the real skill of project and program managers is about 'driving the bus': making good decisions and moving the project forward. Driving and improving, not just monitoring and controlling. Knowing when decisions need to be taken, appreciating the risks, using team wisdom and moving the team along together. (Themes explored extensively in Part 3.)

Different styles, techniques, and tools for problem-solving and decision-making are used for different types of problem: some are open-ended and unstructured (such as brainstorming a risk register or in a value management workshop), while others are more analytical (such as Cause-and-Effect diagrams, Pareto analyses, portfolio analysis tools, estimating wizards, and so on).

Generally, project and program managers are thought to be decisive, implementation types. As such – execution-oriented people (sic) – they would be expected to be 'rational decision-makers', in the Thompson and Tuden sense[35], i.e. if there is agreement on objectives but the means of attaining them may be agreed (then they're 'computational') or may not (then 'judgemental'). Conditions may not allow such rationality, however[36]. Indeed, in the early stages of a project (the front-end, or when starting-up a new contract) 'coalition' and even 'garbage can'[37] may be more appropriate decision-making models. These forms reflect uncertainty and ambiguity and, as with the organic/mechanistic distinction discussed under organisational structure (Chapter 10, pp. 151, 156), they emphasise the need to act in less deterministic modes in the early creative 'planning' stages rather than the heavy 'build' mode of later 'execution'.

Nor need decisions necessarily be taken quickly: many can be delayed[38]. Decisions have their own dynamics (schedule needs) and it is often better practice to avoid making an unnecessary decision. This, too, applies particularly at the strategic level of project and program management.

Group problem-solving is common (obviously) in team situations. Again, beware of the tendency to rush to judgement and watch that strong personalities are not forcing bad decisions. Inclusiveness and care in deliberation, reflection and validation, are good habits in team-based decision-making.

Delegation and empowerment[39]

Projects and programs are generally performed by networks of individuals often working in teams, held together by contracts of varying degrees of formality. In this environment, delegation and empowerment are the natural and largely preferred ways of working.

Expectations should be established from the outset as to decision-making behaviours and limits of authority. Delegate to the lowest level practicable (and distinguish between formal and informal power and authority); but note that upwards delegation is also often necessary. Monitor practice and give feedback on an on-going basis. Avoid meddling. People should be allowed to make mistakes – so long as they learn and the project doesn't suffer.

Motivating

Project people are generally well motivated. Change is stimulating! In Ernst Maslow's hierarchy of needs, relationships, esteem and self-fulfilment represent the top three levels (physiological needs and safety the bottom two): projects are rich in these characteristics. (The same applies for Frederick Herzberg's 'motivation' as opposed to 'hygiene' factors.)

Change is nevertheless difficult, and projects and programs can be stressful. People resent waste, inefficiency and being mucked about! Clarity of purpose and transparency of communication and decision-making (or at least the perception of them) are important. Outcomes, both intrinsic and extrinsic, and perceived fair rewards should clearly relate to performance; and the conditions need to be available to perform effectively. This is 'expectancy theory', where individuals are seen as rational beings who have views about the consequences of their behaviour: where work motivation is seen as depending on the specific needs of the individual and on the expectation of these needs being fulfilled through productive behaviour[40].

Being able to motivate project and program staff is thus an important project management competency. So too is watching out for individuals' health and well-being. Not all motivation is the result of pressure. Indeed, too much stress, poorly managed, can be unhealthy. Good project and program management reflects a duty-of-care to individuals.

Clearly there's more that could be said about people, their role in projects and programs, and how they might be managed, or led. But it's time to move on. The principal topics have been discussed – not just for 'People' but for Levels 1 and 2 as a whole. So let us turn now to Level 3: managing the context – particularly the institutional context – in which projects and programs occur.

References and Endnotes

[1] Bass, B. M. (1985), *Leadership and performance beyond expectation*, Free Press: New York; Briner, W., Hastings, C. and Geddes, M. (1996), *Project leadership*, 2nd edition, Gower: Aldershot; Burns, J. M. (1978), *Leadership*, Harper Row: New York; Fiedler, F. (1967), *A theory of leadership effectiveness*, McGraw-Hill: New York; Bowen, H. K., Clark, K. B., Holloway, C. A. and Wheelwright, S. C. (1994). Make projects the school for leaders, *Harvard Business Review*, 72, 5, pp. 131–140; Gadenken, O. C. (2002), What the United States defense systems management college has learned from ten years of project leadership research, in Slevin, D. P., Cleland, D. I. and Pinto, J. (eds.) *The frontiers of project management research*, Project Management Institute: Newton Square, PA, pp. 97–113; Hersey, P. (1998), *The situational leader*, Center for Leadership Studies: Escondido, CA; Kangis, P. and Lee-Kelley, L. (2000), Project leadership in clinical research organisations, *International Journal of Project Management*,

18, 6, pp. 393–401; Keegan, A. E. and Den Hartog, D. N. (2004), Transformational leadership in a project-based environment: A comparative study of the leadership styles of project managers and line managers, *International Journal of Project Management*, 22, 8, pp. 609–618; Kotter, J. (2000), *A force for change*, Free Press: New York; Kouzes, J. M. and Posner, B. Z. (2003), *Leadership challenge*, 3rd edition, Jossey-Bass: San Francisco, CA; Schein, H. (2004), *Organizational culture and leadership*, Jossey-Bass: San Francisco, CA; Lindgren, M. and Packendorff, J. (2009). Project leadership revisited: Towards distributed leadership perspectives in project research, *International Journal of Project Organisation and Management*, 1, 3, pp. 285–308; Müller, R., Geraldi, J. and Turner, R. (2012). Relationships between leadership and success in different types of project complexities, *IEEE Transactions on Engineering Management*, 59, 1, pp. 77–90.

[2] Yukl, G. (1989), Managerial leadership: A review of theory and research, *Journal of Management*, 15, 2, pp. 251–289

[3] Conger, J. A. and Kanungo, R. N. (eds.) (1998), *Charismatic leadership in organizations*, Sage: Thousand Oaks, CA; Bass, B. M. (1985), *Leadership and performance beyond expectations*, Free Press: New York.

[4] Fiedler, F. E. (1967), *A theory of leadership effectiveness*, McGraw-Hill: New York, NY.

[5] Hersey, P., Blanchard, K. H. and Natemeyer, W. E. (1979), Situational leadership, perception, and the impact of power, *Group Organization Management*, 4, 4, pp. 418–428; Hersey, P. and Blanchard, K. H. (1982), *Management of organizational behavior: Utilizing human resources*, Prentice-Hall: Englewood Cliffs, NJ; Hersey, P. (2009), Situational leaders, *Leadership Excellence*, 26, 2, p. 12.

[6] Pinto, J. K., Thomas, P., Trailer, J., Palmer, T. and Govekar, M. (1998), *Project leadership: From theory to practice*, Project Management Institute: Newtown Square, PA; Pinto, J. K. and Trailer, J. W. (eds.) (1998), *Leadership skills for project managers*, Project Management Institute: Newton Square, PA.

[7] Kotter, J. (1996), *Leading change*, Harvard Business School Press: Boston, MA.

[8] Katzenbach, J. R. and Smith, D. K. (2003), *The wisdom of teams*, Harper Business: New York; Lewis, J. P. (2002), *Working together*, McGraw-Hill: New York; Sense, A. J. (2003), Learning generators: Project teams re-conceptualised, *Project Management Journal*, 34, 3, pp. 4–12; Thamhain, H. J. (2004), Linkages of project environment to performance: Lessons for team leadership, *International Journal of Project Management*, 22, 7, pp. 533–544; Thamhain, H. J. (2004), Team leadership effectiveness in technology-based project environments, *Project Management Journal*, 35, 4, pp. 35–46.

[9] Muethel, M., Hoegl, M. and Gemuenden, H.-G. (2011), Leadership and teamwork in dispersed projects, in Morris, P. W. G., Pinto, J. K. and Soederlund, J. (eds.) *The Oxford handbook of project management*, Oxford University Press: Oxford, pp. 483–499.

[10] Belbin, R. M. (1993), *Team roles at work*, Butterworth-Heinemann: London; Belbin, R. M. (2010), *Management teams: Why they succeed or fail*, Butterworth-Heinemann: London.

[11] Gersick, C. (1989), Marking time: Predictable transitions in task groups, *Academy of Management Journal*, 32, 2, pp. 274–309.

[12] Buchanan, D. A. and Huczynski, A. A. (1985), *Organizational behaviour*, Prentice Hall International: London, page 147.

[13] Thamhain, H. J. and Wilemon, D. L. (1987), Building high performing project teams, *IEEE Transactions on Engineering Management*, 34, 2, pp. 130–142.

[14] Ministry of Defence (MOD) (2004), *The smart acquisition handbook*, Ministry of Defence (MOD): London.

15 Armstrong, S. and Beecham, S. (2008), Studying the interplay between the roles played by stakeholders, requirements and risks in projects, *Project Management Perspectives*, 2008, pp. 46–51; Littau, P., Jujagiri, N. J. and Adlbrecht, G. (2010), 25 years of stakeholder theory in project management literature (1984–2009), *Project Management Journal*, 41, 4, pp. 17–29; Newcombe, R. (2003), From client to project stakeholders: A stakeholder mapping approach, *Construction Management and Economics*, 2, 8, pp. 841–848.

16 Office of Government Commerce (2003), *Managing successful programmes*, TSO: Norwich.

17 Schein, E. H. (1990), Organizational culture, *American Psychologist*, 45, 2, pp. 109–119; Trompenaars, F. and Turner, C. H. (1997), *Riding the waves of culture: Understanding cultural diversity in business*, 2nd edition, Nicholas Brealey: London.

18 BP (2010), *Deepwater Horizon accident investigation report*, BP: UK.

19 Kendra, K. and Taplin, L. J. (2004), Project success: A cultural framework, *Project Management Journal*, 35, 1, pp. 30–45.

20 Murphy, K. (1983), *Macro-project development in the Third World*, Westview Press: Denver, CO.

21 Hofstede, G., Hofstede, G. J. and Minkov, M. (2010), *Cultures and organizations: Software for the mind: Intercultural cooperation and its importance for survival*, 3rd edition, McGraw-Hill Professional: New York; van Lieshout, S. and Steurenthaler, J. (2006), *Effective multi-cultural project management: Bridging the gap between national cultures and conflict*, University of Gävle: Gävle, Sweden.

22 Roberts, A., Kelsey, J., Smyth, H. J., and Wilson, A. (2012), Health and safety maturity in project business cultures, *International Journal of Managing Projects in Business*, 5, pp. 776–803.

23 Hage, J., Aiken, M. and Marrett, C. B. (1971), Organization structure and communications, *American Sociological Review*, 36, pp. 860–871.

24 Wasserman, S. and Faust, K. (1994), *Social network analysis: Methods and applications*, Cambridge University Press: Cambridge.

25 Mayer, R. C., Davis, J. H. and Schoorman, F. D. (1995), An integrative model of organizational trust, *Academy of Management Review*, 20, 3, pp. 709–734; Rousseau, D. M., Sitkin, S. B., Burt, R. S. and Camerer, C. (1998), Not so different after all: A cross-discipline view of trust, *Academy of Management Review*, 23, pp. 393–404.

26 Herzog, V. L. (2001), Trust building on corporate collaborative project teams, *Project Management Journal*, 21, 1, pp. 28–37; Herzog, V. L. (2001), Trust building on corporate collaborative project teams, *Project Management Journal*, 21, 1, pp. 28–37.

27 Goleman, D., Boyatzis, R. and McKee, A. (2002), *Primal leadership: Realizing the power of emotional intelligence*, Harvard Business School Press: Cambridge.

28 Salovey, P., Detweiler-Bedell, B., Detweiler-Bedell, J. and Mayer, J. (2008), Emotional intelligence, in Lewis, M., Haviland-Jones, J. and Barrett, L. (eds.) *Handbook of emotions*, 3rd edition, Guilford Press: New York.

29 Goleman, D., Boyatzis, R. and McKee, A. (2002), *Primal leadership: Learning to lead with emotional intelligence*, Harvard Business School Press: Cambridge.

30 Mersino, A. (2007), *Emotional intelligence for project managers*, American Management Association: New York.

31 Borg, J. (2010), *Persuasion: The art of influencing people*, 3rd edition, Prentice Hall: London; Dent, F. E. and Bent, M. (2006), *Influencing: Skills and techniques for business success*, Palgrave Macmillan: New York; Fisher, R. and Shapiro, D.

(2005), *Beyond reason: Using emotions as you negotiate*, Viking Books: New York; Fisher, R., Ury, W. and Patton, B. (2003), *Getting to yes: The secret to successful negotiation*, Random House: London; Kennedy, G. (2009), *Essential negotiation: An A to Z guide*, 2nd edition, Bloomberg Press: New York.

[32] Giddens, A. (1984), *The constitution of society: Outline of the theory of structuration*, University of California Press: Berkeley and Los Angeles, CA, p. 283.

[33] Pinto, J. K. (1996), *Power and politics in project management*, Project Management Institute: Upper Darby, PA.

[34] Mayer, B. S. (2000), *The dynamics of conflict resolution: A practitioner's guide*, Jossey-Bass Wiley: San Francisco, CA; Rahim, M. A. (2010), *Managing conflict in organisations*, 4th edition, Praeger: Westport, CT.

[35] Thompson, J. D. and Tuden, A. (1959), Strategies, structures and processes of organizational decisions, In: Thompson, J. D., Hammond, P. B. and Hawkes, R. W. (eds.) *Comparative studies in administration*, University of Pittsburgh Press: Pittsburgh, PA, pp. 195–216.

[36] Simon, H. A. (1959), Theories of decision-making in economics and behavioural science, *Administrative Economic Review*, 49, pp. 253–283.

[37] March, J. G. (1978), Bounded rationality, ambiguity, and the engineering of choice, *Bell Journal of Economics*, 9, pp. 587–608.

[38] McCaskey, M. B. (1082), *The executive challenge: Managing change and ambiguity*, Pitman: London

[39] Gemmill, G. R. and Thamhain, H. J. (1973), The effectiveness of different power styles of project managers in gaining project support, *IEEE Transactions on Engineering Management*, EM-20, 2, pp. 38–44; Gemmill, G. R. and Wilemon, D. L. (1970), The power spectrum in project management, *Sloan Management Review*, 12, 4, pp. 15–25; Thamhain, H. J. and Wilemon, D. L. (1975), Conflict management in project life cycles, *Sloan Management Review*, 16, 3, pp. 31–50.

[40] Vroom, V. H. (1964), *Work and motivation*, John Wiley: New York.

Level 3: The Institutional Context

The thrust of our attention so far has been on what managers working in projects need to know in order to develop and deliver them successfully.

In Levels 1 and 2 – what we defined in Chapter 8, pp. 117–9, as the technical and strategic levels of managing projects – managers work in the project, shaping and driving it forward to meet the goals of the sponsor and other stakeholders.

Level 3, on the other hand, is about developing the enterprise's institutional context for projects and programs to enable them to succeed and to enhance their effectiveness. Management at Level 3 is primarily concerned with improving the likelihood of success not of a specific project but of the projects generally falling within the enterprise's own organisational environment; that is, projects:

- either in the parent organisation(s),
- or in the wider environmental context within which the project is located,
- or both (see Figure 8.1).

It is about management *on* or *for* projects as opposed to management *of* or *in* them.

What can management do to improve the context that projects and programs exist in? Chapter 19 will look in some detail at whether and how management can influence context. For the moment, however, we shall orient the discussion around the organisational function that is specifically charged with attending to the (internal) p.m. context – the PMO. (The Project or Program Management [Support] Office, sometimes labelled 'the Centre for Excellence in Project/Program Management'.)

At about this point we need to start discussing portfolio management, that is, the management of the enterprise's collection of projects and programs in terms of the returns they potentially offer, the risks they pose, the

Reconstructing Project Management, First Edition. Peter W.G. Morris.
© 2013 John Wiley & Sons, Ltd. Published 2013 by John Wiley & Sons, Ltd.

resources they need. Portfolio management is a significantly different type of skill (competence) from program or project management. These two are about managing the development of the project (product) through its life-cycle – through time management. Portfolio management is more analytical. Further, while it may be performed in a portfolio management function, some of its activities may alternatively be performed by or in program or project management – for example, assessing the enterprise's long-term p.m. resource needs.

From now on we shall begin using the term p^3m to capture this broadened scope of activity – project, program and portfolio management; or 'mop' to refer to the subject of the management of projects. (See Chapter 4.) We shall often just use the acronym 'mop/p^3m' to refer to the general domain/discipline.

But now to PMOs.

PMOs

PMOs first surfaced around the mid-1990s, initially as monitoring and control units charged with reporting the status of projects and programs across the enterprise's whole portfolio. To this monitoring role ('policing', as someone in Unisys once characterised it) was soon added a 'social serv-ices' help and assistance role. And from this it very soon grew into an in-house centre of expertise (a centre of excellence) in the management of the enterprise's project and programs.

   PMOs take many forms, from administrative support, through centres of excellence, to full organisational functions which may even sometimes be responsible for managing and delivering projects[1]. What seems to be consist-ent across all studies, guides and textbooks on PMOs is that they aim to provide a standing platform of project and program management tools, techniques, expertise and advice serving the enterprise's organisation-wide needs which survives beyond the individual project's transitory lifecycle. They are often located adjacent to the Project/Program Management func-tion (Head of Projects/Programs) and there may well be satellite PMOs in different units (company business units) – but if there are, there should be discipline in ensuring the satellite PMO support and advice is aligned with the enterprise's central PMOs.

Functions of the PMO

In general PMOs cover the following functions.

- Define the enterprise's p.m. standards (methodologies, guidelines, etc.).
- Define the required p.m. competencies.

- Assess staff competence and organisational capability gaps with respect to this tailored best appropriate set of guidelines; that is, assess the enterprise's p.m. 'maturity'.
- Specify and 'hold' central knowledge on project and program management tools and techniques, especially p.m. computer systems, particularly those that are mandated to be used across the enterprise.
- Assess the enterprise's long-term project-related resource needs (particularly of p.m. staff).
- Gather 'lessons learned' from past or on-going projects within the context of a p.m. knowledge management/organisational learning program.
- Arrange periodic project reviews, whether as assurance reviews or project 'assists'.
- Act as the functional owner of p.m. training/learning and development (hence defining its shape and assessing its delivery and efficacy).

This list in fact constitutes a reasonable agenda for us to use in discussing much of the required Level 3 knowledge, which we shall now do.

Project management 'standards' for the enterprise

Given the tendency for projects to go wrong, enterprises have been trying since just about the inception of the discipline to specify their project and program management expectations and requirements (as in the DoD 375 Series – see Chapter 3, p. 42), or have provided written guidance on 'the way we manage projects [and programs] in the ABC Inc. enterprise'. Often such guidance becomes, in effect, a project (or program) management methodology.

Since the 1990s, there have been several initiatives at the national or professional level to provide guidance which can be used generically to improve project and program management performance. As we saw in Part 1, the UK's Office of Government Commerce (OGC) for example produced *Managing Successful Programmes* and *PRINCE2* as well as the Gateway Review process, publications (methodologies) which have proved immensely useful and valuable. PMI's *PMBOK® Guide* is of the same nature: it is in reality more a methodology than a (guide to) a body of knowledge, and as such is considered highly useful by many practitioners.

Prima facie it would seem sensible for such guidance to be published widely, as it has been, even though actual practice will almost certainly vary by sector, by company, by Business Unit possibly, and by project or program. We find however – not surprisingly – that such generic guidance (methodologies, etc.) is most used by those 'on-boarding' and in the early stages of their career[2]. Others think they know it. Maybe they do: that's the aim of 'Certification' to find out. (See p. 223.)

Defining project management competencies for the organisation

The idea of competencies has its origins in psychology – the underlying characteristics of a person to perform effectively in a role[3] – and, as 'core competencies', in corporate strategy – what explains the ability of an enterprise to compete in a consistently superior manner[4].

The terminology is not very mature yet, however. In America, competency tends to be defined as superior performance. In Britain, it is the knowledge, skills and behaviours required simply to perform a role effectively. It is role-specific. Competencies are essentially output, performance focused.

Competency strengthens with experience – knowledge and behaviours anyway; 'skills' depends on which skill.

Competencies are not job descriptions: a job description describes the tasks that will need accomplishing in fulfilling a role. A competency framework shows what knowledge, skills and behaviours are needed by individuals to perform a role effectively. As such, they allow everyone to be clear on what is required. This is helpful in showing the 'development gap' that may exist between the actual and desired levels of competence and that may need closing through training, coaching or other forms of support.

Emanating from HR departments, historically such competency frameworks comprised three elements: professional, behavioural, and company-specific. These days we are seeing a more holistic view being taken of the knowledge, skills and behaviours needed to perform a project or program management role – project control, technical management, and probably some commercial things, as well as a competence in managing people and a broad knowledge of projects and programs and their management – often using the topics from a Body of Knowledge as a starting point or framework – indeed APM has published its own version of a generic project management competency framework[5]. Competencies defined at the enterprise level may need tailoring to meet the needs of individual projects.

By appreciating what competencies are required to be performed effectively, the enterprise is better able to ensure the right people are selected for the job, and that development needs are effectively addressed both at the individual and at the organisational levels*.

Assessing competence and organisational maturity

Logically, before specifying what training or learning and development may be needed, some form of 'Training Needs Assessment' should be carried out – as alluded to just now: assessing the development gap.

The most straightforward technique is benchmarking, as practised for example by IPA. One has to be careful in doing this, however. It's important that differences in the benchmarked set, whether of project type or context, don't make data comparisons invalid or of questionable validity, or even misleading and dangerous.

---

*There is some confusion between the terms 'competencies' and 'capabilities'. The *Oxford English Dictionary* defines them as follows. Capability: the power or ability to do something; Competence: having the necessary ability or knowledge to do something *successfully*. Davies, A. and Hobday, M. (2005), in *The business of projects*, Cambridge University Press: Cambridge, pages 58ff., use competencies for people (knowledge, skills and behaviour) and capabilities for the organisation (structure, roles, systems, processes, procedures, etc.) Though the legitimacy of this distinction is not clear, it sounds sensible.

A broader approach is to perform a maturity gap analysis of the enterprise's competences and capabilities to determine what kind of development and support may be needed. Let us look at this idea for a moment.

Project management maturity

Originally developed, as we saw in Chapter 6, page 101, for software engineering by the Carnegie Mellon Software Engineering Institute[6], the Capability Maturity Model (CMM)* proposes five levels of organisational capability (maturity): *Initial* – chaotic, ad hoc; *Repeatable* – the processes are documented sufficiently such that repeating the same steps may be attempted; *Defined* – the processes are defined as standard business processes. Practices may be taught; *Managed* – the processes are quantitatively managed in accordance with agreed-upon metrics; and *Optimising* – process management includes deliberate process optimisation/improvement.

Since its introduction in 1991, SEI's Capability Maturity Model has proved highly effective in improving the competence of many poorly organised software engineering organisations. The attempts by PMI, OGC and others to imitate this for project and program management ran into several difficulties, however[7].

The most serious is that the competencies needed by an organisation to manage projects or programs effectively are of a much broader range than those required for software engineering or other more narrowly defined disciplines – particularly if one is talking about 'the management of projects' paradigm rather than project management only as execution management (which one should be).

Second is the implicit notion that there is some form of sequential progression ('linear ascent') up the maturity scale. This might not be the case: performance, sadly, sometimes regresses; also it gets harder to improve the better things are. Tellingly, CMM has proved more useful at the lower levels (1–3) than at the higher levels (4 and 5)

Third, and more critically, because progress will generally vary by p.m. function, it can be misleading to portray competence or capability by a single number.

None of these difficulties are insurmountable by any means, but they do represent methodological, and conceptual, challenges.

Maturity gap analysis

The 'maturity' of both the owner organisation and the supplier organisation(s), however this is measured, influences the way project and program management are carried out. Defining what elements should be measured in determining maturity is obviously important and is still subject to considerable debate in the p.m. world – various maturity models of varying completeness

---

*CMMI – the 'I' is for integration – is the successor to CMM. Released in 2002 with upgrades in 2006 and 2010, it is not as restricted to software as was CMM.

currently exist, as we saw in Chapter 6 (p. 101). In general, however, we are talking about organisational structure, processes and roles, and p.m. practices. Some models fold in 'people', that is, personal competencies, but conceptually it is probably cleaner to keep the two sets of data separate although they should be evaluated jointly: actors perform on an organisational stage. The attributes to be measured in a good maturity model ought to be relatively stable, however, and should, one would have thought, align closely with the profession's body of knowledge. At the time of writing there is still work to be done here.

Generically, three issues arise in using maturity models:

- What topics should be assessed?
- How should one assess the topics' relative appropriateness?
- How could the relevant levels of performance be determined?

As we saw in Parts 2 and 1, most of the publicly available maturity models fail to offer an appropriate range of potential topics. The scope is wrong. (Too narrow; too focused on program management.) I would propose therefore that, at least as a starter, the list of p.m. elements discussed in Part 2 be used as the basic list of candidate topics. Different levels of performance maturity can then be defined for each of the relevant topics and the level of performance in practice of each topic then assessed and scored. The relative appropriateness (to the enterprise and/or the project or program) of each topic can then be weighted. Multiplying the score by the weighting will provide a prioritisation: 'not done well but not important' versus 'not done well but important'. Finding data to compare relevant levels of performance essentially boils down to either collecting internal data, if the enterprise is big enough and organised to collect this, or accessing some external body.

Where organisational capability falls short of desired levels, action should be taken. Treat such upgrade work as any project (or program) being especially realistic about estimates

Articulating and 'holding' central project management knowledge

This function applies to the whole range of p.m. practices, tools and techniques but particularly those that are mandated to be used across the enterprise. It involves (1) having expert knowledge on these tools and techniques so that 'Help Desk' enquiries can be readily and fully answered and (2) having a view on the long-term development options and preferences for the enterprise. How hard do we push Earned Value, for example; what should we be doing to integrate, say, Computer-Aided Asset Management, Requirements Management and Project Management systems?

The theories of project control have changed comparatively little in the last few decades, but the functionality of information technology has developed significantly. As a result, the architecting of project information systems is still a major challenge for many institutions. The PMO should be demonstrating leadership in systems design, systems support, and reporting.

Reporting on the project and program portfolio

Not all PMOs are responsible for the systematic reporting of the portfolio's performance, nor is it necessary that they should be. However, in accordance with good governance, the Board ought to have visibility of the status of the complete investment program, including risks, and someone should be integrating and supplying it and the PMO is often seen as the natural home for doing this.

Some degree of audit verification of information supplied by business units may also be necessary.

Try not to leap straight just to Traffic Light reporting on a Probability–Impact grid: it too often masks a wealth of useful information! Chapman and Ward excoriate it[8]!

Assessing the enterprise's long-term project management staffing needs

Most companies are increasingly, if not acutely, aware of a growing shortage of experienced project management staff, as the post-war baby boomers retire, and in many cases there is a sharp rise in the challenges that upcoming projects are posing (see Part 3). The prediction of staff numbers and competencies needed to manage the pipeline of projects and programs is thus becoming a matter of considerable importance.

Doing this well clearly requires a good understanding of the nature of the roles and competencies required for the developing portfolio and good visibility of the probable availability of project personnel. There also needs to be close liaison with the HR department and other senior executives regarding the career development path that is likely to be most suitable for individuals ('talent programs'). Gaps might be filled by training, development programs, or recruiting new talent; or even something more drastic such as buying a company possessing the requisite staff competencies. (Like local presence and knowledge, for instance.)

This function could be viewed as belonging to Portfolio Management but insofar as it lies close to the enterprise's knowledge of project status (across the portfolio) and of competencies, it is often located either in the PMO or adjacent to it as part of the central Project Management function.

Gathering 'lessons learned'

Since at least Peter Drucker's famous characterisation in 1969 of the modern economy as knowledge based[9], companies have become acutely conscious that knowledge is an organisation's hidden asset. "If only IBM knew what IBM knows". This realisation has spawned a whole industry of knowledge management and organisational learning[10].

In theory, the activities needed to drive these functions forward are relatively clear. Knowledge management is essentially the librarian's job: capturing, indexing, filing and retrieving knowledge[11]; organisational learning is how the organisation as an entity learns, and improves, once it gets a hold of this new knowledge[12]. But we immediately run into difficulties[13].

Data are bits and bytes (e.g. R, H, 1). Information is interpreted data (e.g. T, A, B, L, E, can be either something we have dinner off, or an Excel spreadsheet). Knowledge is information in a theoretical context: it has predictive power (e.g. if I slap you on the face your cheek will redden and your pupils dilate [natural science knowledge – totally predictable] and you might just strike me back [neither of us knows, partly due to the mood of the moment [social science knowledge, and in particular knowledge-in-action – the basis of drama].) Wisdom has become popular: for me it is the just evaluation and application of knowledge.

There are several classes of knowledge. We must differentiate natural science knowledge from social science; the latter is not wholly reducible, repeatable or refutable – i.e. wholly reliable as public knowledge (Karl Popper's criteria[14]). Social science is neither context-independent nor value-free and one can never reproduce exactly the same conditions to make prediction entirely secure. There are many theories – epistemologies – of social science ranging from the positivist, which emulates the objectivity of natural science, to interpretivism, of which there are many kinds. And tacit knowledge – that which is experiential and which is 'held' within us and is hard, if not impossible, wholly to explicate, should be distinguished from that which is explicit – 'readily available': written down. And though benefiting from being generally more clearly structured than tacit knowledge (because it's much more visible), explicit knowledge never seems quite as relevant or as valuable. Management knowledge is contextual and is pre-eminently experiential, which is why tacit knowledge in management is so much more valued. Tacit knowledge is generally more focussed and more potent – straight from the horse's mouth! Since tacit knowledge essentially resides in people, people will need to be available as a knowledge source (for reference, training, support, guidance and so on), however formally structured and accessible the explicit knowledge is.

The important point is that project and program management knowledge is a melange, of both tacit and explicit knowledge, and, given the range of topics it involves, from accounting and HSE to risk and people, is predicated on a range of epistemologies, from interpretivist to critical realist to positivist (see also Chapter 18). So, in truth, truth regarding the management of projects is at times a questionable commodity. Management knowledge is difficult to pin down.

Yet learning is central to our ability to manage projects and programs better. The problem is, we don't understand very well the best ways of doing this. The US Army's 'After Action Reviews' is frequently cited as an example of good practice in appraising actions in order to pick out learnings and feed these into the planning and conduct of future operations[15]. The analogous post-project appraisal is promoted as good project management practice.

These techniques work up to a point but they are limited. They tend to be for the benefit of individuals or teams, less for the enterprise as an organisation: how do you deal with learnings drawn from hundreds of projects? Make lists? (Soon soporifically worthless.) And how useful are such lessons when separated from context (e.g. "exercise effective change control")? Searching databases of explicit knowledge has its place but listening to human beings with relevant tacit knowledge is likely to be better, albeit the knowledge may not be so comprehensive, detailed or easy to access. And it may have in-built bias. In any event, we are learning now that we don't have to wait till after the project is finished before we can call a 'lessons-leaned' analysis. Why not do it at a stage-gate? (Actually, anywhere.)

In fact, it is useful to acknowledge that there are many opportunities for learning in projects: start-up meetings; stage-gate reviews, quality reviews, design reviews, progress reviews, or risk or value management reviews; personal appraisals; training; post-completion reviews; and just plain old reflection.

The trouble is, we still don't have very good models of how organisations learn. Single- and double-loop learning has been suggested – 'single' by responding to direct feedback; 'double' by reflecting cognitively a meta level up on one's mental models of what's happening, and adjusting the model(s) to reflect the new insight[16]. A spiral model involving an interchange between tacit and explicit knowledge has been proposed by Ikujiro Nonaka and Hiroaka Takeuschi[17], though this has been criticised on the grounds that tacit and explicit knowledge are really two versions – types – of the knowledge we have and as such cannot be converted one to the other[18]. Nevertheless, the idea is attractive and research we conducted at UCL suggested that tacit–tacit exchange – so-called socialisation – of knowledge seems the most preferred in projects.

Reflection becomes a seminal idea in organisational learning, following Donald Schön's work on how professionals continuously reflect on the established, core 'body of knowledge' and adjust to their experience of recent practice – the surgeon having just completed an operation, the musician just having played a piece of music; the professor just having delivered a lecture[19]. (Anthony Giddens has as an important feature of Structuration the reflexivity – the meta review – of social actors' 'knowledgeablity'[20] – a parallel but much broader and denser concept than Schön's.) Hubert and Stuart Dreyfus argue that true expert professionalism involves intuitive judgment on how to act in a given situation based, obviously, on formal knowledge, and the acceptance of personal responsibility for deviations or mistakes where they occur, and for learning from them[21].

Communities-of-Practice[22] and Subject-Matter Experts are two popular means of generating, maintaining and applying learning at the organisational level. Jean Lave and Etienne Wenger's idea of Communities-of-Practice (CoPs) is attractive and it rings true: a group of *practitioners* (because management is pre-eminently practitioner-based – learning through doing[23]) meet regularly, socialise, form a community, exchange tacit knowledge; then, at some point when the group feels it is comfortable to do so, it 'reifies' – makes 'thing-like'; makes explicit – its tacit knowledge. (It could be about HSE, or risk, or doing business in Vietnam, etc.)

A member of that Community-of-Practice could be a 'Subject-Matter Expert' (SME), although he or she doesn't have to belong to the CoP. The SME is an individual recognised as a company expert on a particular topic. Some companies involve the SMEs in peer assists, workshops, and in coaching and mentoring.

'Owning' project management training/learning and development

Training is a characteristic of a discipline. The PMO should act as the functional owner of p.m. training/learning and development (hence defining its shape and assessing its delivery and efficacy).

Learnings need to be communicated and learnt. Yet many organisations fail to link their knowledge management and lessons-learned programs into training and development.

The PMO should be in the best place in the enterprise to see where the competency gaps are and, probably working with the Learning and Development arm of the enterprise, devising a program to give people any missing knowledge or new skills or provide the means to learn new behaviours. Not all such interventions need be traditional training-type programs: coaching and mentoring are two obvious alternative ways of helping fill competency gaps.

Some organisations take Certification extremely seriously: not just providing a certificate of attainment but the idea that particular classes of project should only be managed by people who have demonstrably proven, via an assessment process, and are certificated as such, that they are competent to do so – a so-called 'Licence to Operate' scheme. (Many have tied this quality level explicitly into professional standards, such as PMI's PMP, APM's APMP or PQ, or IPMA's certification structure[24].) Doubts can be raised of such an approach – e.g. simplistically, is a knowledge test the appropriate assessment of management capability? – but the wiser organisations, like Siemens for example, use a whole battery of assessment techniques which, when combined, are very powerful. So, on balance, better this surely than nothing, or just winging it!

Arranging project reviews

Project reviews will range from assurance reviews – quality audits on behalf of the enterprise to check and verify that things are in order, that the project is in good shape – to 'assists' where external review and support is brought in to review the project and see whether improvements could be made. Ideally there should not be a great need for either but with the wrong people on the project the need increases. 'Assists' obviously are best placed when activities are predominantly in the planning stages; 'reviews' in the execution stages.

Such assists and even reviews may be facilitated and performed by Subject-Matter Experts – for example, by HSE, risk, C&P (Contracts and

Procurement) or technical specialists – and in this case, the PMO acts more as an organiser than performer.

Shaping the external environment

It is in the interests of the project to stabilise its environment and minimise the possibility of external disruption. While it would be dangerously naïve to imagine that one can 'manage' the project's environment it is not impossible that it can be influenced. This is, after all, what politicians try and do all the time. The task here above all is *to shape the context*. Leadership needs to be given. A project champion role would seem attractive. Coalitions may need building to protect the project or program and to assemble the skills and resources that will be needed.

All this may seem a world away from the reality of many project managers working at Levels 1 and 2, and in truth it probably is. Much of the work at Level 3 will be carried out by senior managers and specialist experts – lawyers, financiers, Public Relations people, Corporate Affairs types. Theoretically very little has been written about it.

Stakeholders will, as we've seen, need identifying and assessing, and influencing. Current stakeholder management practice revolves largely around mapping stakeholder influence[25]. There is room to go beyond that, however. Institution theory would seem to offer rich potential[26], as for example with Richard Scott's 'pillars'[27] – the 'regulative, normative and cultural-cognitive' – to study how best to engage stakeholders. Thus Ryan Orr and Richard Scott showed how costs rose on 23 large projects "after failing to comprehend cognitive-cultural, normative, and/or regulative institutions in an unfamiliar societal context"[28].

From an institutional perspective, stakeholder groups may be viewed as stand-alone organisational networks or as clusters in the sense of 'sets' or 'fields'[29]. Groups such as banks or regulators, although external to individual projects or programs, may exert great influence over such project groups. Others will relate more directly to them, for example suppliers in framework agreements, either already existing[30], or being formed through economic or other stimuli [31], or being specially created either by the project (Level 2) or by the enterprise specifically for the project or program (Level 3).

Each of these groups exists in an organisational context – an environment – which may influence the project significantly. The potential for management to shape the interactions and contexts within which such groups operate is only beginning to be appreciated and researched comprehensively and systematically[32]. It is the subject of Chapter 19.

Clearing the Decks for Reconstruction

These then are the elements of the discipline of managing projects. The central task of the discipline is to deliver projects successfully. (Which, we've seen, includes development in the front-end stages[33].) Doing this involves, at a minimum (Figure 16.1):

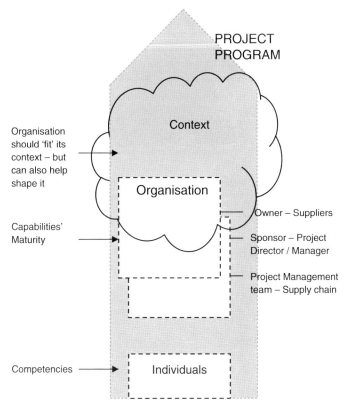

Figure 16.1   The elements of project management capability.
Note: Getting the project, or program, competencies (people) and capabilities (organisational) to match its need can be complex and difficult. They need to be appropriate to the characteristics of the project/program, and to the environmental context that the project is to operate in. And while capabilities can be slowly built up, they are rarely very flexible. People, meanwhile, are a mobile resource and can leave the project, either as the result of a portfolio decision or their own volition.
Source: Author's own.

- having an organisational platform or platforms appropriately mature in terms of the infrastructure – systems, processes, structures, support, etc. – on which project personnel can perform effectively;
- staffing the project, or program, with people who have competencies appropriate to its characteristics (its strategic needs, technological characteristics, commercial platforms, etc.) and the environment within which the project or program is based.

And it requires actions and resources to be deployed effectively, for example as Chapter 20 argues, actively seeking out ways of enhancing value for the sponsor.

Addressing how to do this effectively will occupy the rest of this book.

References and Endnotes

[1] Aubry, M., Hobbs, B. and Thuillier, D. (2008), Organisational project management: An historical approach to the study of PMOs. *International Journal of Project Management*, 26, 1, pp. 38–43; Bolles, D. (2002), *Building project-management centers of excellence*, American Management Association: New York; Englund, R. L., Graham, R. J. and Dinsmore, P. C. (2003), *Creating the project office: A manager's guide to leading organizational change*, Jossey-Bass: San Francisco, CA; Hill, G. M. (2003), *The complete project management office handbook*, Auerbach Publications: Boca Raton, FL; Hobbs, B. and Aubry, M. (2008), An empirically grounded search for a typology of project management offices, *Project Management Journal*, 39, 1, pp. 569–582; Marsh, D. (2000), *Project and programme support office handbook, vols. 1 and 2: Foundation and advanced*, Project Manager Today Publications: Branshill, Hants.

[2] Wells, H. (2011), *An exploratory investigation into the contribution of project management methodologies to the successful management of IT/IS projects in practice*, PhD thesis, UCL.

[3] Boyatzis, R. E. (1982), *The competent manager: A model for effective performance*, John Wiley: New York.

[4] Hamel, G. and Prahalad, C. K. (1994), *Competing for the future: Breakthrough strategies for seizing control of industry and creating markets for tomorrow*, Harvard University Press: Cambridge, MA.

[5] APM Knowledge (2008), *APM competence framework*, Association for Project Management: High Wycombe.

[6] Pennypacker, J. S. and Grant, K.P. (2003), Project management maturity: An industry benchmark, *Project Management Journal*, 34, 1, pp. 4–11; PMI (2003), *Organizational project management maturity model (OPM3)*, Project Management Institute: Newton Square, PN.

[7] Thomas, J. and Jugdev, K. (2002), Project management maturity models: The silver bullets of competitive advantage? *Project Management Journal*, 33, 4, pp. 4–14.

[8] Chapman, C. and Ward, S. (2011), *How to manage project opportunity*, John Wiley & Sons: Chichester, pp. 49–51.

[9] Drucker, P. F. (1969), *The age of discontinuity*, Harper: New York.

[10] Argyris, C. and Schön, D. A. (1978), *Organizational learning: A theory of action perspective*, Addison-Wesley: Reading, MA; Bresnen, M., Edelman, L., Newell, S., Scarbrough, H. and Swan J. (2003), Social practices and the management of knowledge in project environments, *International Journal of Project Management*, 21, 3, pp. 157–166; Collison, C. and Parcell, G. (2001), *Learning to fly*, Capstone: Oxford; Fernie, S., Green, S. D., Weller, S. J. and Newcombe, R. (2002), Knowledge sharing: Context, confusion and controversy, *International Journal of Project Management*, 21, 2, pp. 177–188; Huang, J. C. and Newell, S. (2003), Knowledge integration processes and dynamics within the context of cross-functional projects, *International Journal of Project Management*, 21, 3, pp. 167–176; Morris, P. W. G. and Loch, I. C. A. (2004), Knowledge creation and dissemination (organizational learning) in project-based organizations, in Slevin, D. P., Cleland, D. I. and Pinto, J. K. (eds.) *Innovations: Project management research*, Project Management Institute: Newton Square, PA; Garrety, K., Robertson, P. L. and Badham, R. (2004), Integrating communities of practice in technology development projects, *International Journal of Project Management*, 22, 5, pp. 351–358; Nonaka, I. and Takeuchi, H. (1995), *The knowledge creating company*, Oxford University Press: New York; Schön, D. A. (1983), *The reflective*

practitioner, Basic Books: New York; Schindler, M. and Eppler, M. J. (2003), Harvesting project knowledge: A review of project learning methods and success factors, *International Journal of Project Management*, 21, 3, pp. 219–228; Senge, P. M. (1990), *The fifth discipline*, Currency Doubleday: New York; Sense, A. J. (2003), Learning generators: Project teams re-conceptualised, *Project Management Journal*, 34, 3, pp. 4–12; Wenger, E., McDermott, R. and Snyder, W. M. (2002), *Cultivating communities of practice: A guide to managing knowledge*, Harvard Business School Press: Cambridge.

[11] Schindler, M. and Eppler, M. J. (2003), *op. cit.*: 10, pp. 219–228; Williams, T. (2008), How do organizations learn lessons from projects – And do they? *IEEE Transactions on Engineering Management*, 55, 2, pp. 248–266.

[12] Learning Organization is an organization that learns in relation to goals. It is teleological. Organizational Learning need not be goal oriented. See Easterby-Smith, M., Araujo, L. M. and Burgoyne, J. (1999), *Organizational learning and the learning organization: Developments in theory and practice*, Sage: London.

[13] Swan, J., Newell, S. and Scarbrough, H. (2010), Why don't (or do) organizations learn from projects, *Management Learning*, 41, pp. 325–344.

[14] Popper, K. (1959), *The logic of scientific discovery*, Basic Books: New York.

[15] Collison, C. and Parcell, G. (2001), *op. cit.*: 10.

[16] Argyris, C. and Schön, D. A. (1974), *Theory in practice: Increasing professional effectiveness*, Jossey Bass: San Francisco, CA.

[17] Nonaka, I. and Takeuchi, H. (1995), *op. cit.*: 10.

[18] Tsoukas, H. (2003), New times, new challenges: Reflections on the past and the future of organization theory, Chapter 23, in Tsoukas, H. and Knudsen, C. (eds.) *The Oxford handbook of organization theory*, Oxford University Press: Oxford, p. 615.

[19] Schön, D. A. (1983), *op. cit.*: 10.

[20] Giddens, A. (1984), *The constitution of society: Outline of the theory of structuration*, University of California Press: Berkeley and Los Angeles, CA, p. 283.

[21] Dreyfus, H. L. and Dreyfus, S. E. (2005), Expertise in real world contexts, *Organization Studies*, 26, 5, pp. 779–792.

[22] Wenger, E., McDermott, R. and Snyder, W. (2002), *op. cit.*: 11.

[23] Cook, S. D. N. and Brown, J. S. (1991), Bridging epistemologies: The generative dance between organizational knowledge and organizational knowing, *Organization Science*, 10, 4, pp. 381–400.

[24] Crawford, L. (2004), Global body of project management knowledge and standards, in Morris, P. W. G. and Pinto, J. K. (eds.) *The Wiley guide to managing projects*, Wiley: Hoboken, NJ.

[25] Littau, P., Jujagiri, N. J. and Adlbrecht, G. (2010), 25 years of stakeholder theory in project management literature (1984–2009), *Project Management Journal*, 41, 4, pp. 17–29.

[26] See for example Henisz, W. J., Levitt, R. E. and Scott, W. R. (2012), Toward a unified theory of project governance: Economic, sociological and psychological supports for relational contracting, *The Engineering Project Organization Journal*, 2, pp. 37–55.

[27] Scott, W. R. (1995), *Institutions and organizations*, Sage: Thousand Oaks, CA.

[28] Orr, R. J. and Scott, W. R. (2008), Institutional exceptions on global projects: A process model, *Journal of International Business Studies*, 39, pp. 562–588.

[29] Meyer, J. W. and Rowan, B. (1977), Institutional organizations: Formal structure as myth and ceremony, *American Journal of Sociology*, 83, pp. 340–363; DiMaggio, P. J. and Powell, W. W. (1983), The iron cage revisited: Institutional

isomorphism and collective rationality in organizational fields, *American Sociological Review*, 48, 2, pp. 147–160.

[30] As in the case of the Hollywood film cluster (DeFillippi and Arthur, 1998): DeFillippi, R. J. and Arthur, M. B. (1998), Paradox in project-based enterprise: The case of film making, *California Management Review*, 40, 2, pp. 125–139.

[31] Manning, S., Ricart, J. E., Rosatti Rique, M. S. and Lewin, A. Y. (2010), From blind spots to hotspots: How knowledge services clusters develop and attract foreign investment, *Journal of International Management*, 16, 4, pp. 369–382.

[32] Mahalingam, A. and Levitt, R. E. (2007), Institutional theory as a framework for analyzing conflicts on global projects, *Journal of Construction Engineering and Management*, 133, p. 517.

[33] For example, Miller, R. and Lessard, D. R. (2000), *The strategic management of large engineering projects*, MIT Press: Cambridge.

Part 3

Reconstructing Project Management

Its futures reconstructed

Reconstructing Project Management, First Edition. Peter W.G. Morris.
© 2013 John Wiley & Sons, Ltd. Published 2013 by John Wiley & Sons, Ltd.

Introduction to Part 3

Is project management a discipline or a domain? If it's a domain, we can sit back and ruminate at our leisure. If it's a discipline, things get tougher: there has to be a logic – a discipline – to the way its knowledge and its actions are deployed. And people need to be trained in it and practise it. This is the subject of Part 3: how should the elements of project and program management be re-assembled – reconstructed – so that the discipline adds value, given today's and tomorrow's social and business needs?

Part 2 presented the elements of the subject. Part 3 looks at the application of these elements but from three new perspectives – managing context, adding sponsor value, and addressing society's needs. It asks, in what way can the discipline be best deployed to meet such challenges as these?

A Discipline

So, is project management a discipline or a domain*? A discipline implies ownership of some specialised body of knowledge and the exercise of judgement in applying this knowledge, both with regard to what aspects, when, how and by whom[1]. Paul Griseri, in his book, *Management Knowledge*[2], says: "a discipline suggests . . . that there must be a shared understanding of the key issues and the key ways of investigating these." He goes on: "There may be controversy and disagreement about specifics . . . [but] . . . there is a general agreement about the overall range of subjects . . . about the problems, presumptions, paradigms or methodologies of investigation." We have seen in Part 2, and Part 1, that while for many people there is such

---

*Answer: obviously, both. The term 'project management' is of course, as we've seen, elastic, socially constructed and subject to interpretation: what one person may mean by project management may be different from what another may mean.

Reconstructing Project Management, First Edition. Peter W.G. Morris.
© 2013 John Wiley & Sons, Ltd. Published 2013 by John Wiley & Sons, Ltd.

a shared discipline, not everyone agrees about the paradigms, nor even is there is an agreed body of knowledge. Further, it can be charged, the commonly accepted presumptions are not always either sufficient or adequate for doing what the discipline is meant to do, namely deliver projects successfully.

Knowledge is core to a professional discipline. Other professional disciplines such as engineering or law have relatively robust, normative cores (the laws of engineering, the legal statutes, etc., often in a disassembled manner as in Part 2 of this book, under the intellectual guardianship of academia[3]) but the scope of the management of projects – the real scope of what it takes to develop and deliver projects effectively, as we've seen it in Parts 1 and 2 – is both very large and resides in several theoretical geni having different epistemologies (see Chapter 18, p. 237ff.). Is it reasonable to expect a professional in the field to master the whole domain? Is it too large for project management professionals to constantly reflect, as Donald Schön proposed, on how their practical experience shapes and modifies the discipline's central 'body of knowledge'? Or don't we tailor our own version anyway, reflecting our role and the characteristics of the project/program, rather as I argue in Chapter 19?

A discipline also implies application. Doing. Practice. Medical procedures; processes of law. (Again, Schön's *Reflective Practitioner*.) The *PMBOK® Guide* is strong on process – it is virtually a methodology – but the Bodies of Knowledge and guides reflecting the broader conceptions of the subject, such as the APMBOK, do not concern themselves much with what would be a useful or appropriate sequence of application.

What would it matter if the management of projects was *not* a discipline? After all, countless numbers of projects got designed and built before project management was invented. The argument is that by deploying the discipline of mop/p³m, we can do better. Unfortunately, it is not irrefutably clear that it does do it better, or at least to a standard that is unquestionably consistently appropriate. Unacceptable mistakes are still made; performance still has too many shortcomings (see Appendix 1). Is there compelling evidence for a group of engineers and contract managers, say, to adopt modern project management, as described in Part 2, because to do otherwise would most likely lead to inadequate performance? No, I don't think there is.

My concern is not so much that individuals may make mistakes ('to err is human') but that, at the most fundamental level, there is still too much disagreement about what 'it' (the discipline, the domain) is, and how it should be applied; what the scope and content – what I shall in a moment call the ontology – of the discipline should be. What the purpose of the discipline should be.

A Knowledge Domain

There are additional concerns regarding project management as a discipline. For the sceptics, as we saw right back in the Introduction, given the effects of context, let alone of individual actors, there have to be doubts as to whether prescriptive, normative guidance can validly be formulated and mandated for such a large, complex area of practice. For such sceptics, it makes more sense to approach the field predominantly as a domain of knowledge and enquiry than as a discipline, their interest being more on projects as organizational phenomena[4]: less in instrumental advice than in underlying questions and theoretical issues; attracted more to discussion of the lenses through which issues may be viewed than with what to do about the issues themselves; with means rather than ends.

Foundations for the Future

Both approaches – the looking at projects as organizations and managing them to achieve their targets – are, of course, valid. What I argue for is indeed a bit of each: that the unit of analysis should be the project, rather than project management processes or functions, but that, in addition to studying projects as organizational phenomena, we should be looking at how their conception, development, execution and handover can, and should, best be managed.

And that we should be thinking of the purpose for which they should be being managed. They should add value to the realisation of the sponsor's aims and strategies, should 'fit' the context in which they and their sponsor will be operating, and should be alert to, and address, the changes in business and societal conditions that we can see coming down the pike.

To these ends, Part 3 begins by revisiting the intellectual foundations of the discipline. We then look in practical detail at how management can shape projects to fit their environment, and how it can provide value to the project's sponsor. Finally, Part 3 looks at changing trajectories in our world over the next, say, 30–60 years and what the implications might be to the discipline. (Which, as it happens, is just about the length of time it has, at the time of writing, been formally in existence!)

References and Endnotes

[1] Hodgson, D. and Muzio, D. (2011), Prospects for professionalism in project management, in Morris, P. W. G., Pinto, J. K. and Söderlund, J. (eds.) *The Oxford handbook of project management*, Oxford University Press: Oxford.

[2] Griseri, P. (2002), *Management knowledge: A critical view*, Palgrave Macmillan: Basingstoke.

[3] Abbott, A. (1992), *The system of professions*, University of Chicago Press: Chicago.

[4] See Bredin, K. and Soderlund, J. (2001), *Human resource management in project-based organizations*, Palgrave Macmillan: London, for a summary of the contemporary view of projects as temporary organizations, *viz.* non-routine, a pre-determined end point, performance individually assessed and requiring special management to deal with their uncertainty and complexity (p. 8).

The Character of our PM Knowledge

If the elements discussed in Part 2 are on the whole relatively clear, there is less of a consensus on the way they might be combined and represented as a higher 'whole'. This chapter engages with some of the philosophical issues that affect what we can say about the nature of the overall subject: its essence, its robustness. For as future events develop and impact it, so its character will be subject to change. It should help, therefore, to have our thinking as straight as possible with regard to its nature.

Maybe 'philosophical' is too ambitious a word. Really, I am just someone trying to work-out (a) what knowledge is needed in order to manage projects successfully, and (b) whether there is a discipline of managing projects. Many of the things I am trying to 'essay' in this chapter are relevant to our understanding of the potential answers to these two aims. Methodology and epistemology are. Ontology is harder – partly because it's conceptually more difficult, partly because I have almost certainly been too simplistic in what follows.

We begin by defining what we are taking to mean by the principle subdivisions of the discipline, for there is overlapping terminology and plenty of opportunity for confusion. We are like the blind men trying to describe an elephant: we can describe the parts we know but not the thing itself.

Terminology

Programs are collections of projects having shared goals and objectives and perhaps resources, all of whose benefits must be realised for the overall program to work. *Program Management* is the management of the collective set of such interdependent projects. There are many who claim, particularly in the United Kingdom (but also PMI in its Program Management 'stand-

Reconstructing Project Management, First Edition. Peter W.G. Morris.
© 2013 John Wiley & Sons, Ltd. Published 2013 by John Wiley & Sons, Ltd.

ard'), that program management is qualitatively very different from project management: that it is about organisational or strategic change. It can be but it is also used without this emphasis in product development and multi-project management[1].

All *projects* follow the same development life-cycle, going from the 'front-end' to Operations and Maintenance (or equivalent terminology; see p. 150). Major projects are those which are large, complex and difficult and which exceed the normal experience and/or capability of the organisations executing them[2]. The differences between major or mega projects and programs can be hard to identify: a big infrastructure project can, for example, be broken down into a number of very large sub-projects. The ensemble will have one or more shared goals, shared resources and shared benefits – the criteria for program management? Does this mean it really is a program? Maybe. London's £15 bn. new railway, Crossrail, would say yes; Heathrow's Terminal 5, no. (Because that's how they labelled them.)

Project Management is either the generic term for the knowledge domain or for the discipline of managing the development and delivery of projects. In practice, in many organisations, the term is also used to refer only to the management of project execution (after requirements have been identified). If this is the case, then we need to ask, what is the discipline that is responsible for managing the front-end stage of the life-cycle – development management? (To me, it would seem best to extend project management to include this activity.)

Agile is essentially an approach to delivering at least some functionality reliably within a time period (or budget – or even scope constraint), but since there is often little respect for the *overall* project life-cycle, and since the iron triangle (time, cost, technical performance) no longer necessarily holds, is it really project management?

Portfolio Management is the management of a portfolio of assets (or liabilities) – here projects or programs – in terms of the risk they pose, resources they need, and returns they offer. As a function, it is qualitatively different from project and program management. Unlike them, it does not focus on driving the emerging product (output) through its development life-cycle.

The Management of Projects covers both programs and projects. Its unit of analysis is the project. The project is defined, pre-eminently, by its development life-cycle. *The Management of Projects* is as concerned with managing the front-end as with down-stream execution. ('Front-end' is defined as either the period prior to definition of the project's, or program's, requirements – or the period prior to 'sanction execution' being given.) The Management of Projects stresses the need for the project's strategy to flow from the sponsor's. It emphasises the creation of value for the sponsor. It includes proactive stakeholder management. It recognises the challenges frequently found in technology and innovation and the importance of having an appropriate and well-managed commercial platform to work off. And it acknowledges that people are central to projects, and programs. 'Projects are built by people, for people, through people'.

As I've already said, I shall now be calling the general area 'mop/p³m'.

Since there are several quite different members of the mop/p³m family, what can we find to say that characterises their essence? This is the question that their ontology addresses.

Ontology

Ontology is the branch of philosophy that refers to the essence of a subject: its existence: the nature of its being; how it relates to other things; and what it should be addressing and to what purpose[3].

Discussion of ontology in the field of project management has tended to focus on how project management is perceived, distinguishing traditional conceptions of project management, which adopt a predominantly realist perspective, from post-modern ones where projects are seen primarily as temporary endeavours (theatre for example), to hyper-modern views (for example, considering whether its essence is not so much about 'being' as 'becoming'[4]).

Realists believe that things, such as projects, exist independently of our thinking about them: they have a reality of their own. An alternative to realism is nominalism[5]. Nominalists believe that the social world external to individual cognition is made up only of names, concepts and labels which are used to structure our perception of reality; there is not any concrete 'real' structure to the world which is external to our minds. Words define reality.

Realism seems natural to most project people, particularly if their world is execution and their projects have a strong engineering base. Of course, there is a reality: we're building it: you can touch it! If we enlarge the scope of the subject to include the management of the project front-end, would this change its realist ontology? Yes. Decisions about purpose and scope, ethos and definition, change the 'essence' of the discipline. Even billion-dollar engineering projects become less obviously just real, even more 'becoming' than 'is', at the front-end. What was the essence of the US SST: was it a necessary 'real' project or was it a piece of imagination conjured by a name*? Is the UK's proposed new London–Birmingham (and beyond) High-Speed Railway an environmentally damaging abomination or a vital new piece of infrastructure? Both claims are made. And in a way, the project can be both. In a Parmenidian sense†, its essence is shaped by everything that is said about it.

So if a project, or program, can, as it develops, experience different core characteristics of being, so may the discipline of managing projects and

---

*Was President Reagan's Star Wars program, SDI, the Strategic Defense Initiative, which essentially outbid the USSR's nuclear pretensions and led President Gorbachev to Reykjavik in 1986 and to the end of the Cold War, 'real' or was the reality only in the words used to describe it?

†Parmenides (c. 515–450 BC) had a 'being' ontology. Just the act of discussing something means that in some ways it exists. A colleague put it in an unguarded moment, nicely if not wholly accurately, 'what's 'is' is 'is''. Heraclitus (c. 540–480 BC), on the other hand, had a 'becoming' ontology: 'you never bathe in the same river twice'.

programs. The broader 'management of projects' perspective is going to have an altogether more complex and richer ontology than does project management as and when viewed as a predominantly execution-oriented discipline. Its ontology will, it seems to me, should such a thing be conceptually acceptable, exhibit in a project both a shift from a predominantly nominalist position to a realist one, and simultaneously a move from a 'becoming' to a 'being' ontology, entirely as a consequence of following the project life-cycle, which is of course the single differentiator between projects and non-projects. Gardening shares this ontological shift based on the project life-cycle, as does cooking. Chefs and gardeners (orchestra conductors and theatre directors) all make good project managers. The reverse is not necessarily true.

What practical impact does this philosophising have, asks the practitioner. Well, philosophy needn't and often doesn't have a big impact on practice but to the extent it does, it is generally because it helps us see things differently, and better. And here? One, we recognise the broader, more varied knowledge and skill base inherent in taking the 'management of projects' paradigm, particularly at the front-end, and all that comes with this instead of the more narrowly focussed 'execution' one. Two, we should appreciate the changing nature of projects as they develop, the early stages being much more about creating visions – through description – than touching a hard reality: the former is based on intangibles, on promises and imaginings; these shape the realism that execution then produces. Three, Third Wave, institutional management of projects will need to straddle both types of ontology and various epistemologies.

Epistemology and Theories of Project Management

How robust can we be about what we say we know about mop/p³m? Partly this depends on the research methodology we have employed (see below) but partly it depends too on the kind of knowledge we are working with. With project management knowledge, we are nearly always operating with the social sciences rather than with natural science. Natural science-based knowledge is independent of context and of value-systems (except where context is part of the science, e.g. things change under different environmental conditions). Knowledge of the social world is not independent of context and of value-systems. Nevertheless, positivism is an epistemology that seeks to bring to social science the same certainty of value- and context-free knowledge that we find in natural science. But this is difficult when humans become involved: people bring ideas, values and sometimes unexpected behaviours. Our knowledge needs to become more interpretive.

Culture becomes important in that, should the culture require it, certain things about managing projects may need to be addressed consistently regardless of local conditions or opinions. Requirements about financial disbursements perhaps, employment conditions, health and safety, or engineering design, document management or change control. Guidance or rules may need to be given in a normative manner, and in doing so it may seem to be using a positivist perspective: at the project level, the knowledge being followed may seem significantly context-independent and requirements, or our advice, stand regardless of our disposition. Underpinning this positivist, normative advice, however, will be the norms and values of the enterprise and of society. These can be so strong that they seem to render the knowledge virtually absolute. Health and Safety standards (requirements, guidance, knowledge) for example may seem to reflect a strongly positivist epistemology, but ultimately only because ethically that is how our society expects it to be. If we were Romans (or Victorians, or Napoleon) we would not see such rules as positivist knowledge akin to nature's's laws.

In reality, many project and program managers positively (if I can use the word) seek guidance and their managers feel that normative guidance must surely be helpful for the enterprise. They are, after all, supposed to be responsible for delivering successful outcomes, and one obvious route is to impose some rules. Indeed, without rules of some kind, we'd be lost. This is the thesis proposed by Tom Burns and Helena Flam, discussed later (p. 246), that rules are often essential, whatever critical theory has to say*: one can go to jail if certain rules have been disregarded, for corporate manslaughter for example, on breaches of health and safety legislation. But we should take care: the limitations of norms, standards and so-called truths need acknowledging. 'Best Practice', popular though the concept is, is to some extent a chimera. (How can we ever know what 'Best' is? Best in the world? Best in Bangkok? Best in tunnelling projects?) Project management needs to be sensitive to the interplay between normative recommendation and contextual adjustment, between positivism and interpretavism.

Different epistemologies may seem appropriate (or inappropriate) to different undertakings, but close examination may still undermine the appropriateness to project management of positivism. Guidance which is 'method-oriented' such as *PRINCE2* or *PMBOK* is typically presented in a positivist character: predictive, independent of context and of value-systems. But this is rarely really the case in practice. While it might seem to work to some extent for such arguably mechanistic topics, say, as scheduling, the effect of human behaviour on most p.m. knowledge areas – say, estimating (*vide* Kahneman[6]) or even scheduling itself (Critical Chain's motivational dimensions[7]), let alone leadership and the other 'people' topics – would suggest that more interpretative epistemologies are needed here.

---

*Critical management, as we saw above, argues that there is no such thing as value-free knowledge. Instead "knowledge production is driven by three cognitive interests: a technical interest in production and control (the main concern of positivism); a practical interest in mutual understanding (the chief preoccupation of interpretivism); and an interest in emancipation (the focus of critical social science)". Tsoukas, H. and Knudsen, C. (2003), *The Oxford handbook of organization theory*, p. 18. Critical Management writing about project management tends to be excessively concerned with the negative – often putative negative – aspects of project management rules. See for example Cicmil, S. and Hodgson, D. (2006), *Making projects critical*, New York: Palgrave.

So what is an appropriate epistemology for mop/p^{3}m? Personally I favour Critical Realism. Critical Realism, while recognising the value-laden and interpretative nature of much of our knowledge, incorporates a normative viewpoint[8]. It is particularly attractive in addressing the question of how certain we can be that our knowledge is properly representative (for example, of what we believe good practice to be) and applicable ('as is' or does it need tailoring?).

How can we be sure that the knowledge we are articulating is representative of the domain; how do we know that the domain knowledge is true; how confident are we about generalising from an unknown universe of projects (what's a valid sample size, how do we know there isn't a black swan lurking to destroy our inferences); how do we know that what we believe is valid knowledge (normative guidance say) is applicable in such-and-such a situation? These are all serious questions. Critical Realism addresses such questions by saying that "reality is stratified: there may indeed be causal relationships (laws, event sequences, etc.) discernible at a level of observation but these are just subsets of what can be observed, and what can be observed is itself a subset of what, at a deeper level of reality, in fact exists."[9]

The problem is, how to tailor what at a high level might seem almost platitudinous generalities into something which at the operational level is more focussed and meaningful. Take the guidance summarised in Table 18.1, for example. This is an attempt to outline some best practice principles on the management of projects. I would contend that it is conceptually valid – "*what, at a deeper level of reality, in fact exists*". However, I acknowledge that, at this level of generality, it doesn't say too much that is of value. (It is boring.) To give it more bite, we need to go down a few organisational levels. But then, what may be true for Organisation 1 in context B might not apply so well for Organisation 2 or context X.

Do the generic principles guidance outlined in Table 18.1, recognising that the language is highly abbreviated, represent a theory of project management? Does it have predictive power? (Yes.) Is there a conceptual explanation of why this predictive power works? (Yes.) Is the above transparent? (Yes.) Is it reducible, repeatable and refutable in the way Karl Popper proposed that scientific – public – knowledge needs to be[10]? (Yes.) Does this all constitute a theory? Well probably it does, albeit a very weak theory. It is at best only an outline of a theoretical framework. Is the theory sufficient to explain the domain? No. But like many theories, it works in parts[11].

Even then, remember: "whilst there are plenty of theories in management, there are no laws"[12].

As there is a range of topics that together make up the project management body of knowledge (using the term in its widest sense), and as can be seen from the range of guidance areas shown in Table 18.1, we can conclude that *there is no single 'theory of project management'*[13]. Instead there are several theoretical frameworks that might be appropriate for different aspects of the field. Project management knowledge is pluralistic.

Table 18.1   Generic principles in the management of projects

Governance

- Set, monitor and maintain values, objectives, strategy, assurance mechanism, risk/return expectations

Definition

- Align the project strategy with the sponsor's, including periodic reviews
- Define the requirements (in a testable manner) so that specifications and solutions can be verified
- Manage design and technology so that innovation are thoroughly examined before proceeding to full project commitment

Control

- Define and manage the project scope, schedule, resource, and budget (optimal financing), including limiting changes once the design has been agreed (design freeze), integrating cost, schedule and scope measurement, and conducting trend analyses on anticipated out-turn cost. Ensure realistic contingencies

Resourcing

- Procure/induct resources into the project in as cost-effective/value-enhancing manner as possible
- Build effective project teams
- Exercise leadership

Performing

- Ensure appropriate people and organisational maturity. Enable instinct. Effect decision-making and communications. Influence stakeholders. Shape context. Put a premium on people.
- Seek out value

Learning

- Review lessons learned after and during the project; peer review sessions; link to competency development

Source: Author's own.

Methodology

If epistemology is about how we know what we know, methodology is about how we develop knowledge in a way that is rigorous so that the new knowledge is valid. Just about every postgraduate dissertation has a section on methodology but we still tend to do it poorly. As can be seen in Appendix 2, sometimes some quite prestigious reports are based on extremely weak methodology: one has to keep one's wits awake. My colleague at UCL, Hedley Smyth, and I reviewed a sample of *International Journal of Project Management* papers published in 2005. We found that over 90% of the authors did not make their methodology explicit. Further, many used the term 'methodology' to mean 'method'. As a result, methodology was mislabelled as tools and techniques, thereby obscuring the epistemological requirements of addressing context, causality and general–particular explanations[14].

Scholarship has to be based on sound methodology. Our conclusions and assertions should be substantiated by evidence. Generally this will be research data or reference to previous research (for example, as quoted in the literature) that is representative of the issue we are investigating and that is as free from bias as possible. We need to be conscious of the limits of reliability and the credibility of empirical data and we need to be careful about generalising too widely, ensuring that we have sufficiently robust evidence to serve as the basis of our proffered knowledge. All this is essential when using such data, combined with theory, to postulate guidance or to predict future conditions. And when things change in the future, as they invariably will, so our data and theoretical models most likely will as well.

Such are the foundations of scholarship. Yet one has to be careful even with the best in the field. It's surprising how statements – theories even – are made on the basis of weak methodology. Take two examples.

Aaron Shenhar and Dov Dvir's book *Reinventing Project Management*, which aims to show how project management needs to adapt in form and application according to four parameters – novelty, technology, complexity, and pace – is subtitled *The Diamond Approach to Successful Growth and Innovation*. Why 'diamond'? Because a diamond shape emerges when points – scores – along the four axes are joined up. But where do the four dimensions come from? Originally Shenhar and Dvir said (p. 41): "Drawing on classic contingency theory, we concluded that we can define three dimensions that characterize each project: uncertainty, complexity, and pace", and we are then referred to appendix 3 for further information – where there is no discussion as to why these three and no other dimensions were selected.

But what about the four dimensions? All we are told, a little later on the same page is: "In practice, we have found it helpful to expand this model, recognizing that there are really two major sources of uncertainty: market (or goal) uncertainty and technological (or task) uncertainty. Thus the NTCP ('novelty, technology, complexity, and pace') model emerged" (p. 41). And that is the sole basis on which the model, with these as the determining four parameters for deciding how to adapt project management to different contexts, is justified! The framework is built, the selection made, essentially *ex nihilo*, or at least *ex cathedra*.

Also, we should be careful in our use of terms. Shenhar and Dvir define technology by four relatively crude terms: 'super high tech', 'high tech', 'medium tech', and 'low tech' (where, in the latter, sits Dunce-like, poor old construction [appendix 5B]). Meanwhile, 'novelty' uses similarly blunt measures dragged straight from Wheelwright and Clark (p. 108ff.), namely 'derivative', 'platform', and 'breakthrough' projects – technically they are really 'programs'. There may nothing wrong with these definitions so long as we acknowledge their bluntness and limitations.

Shenhar and Dvir's contingency determinants – ntcp – are furthermore only characteristics of the project. Unlike the original contingency theory they say nothing about the characteristics of the environment the project finds itself in – a mistake avoided by GAPPS in its measure of complexity. But this too, despite its eminent provenance, has problems.

GAPPS – the Global Alliance for Project Performance Standards (see above, Chapter 6, p. 103) – runs into difficulty over its measures of complexity. Forty eight factors are collapsed into seven measures. These, despite GAPPS' exhaustive international discussions, end up being not so much a measure of project complexity *per se* as of environmental challenge, size, importance, impact, and indeed project interface complexity* (a conception very close to that given at the end of Chapter 16, Figure 16.1). The ideas seem right but the framing and use of terms seem questionable. Complexity is not 'what it says on the tin'!

Or take attempts to conclude that the Manhattan Project's pursuit of parallel technology development routes – the Fat Man (subcritical material imploded) and the Thin Man (fissionable material shot at subcritical material) – suggest a lesson for project management today[15]. Well, while there may indeed be a hypothesis to be explored – that innovative development strategies should be considered in the light of a project's strategic aims, for example – inferring a general rule for the discipline on the basis of one, highly unusual, project, in which technical uncertainty and schedule pressure were massive but which had no funding limitations, and which furthermore did not apply project management as a discipline (because it had not yet been invented), is methodologically weak, to say the least.

Or the Standish (CHAOS) reports (see Chapter 5, pp. 87–8): not available publicly, with no discussion of methodology, let alone a critique as would be found in any half-decent university dissertation, these reports have achieved a notoriety totally unsupported by their methodological rigour, whether in the portrayed shocking-ness of the reported project performance or their recommendations, for example to chunk projects up into smaller units.

Flexible, agile, innovative project and program structures are almost certainly forms of the coming reconstructed project management landscape, but our justification for them and our knowledge about them would be more reliable if it were based on solid social science methodology: on sound empirical data, rigorously and transparently analysed: collected in an unbiased manner within a clear methodology, be that inductive ethnographic, exploratory analyses of statistical correlations, case studies, or deductive testing of hypotheses; using data which have been analysed objectively and critiqued acutely and presented with transparent clarity. There should be a clear line of sight between aims, theory, data collection and analysis, findings and conclusions – and implications.

---

*GAPPS' seven measures of project complexity are: (1) stability of the overall project context; (2) number of distinct disciplines, methods, or approaches involved in performing the project; (3) magnitude of legal, social, or environmental implications from performing the project; (4) overall expected financial impact (positive or negative) on the project's stakeholders; (5) strategic importance of the project to the organisation or organisations involved; (6) stakeholder cohesion regarding the characteristics of the product of the project; and (7) number and variety of interfaces between the project and other organisational entities: GAPPS (2007), *A framework for performance-based competency standards for global level 1 and 2 project managers*, Sydney: Global Alliance for Project Performance Standards.

Does all this matter? Can't we just get on and 'do it', as the Nike advertisement famously said? Yes, but the way we define the subject will inevitably, obviously, influence the way people perceive it and hence how they do 'do it'. Words matter: they communicate meaning and this affects, significantly, how people act[16]. (Nominalism.) Projects and project management are, to repeat, 'invented not found'. It is us who name things, who shape the discipline.

We therefore need to think carefully about 'standards': how fixed are they, who defined them, are they valid? The professional community will coalesce around certain conceptions and definitions, but that doesn't mean to say they are always right. With new times come new needs and new perceptions. Which is why we need to be as clear as possible regarding the definition of what we take the discipline to be.

But we should recognise that project people generally have limited thirst for such philosophical discussion: they are typically action-oriented people, practical and goal-driven. Yet they and the field are teleological and it is worth testing their patience a shade longer to reflect on what this means to our quest to describe the knowledge needed to manage projects.

Teleology

All projects and programs have, as we've seen, as a fundamental characteristic, the development life-cycle. A second characteristic, however, is their strong goal orientation: they, like project and program managers, are teleological.

Teleology is concerned with 'end purpose': aims, goals, functions etc. As an area of philosophy and natural science, it is relatively unfashionable today but it is recognised as relevant in some areas of management and science, for example in feedback and control in cybernetics; which of course is the core of (Level 1) project management. It also resonates with those who have approached the study of organisations from an 'instrumental rationality' perspective: "rational action oriented to practical goals" as portrayed by Max Weber[17] and James Thompson[18], who saw the task of management as being to control uncertainty and achieve planned results – a view dominated by rationality of purpose and decision-making: exactly the perspective of project management!

There has been little or no reference to teleology in project management research as yet. While Andersen, Grude and Haug's *Goal Directed Project Management*[19] sounds like a candidate, in my mind it isn't. Clegg *et al.*'s exploration of the 'future perfect' idea on the other hand is an example. The future perfect is, for them, "the simultaneity of a notion of something being projected into the future and being thought of at the same time as if it were already complete"[20], their thesis being that detailed project planning is not so necessary if the project team has a shared visualisation of what it is that will have been accomplished when the project is finished. The irony is that Clegg is a leading hyper-modernist in organisation theory, and as such, one would have thought, inherently suspicious of goal-directedness. Andersen *et al.*'s book on the other hand is really a methodology with little reflection on the philosophical implications of being goal-directed.

One consequence of the teleological orientation of the domain is to encourage a pragmatic disposition to accepting normative and instrumental advice, so long as the advice is good – if it leads to the goals being achieved. (We still need to be rigorous in making sure such advice is correct and alert to its generic transferability.)

This brings us back to the issue raised in the Introduction: whether it is even possible to offer such generalised advice in a context-dependent area such as project management. I maintain that some level of offering is possible so long as we are conscious of the ontological framework and the epistemo-logical and methodological robustness of the advice is understood.

The high-goal orientation of projects and programs and of those working on them influences their behaviour, often very strongly. Project practitioners have, as we've seen, a tendency to get on with things: they have a bias for action, are generally very positive and altogether are rewarding people to work with. Few of them would know, or care to know, that this is in part because they, their work environment, and their project colleagues are typi-cally teleological. Form fits philosophy!

Typology

We have, of course, recognised for several decades that project management guidance and application needs to be tailored to its context. This insight has led to several initiatives, none of them in my view wholly successful, to attempt to characterise the context within which projects occur so that one can then apply the required tailoring[21].

Such attempts fail because while the environment certainly does modify the character of project management deployment, so do the characteristics of the project. I have long held that the deployment of project management practices depends on: (1) the character and expectations of the industry sector – say, defence/aerospace versus big pharma drug development versus oil and gas; (2) the character, culture and expectations of the enterprise; (3) ditto of the Business Unit; and (4) ditto possibly of the program or project. There is no real research to back this up, however.

But in addition, valid though this might be, forces operating at the project level *also* affect the deployment of project and program management, as we discuss in Chapter 19 – factors varying from strategic intent to 'novelty, technology, uncertainty and pace'[22] (if you have not allowed methodological doubts to undermine your trust in this model), or size, speed and complex-ity[23], or equivalent.

Because of the power of these project-level factors in shaping the nature of the project management tailoring, I consider attempts to develop a robust, predictive view of the influence of typology to be of only limited value at best.

The Character of the Field's Substantive Knowledge

Creating the foundations of the knowledge that we need in order to manage projects and programs effectively obviously requires not just foundational knowledge of the type we have just been discussing but substantive subject knowledge too. There are several different kinds of knowledge needed on projects, each having its own character affecting the way we think. In practice, I'd suggest that most project people think either as technical specialists, from a commercial perspective, and/or from a project controls view. Organisation theory and control theory (cybernetics) provide conceptual frameworks which both explain the need for, and the workings of, project management and help us to understand better how to structure enterprises and peoples' behaviours and to control progress in accomplishing the project. Some strategy is necessary for some working in the management of projects, and other branches of knowledge such as HSSE also provide distinctive areas of project knowledge. But does this add up to a distinctive core 'knowledge set' for the discipline of the management of projects as a whole? None of these on their own are sufficient. And importantly they miss a – possibly 'the' – distinctive core skill: interdisciplinarity, which we discuss at the end of this chapter.

Strategy

Strategy clearly has a place in the management of projects – not so much in conceiving and executing some grand design as in shaping the project in the development stages, emphasising issues of implementation and ways of enhancing value[24]. Strategists are more likely to be nominalists and interpretivists, which is not to say they won't be concerned with reality and empirical evidence.

Technology

A technical knowledge base is generally fundamental in the development and delivery of projects. Most projects generally involve, if only in part, something being created, something involving a technical element. This is true in my personal experience of aerospace, building, civil engineering, development aid, petrochem, pharmaceuticals, power, oil and gas, software, systems and telecommunications. The project manager becomes involved in making decisions of substance, and these decisions are inevitably cast in the technical milieu of the project. In short, project management decisions are often unavoidably technical. That's why I find it non-credible to conjure up the vision of a project manager who would not have an interest in the technological issues and dimensions of the project. This is true even in the early stages of a project.

Project people are often strong on engineering and technology. Engineers tend to the realist and positivist outlook. Issues of organisation and people are acknowledged as important and perhaps fascinating, but the extent of intellectual involvement in others' conceptual worlds – organisation theory's, for example – is generally not huge*.

---

*This is a statement based on personal experience. It has no research data gathered to support or critique it that I am aware of.

Commercial

The commercial area covers subjects such as economics, finance, legal, joint ventures, supplier engagement, contracts and procurement, sales and marketing. As a discipline, it is nowhere near as well studied as engineering and technology. David Lowe and Roine Leiringer of the universities of Manchester and Reading, respectively, attempted a reflection with their 2007 book *Commercial Management of Projects: Defining the Discipline*[25], but the book lacks the breadth one would expect – economics is completely missing, for example, despite the currency of many of its concepts[26], as is law.

Commercial people are likely to be less realist and positivist than engineers: more inclined to a nominalist, 'becoming' ontology; more likely to follow, probably unawares, an interpretivist epistemology.

Organisation theory

Organisation Theory is more used to self-reflection. (Some might say that it does little else: meta-analysis – 'reflexivity' – "aims not so much at generating theory about particular organizational topics but to make the generation of theory itself an object of analysis".[27])

Historically people and organisation structure have been treated separately in Organisation Theory, but a feature of recent thinking has been the way people (actors) can shape their environment rather than just adapt to fit it, and how actors and process mutually influence each other*. Thus, for Anthony Giddens at the London School of Economics, structure is both the product and the medium of action. Giddens emphasises the "complex skills which actors have in coordinating the contexts of their day-to-day behaviour; and [also] the need for the researcher to be sensitive to the time-space constitution of social life"[28]. Tom Burns and Helena Flam of Uppsala University have shown how social rules – localised by region, country, professional or trade group, family, etc. – may adapt as a consequence of the interaction between actors and context[29†].

Most significantly for projects – or more especially, as I'll argue in Chapter 21, for programs – Frank Geels of the universities of Eindhoven, Sussex and now Manchester has studied how actors and organisations interact to shape the future. Geels' interest is in innovation and the way it develops across different institutional levels – micro (niche), meso (regime) and macro (landscape)[30]. His work has included looking at how change and innovation can be engineered in society and we will pick up this aspect of it again in Chapter 21. Meanwhile Geels' multilevel, personal–institutional framework has been

---

*Organisation theory is a sub-area within sociology. We might note in passing that surprisingly little research has been published in the mop/p³m field on the impact of contemporary trends in society – race and gender; poverty and displacement; anger and retribution; the impact of technology (e.g. ICT); and the new pervasiveness of social networks (social media).

†Christophe Bredillet suggests 'conventions': Bredillet, C. (2010) Blowing hot and cold on project management, *Project Management Journal*, 41, pp. 4–20.

taken forward in the management of projects by Gernot Grabher and Oliver Ibert of Hamburg University and the Free University of Berlin, respectively, with their concept of 'project ecology'[31].

Organisation theorists could, and do, cover the spectrum of ontologies and epistemologies. Organisation theorists are pre-eminently analysts or facilitators and are rarely, if ever, in the front-line of project decision-making.

Control

The conceptual base of project control is fundamentally not complex. At its most sophisticated, it is cybernetics and systems theory. More prosaically, it is cost and schedule management, as described in Part 2.

Project controls personnel are likely to believe they are realists and positivists. In fact, they live in a much less certain world where reality is elusive and knowledge is interpretivist.

Summary: Interdisciplinarity and empowerment

So, what does all this tell us about the nature of the knowledge we need for the management of projects? There are several contributions.

1. The emphasis on our ability to influence context, as well as to be sensitive to it, is important in the new world of (reconstructed) project and program management. (This is a feature beginning to be felt in several fields, for example history[32] and geography[33].) We shall explore the role of project management in shaping context in the next chapter.
2. Actors – us: people – can act and make a difference. We can shape our future. We can and we should. This is the theme of Chapter 21.
3. In addition to the core elements of formal knowledge, as reviewed in Part 2 of this book, a distinctive characteristic of the management of projects and programs effectively is its interdisciplinarity.

We are well past the critique that says that project management professionals need to move beyond the mechanical application of practices and techniques. There *is* method behind the application of mop/p^{3}m, but there needs additionally to be understanding in some depth of the different disciplines and knowledge domains that are called on in creating and delivering projects and programs effectively. The ensemble needs to be applied in an ordered way: moving from the business case and strategy; through technical and commercial development; building an organisation; exercising control; making decisions; and managing people. To an extent, doing this can be instinctive, but issues invariably arise calling for the project manager to engage at some depth with the language, concepts and epistemologies of the specialists working on the project. The task of the project manager is to integrate the work of these different specialists to achieve the sponsor's goals.

Lawrence and Lorsch developed the management function of integration to coordinate and control, to bring together, organisational entities that need bridging but which have different characteristics. Project managers provide integration, but they may not see that their work requires "the mutual recognition of organizing concepts, methodologies, procedures, epistemologies and terminologies"[34] that characterises interdisciplinarity. Yet that is what may well be required. Project management is inherently interdisciplinary.

Learning to be an effective manager of projects thus involves:

- Formal knowledge of the discipline of the management of projects as regards its practices, processes and concepts as a collective in its own right.
- Familiarity with the formal knowledge disciplines used on the project or program – strategy, engineering, commercial management, organisation, behaviours/people, etc.
- The exercise of interdisciplinary leadership and integrative skills.

Historically, academia has been central to developing professions' knowledge. "Most professional education occurs in universities . . . because professions rest on knowledge and universities are the seat of knowledge in modern societies"[35]. But in reality this is rarely the case in today's world of project management as a whole, as a summative discipline – largely because *we have very few academics in project management who have ongoing experience of the reality of managing projects: very few would be called on by industry to manage or to advise on the management of projects.* They are not the seat of project management knowledge: practitioners are. Instead, academics disaggregate the subject and research and teach its elements, generally from one particular theoretical stance. And interdisciplinarity generally doesn't even come onto the curriculum.

Developing a semi-instinctive management capability has been touted as a reasonable aim for the project management professional – as, for example, in the Dreyfus and Dreyfus schema: "[since] expertise is based on the making of immediate, unreflective situational responses; intuitive judgement is the hallmark of expertise"[36]. But as Daniel Kahneman, the psychologist and proponent of behavioural economics, has shown, our instinct for 'quick thinking', while in certain times and places is essential, in others results is the wrong answer*: we need to think about some things more carefully, allowing cognition to work-out the answers. Cognition is shaped more by education than is instinct, which is shaped more by one's genes, upbringing and training. Managers are going to require re-education if we want to create new, interdisciplinary ways of thinking.

Knowledge is far from straightforward. We need to recognise its many types and its context. But the task – the challenge – is for academics and practi-

---

*His most famous 'quick' example is "if the bat costs a dollar more than the ball, and the bat and ball together cost $1.10, how much does the ball cost?" The answer is not 10¢.

tioners alike to seize the agenda and make a difference. The next three chapters address ways of doing this. First is the subject of shaping context and contingency. Then in Chapter 20 we look at value creation. And in Chapter 21 we ask how mop/p³m can connect better with business performance and how it might contribute to, and interact with, our changing world.

References and Endnotes

[1] Sapolsky, H. (1972), *The Polaris system development: Bureaucratic and programmatic success in government*, Harvard University Press: Cambridge, MA; Midler, C. and Navarre, C. (2004), Project management in the automotive industry, In: Morris, P. W. G. and Pinto, J. K. (eds.) *The Wiley guide to managing projects*, Wiley: Hoboken, NJ.

[2] This is the definition used by the UK Major Projects Association (MPA), http://www.majorprojects.org.

[3] Blomquist, T. and Lundin, R. A. (2010), Projects – real, virtual of what? *International Journal of Managing Projects in Business*, 3, 1, pp. 10–21.

[4] Cicmil, S. and Hodgson, D. (2006), *Making projects critical*, Palgrave: New York, pp. 1–25; Chia, R. (1995), From modern to postmodern organizational analysis, *Organization Studies*, 16, 4, pp. 579–604.

[5] Burrell, G. and Morgan, G. (1979), *Sociological paradigms and organizational analysis*, Heinemann: London.

[6] Tversky, A. and Kahneman, D. (1974), Judgement under uncertainty: Heuristics and biases, *Science*, 185, pp. 1124–1131; Jacowitz, K. and Kahneman, D. (1995), Measures of anchoring in estimation tasks, *Personality and Social Psychology Bulletin*, 21, pp. 1161–1166.

[7] Goldratt, E. M. (1990), *Theory of constraints*, North River Press: Great Barrington, MA; Goldratt, E. M. (1997), *Critical chain*, North River Press: Great Barrington, MA.

[8] Bhaskar, R. (1975), *A realist theory of science*, Leeds Books: Leeds; Bhaskar, R. (1979), *The possibility of naturalism*, Harvester Press: Hassocks; Harré, R. (1972), *The philosophies of science*, Oxford University Press: Oxford; Harré, R. (1979), *Social being*, Blackwell: Oxford; Sayer, R. A. (2000), *Realism and social science*, Sage: London.

[9] Morris, P. W. G. (2004), The validity of knowledge in project management, in Morris, P. W. G. and Pinto, J. K. (eds.) *op. cit.*: 1: p. 1142.

[10] Popper, K. (1959), *The logic of scientific discovery*, Basic Books: New York.

[11] Note that the questions it asks are different from others who have written on project management's theories, e.g. Soderlund, J. (2004), Building theories of project management: Past research, questions for the future, *International Journal of Project Management*, 22, 3, pp. 183–191; Koskela, L. and Howell, G. (2002), The underlying theory of project management is obsolete, *Conference Proceedings of the 2002 PMI Research Conference*, Seattle, USA. Project Management Institute: Newton Square, PA.

[12] Griseri, P. (2000), *Management knowledge: A critical view*, Palgrave: London, p. 42.

[13] Koskela, L. and Howell, G. (2002), *op. cit.*: 11.

[14] Smyth, H. and Morris, P. W. G. (2007), An epistemological evaluation of research into projects and their management: Methodological issues, *International Journal of Project Management*, 25, pp. 423–436.

15 Lenfle, S. and Loch, C. (2010), Lost roots: How project management came to emphasise control over flexibility and novelty, *California Management Review*, 53, 1, pp. 32–55.

16 Pellegrinelli, S. (2011), What's in a name: project or programme, *International Journal of Project Management*, 29, pp. 232–240.

17 Weber, M. (1948), *The methodology of social science*, Free Press: New York.

18 Thompson, J. D. (1967/2003), *Organizations in action*, McGraw-Hill/New Brunswick: New York.

19 Andersen, E.S., Grude, K.V. and Haug, T. (2009) *Goal Directed Project Management* (4th edition) Kogan Page, London

20 Clegg, S. R., Ptsis, T. S., Marosszeky, M. and Rura-Polley, T. (2006), Making the future perfect: Constructing the Olympic dream, in Hodgson, D. and Cicmil, S. (eds.) *Making projects critical*, Palgrave Macmillan: Basingstoke, p. 2, 67.

21 Hobbs, B. and Aubry, M. (2008), An empirically grounded search for a typology of project management offices, *Project Management Journal*, 39, 1, pp. 569–582.

22 Shenhar, A. J. and Dvir, D. (2007), *Re-inventing project management*, Harvard Business School Press: Cambridge, MA.

23 Morris, P. W. G. (1972), An organizational analysis of project management in the building industry, *Build International*, 6, 6, pp. 595–616; Morris, P. W. G. (1974), Systems study of project management, *Building*, 226, pp. 6816–6817, pp. 75–80, 83–88.

24 Morris, P. W. G. (2009), Implementing business strategy via project management, Chapter 2 in: Williams, T. M., Samset, K. and Sunnevåg, K. J. (eds.) *Making essential choices with scant information: Front-end decision-making in major projects*, Palgrave Macmillan: Basingstoke.

25 Lowe, D. and Leiringer, R. (2007), *Commercial management of projects: Defining the discipline*, Blackwell Publishing: Oxford.

26 For example Schumpeter's theory of 'creative destruction' (Schumpeter, J. A. [1942/1975], *Capitalism, socialism and democracy*, Harper: New York; Nelson and Winter's analysis of economic change in terms of technology and routines (Nelson, R. R. and Winter, S. G. [1982], *An evolutionary theory of economic change*, Harvard University Press: Cambridge, MA; Penrose's work on resourcing (Penrose, E. [1959], *The theory of the growth of the firm*, John Wiley and Sons: New York; and the work on transaction costs by Cheung, Coase and Williamson (Cheung, S. N. S. [1987], Economic organization and transaction costs. *The New Palgrave: A Dictionary of Economics*, 2, pp. 55–58; Coase, R. [1960], The problem of social cost, *Journal of Law and Economics*, 3, pp. 1–44; Williamson, O. E. [1981], The economics of organization: The transaction cost approach, *The American Journal of Sociology*, 87, 3, pp. 548–577; Williamson, O. E. [1985], *The economic institutions of capitalism: Firms, markets, relational contracting*, Free Press: New York; Williamson, O. E. [1996], *The mechanisms of governance*, Oxford University Press: Oxford; Williamson, O. E. [2002], The theory of the firm as governance structure: From choice to contract, *Journal of Economic Perspectives*, 16, 3, pp. 171–195.) And also, the economics and financing of public sector infrastructure and other capital projects, whether in Keynesian terms or the structuring and efficiency of PFI/BOT and its place in budgetary and fiscal management.

27 Tsoukas, H. and Knudsen, C. (2003), Introduction: The need for meta-theoretical reflection in organization theory, in *The Oxford handbook of organization theory*, Oxford University Press: Oxford.

[28] Giddens, A. (1984), *The constitution of society: Outline of the theory of structuration*, University of California Press: Berkeley and Los Angeles, CA.

[29] Burns, T. E. and Flam, H. (1987), *The shaping of social organization: Social rule system theory with applications*, Sage Publications: London.

[30] See for example Geels, F. W. (2004), From sectoral systems of innovation to socio-technical systems: Insights about dynamics and change from sociology and institutional theory, *Research Policy*, 33, 6/7, pp. 897–920; Schot, J. W. and Geels, F. W. (2008), Strategic niche management and sustainable innovation journeys: Theory, findings, research agenda and policy, *Technology Analysis & Strategic Management*, 20, 5, pp. 537–554.

[31] Grabher, G. and Ibert, O. (2011), Project ecologies: A contextual view on temporary organisations, in Morris, P. W. G., Pinto, J. K. and Söderlund, J. (eds.) *The Oxford handbook of project management*, Oxford University Press: Oxford.

[32] Influencing one's environment rather than just being influenced by it is a theme in contemporary historical studies. "For the French Annalistes, notably Fernand Braudel, the environment exercised a formative and even deterministic influence on the evolution of societies that occupied it. . . . Now, by contrast, the landscape is conventionally understood as a cultural construction. Geographers, archaeologists, and anthropologists have taught us to regard it as the biography or autobiography of society". (Walsham, A. (ed.) [2011], *The reformation of the landscape*, Oxford University Press: Oxford.)

[33] Giddens puts the recognition of place, and space, as a defining feature of modern (post-1960s) geography, a principle example being the recognition of regionalisation, as for example, again, in Braudel (*ibid.*: pp. 364–368).

[34] OECD (1972), *Interdisciplinary problems of teaching and research in universities*, OECD: Paris, France, pp. 25–26.

[35] Abbott, A. (1992), *The system of professions*, University of Chicago Press: Chicago.

[36] Dreyfus, H. L. and Dreyfus, S. E. (2005), Expertise in real world contexts, *Organization Studies*, 26, 5, pp. 779–792.

Managing Context

Battles are fought in terms of objectives, resources, ground conditions, spirit of the troops and so on; this applies whether you are a general or a private. Management battles may be less dramatic but the same logic applies. Management knowledge is similarly contextual[1]. It is affected by industry practices, technology, environment, personalities, and so on[2].

Systems theory has shown how organisations as 'open systems' react to their environment to survive, adapting their structure (morphogenesis) and their patterns of dependence. Overall, this open systems, interactive view of organisation can be summarised by its title, 'contingency theory' – a theory where the organisation's structure is contingent on its technology (particularly its routine-ness – organic/mechanic[3], etc.) and the rate of change in the environment around it[4]. (Interestingly, in the open systems view of organisation, adaptation of the organisation is the responsibility of boundary management functions[5], which is precisely what project management is.)

What Shenhar and Dvir did with their 'ntcp' model, and I did when I identified size, speed and complexity as principle factors shaping the organisational characteristics of projects, and influencing the consequences to management (Chapter 18, pp. 240–1), was in effect to position these project characteristics as independent variables. But sometimes some of them may be potentially malleable. Management can mitigate, influence or even utilise them. Thus, for example, innovation (novelty) can be chunked-up and parts managed discreetly; urgency (speed, pace) may be modified through phasing, fast-tracking or concurrent engineering; size or complexity can be reduced by breaking the project into parts and simplifying. All these are partial solutions which management can leverage to modify what have previously been presented as contextual 'givens' to which the project management response must adapt. Management doesn't have to be that passive and supine: it can shape context.

Reconstructing Project Management, First Edition. Peter W.G. Morris.
© 2013 John Wiley & Sons, Ltd. Published 2013 by John Wiley & Sons, Ltd.

So instead of the more traditional approach to contingency management, which posits management as responding to given environmental and project characteristics, the following analysis illustrates the kind of options available to management in responding to independent or semi-independent 'givens' and in deploying shaping actions. (The sequence follows the similar sequences of actions developed in Part 2).

We begin with a group of independent and semi-independent variables that significantly determine the context, and hence the shape and trajectory, of the project or program.

Independent (or Semi-Independent) Variables

First is the sponsor's, and hence the project's, 'strategic intent'[6]: what strategically is the sponsor trying to achieve; what are his aims, objectives, goals; why; what are the strategic benefits and advantages? To go to the Moon and back, thereby achieving – what? – political kudos on Earth? (So keep the TV cameras at all costs.) To develop a commercially successful, revolutionary Web-enabled, information-connecting, handheld hub based on a smart telephone, thereby revolutionising current ICT paradigms of usage and technology? (Emphasise slickness of design, functionality, weight, and use, not just technological sophistication.) To relocate the government's intelligence-gathering HQ, thereby facilitating new ways of working appropriate to post-Cold War threats? *Etcetera*. Each of these very different strategic imperatives sets the project off on a unique trajectory which is then articulated in its strategy.

Second are the 'requirements' (user, system, other – business, functional) that the project will need to satisfy. Are they clearly captured? Are they the right ones? Are they really independent variables? Might they not be modified? Well they might, but in the first iteration, they surely have to be treated essentially as 'given': these are what the users need. Yet on examination, some may have to be modified – and this is precisely one of my arguments for management being engaged in the requirements-elicitation process rather than just rather vaguely 'identifying requirements' (Chapter 12). But initially, and fundamentally, they stand independent of the project response.

Third is the effect of the 'environment' – the project's milieu – on the project's business proposition/strategic intent. The environment covers not just the physical environment (sustainability, carbon emissions, threat of flooding, pollution, physical isolation, etc.) but – and in particular – the political environment (support for the project, political stability [change of government], legislation, fiscal regime, regulation, employment rules, licences, etc.); the economic environment (inflation, product or service demand forecasts, foreign exchange issues, funding constraints [availability, cost of borrowing]); and the social environment (community support/opposition, availability of labour, etc.). All these are *prima facie* independent of the project, though, like requirements, in practice there may be some degree of influencing and modification that the project's management might be able to achieve to mitigate any negative influences on its chances of successfully achieving its objectives and goals, as we discussed at the end of Part 2 regarding institution theory. As (largely) independent variables, they will influence the shape of the project and the work of management.

Dependent Variables

Fourth, but switching now to management's responses to the project's strategic intent as characterised by its requirements and constrained by its environment, is 'funding'. Before proceeding too far into concept and design development we need to know on what terms finance and insurances will be available for the project, and how this fits with the business case. To answer this clearly requires an estimate and an estimate requires a design. So 'funding' iterates with 'solution development'. Funding requirements may then result in re-scoping or re-design, in de-risking, in choosing specific suppliers, in changing the pace of development (e.g. to secure income before drawing down full funding), etc.

Fifth, then, is 'solutions development'. Following the elicitation of user, system and other requirements, the basic technologies need choosing. Their stability, maturity, and risk all need assessing both in absolute terms and in terms of the benefit gained: is it worth accepting the challenge of the proposed technological innovation? What strategies can we deploy to develop and verify technology solutions of an acceptable risk profile? What are the benefits of taking this risk?

Requirements and acceptable technology lead to specifications against which designs are developed. (We saw in Chapter 12, pp. 169–70, that there are industry sector variations on this sequencing.) How should we best transfer from design to build, code, manufacture, assemble, etc? Are there implications for testing, verification and validation, and commissioning?

All this requires, and iterates with, resourcing. *Sixth* therefore is contracting, procuring and organising the resources needed to do all this. This may be a minor task if the project is small and doesn't require specialist resources, or it could be significant. Again management can shape the response: options amongst project, program, matrix or network organisational forms will need choosing; different contracting strategies will radically affect the project's staffing needs and responsibilities. Individuals and teams will need managing to get the best from them.

And *seventh* is planning and controlling and active decision-making to pull everything together and to move the project, or program, forward. Governance arrangements might sit here; in any event, they will set the tone for the way the project is to be managed. Schedule urgency will reflect the project's 'strategic intent', modified by items two to six. Budgets and risks ditto. And the constant aim of the project team should, I contend, be that of improving value while managing risk – a theme we shall explore in the next chapter.

Program management

To the program management community, this seven-stage model may seem too project oriented. The scope of the field of play in program management is usually broader than in projects, and hence the opportunities to modify or leverage dependent characteristics are greater. Program management has more opportunity to take advantage of constraints in planning, particularly in assessing options, in resourcing and in assessing benefit delivery. One of the few works I know that directly suggests actions that might manage context is Sergio Pellegrinelli's *Thinking and Acting like a Great Programme Manager* (2008)[7], although his suggestions are really more like Level 2 ones (in the Morris, Geraldi, Parsons schema) than Level 3*.

Organisation design tools

A parallel but more focussed stream of work centred on shaping projects' fit with their environment is that coming largely from the work of Ray Levitt[8] of Stanford University and Richard Burton[9] of Duke University using contingency-based organisation design tools. These are largely based on Galbraith's information processing ideas (see p. 58). The tools begin by diagnosing the independent variables, much as we did above – strategic imperative, environment, tasks required, etc. They then analyse coordination needs and identify opportunities for organisational rationalisation.

Whether such computer-based assistance is materially helpful is, to my mind, still an open question: I am aware of little formal evaluation of their utility. In any case, they are, at this stage, more products of the research laboratory than of the practical world of action, in which the rest of this chapter has been positioned. Nevertheless we both share the same conceptual stables, and have a similar vision: ". . . tools are needed which are future oriented to understand better the world of what might be. . . . The world of what might be can begin with project managers thinking through the future possibilities. . . ."[10] In this respect we should note the potential impact that changes in manufacturing technology, proposed by some as the Third Industrial Revolution, could have on project management.

Advances in computation are leading to potentially significant restructurings of the product development life-cycle, particularly across the key design–production interface. So-called 'additive manufacturing' is enabling the 'printing' of products directly from the design to 3-D printers[11]. (This

---

*They include such actions as initiating or improving communications; changing the composition of committees – i.e. decision-makers; lobbying for permits; briefing relevant committees; updating resource forecasts; seeking increased stakeholder participation; making benefits clearer to those who will benefit from them; cutting scope to speed schedule completion; and revisiting and, if necessary, revising the business case, improving NPV (lower costs, higher revenue) and seeking alternative funding arrangements.

is quite different from CAD/CAM, where the build is component led; here it is slice driven.) Combine 3-D printing with new strong materials and the linkage between design and manufacturing can become seamless. Although still largely at the R&D stage, this new technology has the potential to fundamentally alter the interdependencies not just between design and manufacturing but between users (as requirements owners), designers, manufacturers (suppliers and installers) and operators (possibly users again) – and as a result, project management's role in providing integration across the life-cycle will change. Organisation design tools that address the impact of these developments on project management are an obvious next step. Agile project management will take on a whole new meaning.

In summary, the way we deploy the elements of the management of projects, given the constraints of the project's strategic intent, environment and requirements, can shape and modify the project's characteristics. So much so in fact, that nearly all of the characteristics that are said to influence the way we should manage projects turn out to be not wholly 'given' but may instead be susceptible to modification. Changes in technology are likely to make the case for doing this even stronger. Often the project organisation can be enhanced, with respect to its strategic objectives, its context and its characteristics, to such an extent that the value of the offering is significantly improved. Which brings us to the next stage of our argument: enhancing sponsor value.

References and Endnotes

[1] Griseri, P. (2002), *Management knowledge: A critical view*, Palgrave Macmillan: Basingstoke.
[2] *Ibid.*
[3] Burns, T. and Stalker, G. M. (1961), *The management of innovation*, Tavistock: London; Woodward, J. (1958), *Management and technology*, HSMO: London.
[4] Emery, F. E. and Trist, E. L. (1965), The causal texture of organizational environments, *Human Relations*, 18, pp. 21–32.
[5] Burns, T. and Stalker, G. M. (1961), *op. cit.*: 3.
[6] Ansoff, I. (1969), *Corporate strategy*, Penguin: Harmondsworth.
[7] Pellegrinelli, S. (2008), *Thinking and acting as a great programme manager*, Palgrave Macmillan: Basingstoke.
[8] Levitt, R. E., Thomsen, J., Christiansen, T. R., Kunz, J., Jin, T. and Nass, C. (1999), Simulating project work processes and organizations: Toward a micro-contingency theory of organizational design, *Management Science*, 45, 11, pp. 1479–1495.
[9] Burton, R. M. and Obel, B. (2011), Computational modelling for what-is, what-might-be, and what-should-be studies – and triangulation, *Organization Science*, 22, 5, pp. 1195–1202.
[10] Caroll, T. N. and Burton, R. M. (2012), A contingency approach to designing project organizations: Theory and tools, *Engineering Project Organization Journal*, 2, 1–2, p. 13.
[11] Markillie, P. (2012), A third industrial revolution, Special Report, *The Economist* April 21, 2012.

Ethos: Building Sponsor Value

Managing projects effectively is not just a question of having the right people organised in a manner appropriate to the characteristics of the project and the environment it is in. Nor is it just a question of using project management tools and techniques. There is something about the 'why' and 'how' one does projects; the 'to what end?', I've called this 'the project management ethos'.

Questions of Purpose

What should be the driving ethos of the discipline? Consider the following quotes:

(a) "Project management can be defined as the discipline of managing project successfully"[1].

(b) "What is the professional remit of project management? Put simply, is it to deliver 'on time, in budget, to scope', or is it to deliver projects successfully to the requirements of the project customer/sponsor? In essence it has to be the latter, because if it is not, project management is an inward looking profession that in the long-term few serious managers are going to get very excited about. What managers in government, business, academia – just about everywhere in fact – are concerned about is that their projects are managed effectively and efficiently; that they represent value-for-money and meet or exceed their strategic objectives"[2].

(c) "Those shaping and executing the projects which flow from an enterprise's strategy can, and should, challenge and contribute to the unfolding of that strategy...project management has more of a contribution to make to strategy implementation than simply ensuring alignment and being efficient in execution. It can add value to the emerging strategy and ensure that benefits are reaped from its realisation"[3].

(d)   "A lot of the real skill of project and program managers is about 'driving the bus': making good decisions and moving the project forward. Driving and improving, not just monitoring and controlling. Knowing when decisions need to be taken, managing the risks".

Quote (a) seems reasonable, surely. Presumably someone – some discipline – needs to be responsible for the successful development and delivery of projects and programs. Curiously, though, there is not a lot of evidence of either theoreticians or practitioners accepting or contesting the statement. Yet if, as seems reasonable, it is accepted, then the implications are quite severe. For we then find that many of the causes of failure are to be found pre-execution, in the project front-end (as we see from Appendix 1).

Effectiveness

Further, we are drawn right from the outset into asking what we mean by 'success' and its antithesis, 'failure'. (Topics we have discussed in Part 1.) Among the contributions to this debate is the work of Miller and Lessard (2000) who, in the IMEC study of 60 large ($1 bn+) energy and transport projects, found that while only approximately 75% met their cost and schedule targets, only about 40% met their sponsors' objectives (Chapter 5, pp. 75–6). The former they termed 'efficiency' targets; the latter 'effectiveness'. Whatever one's thoughts about the percentages, one thing stands out: it's amazing surely that 'value for the sponsor' – effectiveness – is so much lower than the efficiency number, for achieving value for the sponsor is surely the reason the project is being done. Hence statement (b).

But managers can serve their customers better than just "meeting their requirements"! That's not the Total Quality Management approach where the customer comes first; it's not what has driven the world's most successful companies such as Singapore Airlines, Tesco, Toyota, etc. Projects can be developed to exceed their sponsors' expectations. Let's call it for the moment 'Building Customer Value'. This is the essence of statement (c).

Enhancing Sponsor Value

To do this requires a more proactive style of management[4]. Here the people whose profession it is to practise the business of managing projects – the project management team – serve the sponsor by deliberately seeking, where and as appropriate, actions to improve the value that the emerging project is offering the sponsor. This is what is meant by statement (d): that project management, using the term in its widest sense, has a proactive, driving role; one which pre-eminently is about developing and delivering value to

ADDING VALUE

MONITORING & CONTROL

1. Develop the project strategy (in line with Stakeholders' objectives and goals.
2. Establish the scope of the project.
3. Plan the schedule (if not in detail then in outline or progressive detail).
4. Allocate resources.
5. Allow for risks.
6. Allocate contingencies; agree the budget.
7. Monitor and Control that the project keeps to these planned targets.

1. Establish the business case for the project and the proposed strategy for its development and delivery to marry into this.
2. Establish the scope of the project, plan the schedule (if not in detail then in outline or progressive detail); allocate resources; allow for risk, allocate contingencies; agree the budget.
3. Manage the work on the emerging project definition (a major piece of which will be the pre-sanction product definition) and on preparing for downstream implementation.
4. Establish the commercial platform upon which the project work will be done.
5. Enhance performance; build value, harvest benefits, manage risk, control performance and drive progress.
6. Do all this remembering that projects are done by, with and through (and for) people- that is, manage the people involved in the project.
7. Learn and improve.

Figure 20.1   The 'traditional' versus a 'value-driven' approach to the management of projects.
Source: Author's own.

its clients* – the project's sponsor and other stakeholders (including suppliers' sponsors [bosses]) – while managing the risk latent in the project[†]. This is a management activity which is called for from the early front-end stages of the project as well as in the more traditional execution stages.

This ethos is substantially different from what many would consider the norm for project management: namely, that project management is essentially about monitoring and control: that one first plans the project and then monitors implementation, taking action as required to keep it on plan. *This is the cruise-control view of project management.* The PMBOK® Guide and ISO 21500 both reflect this approach. Figure 20.1 illustrates the difference between this 'traditional' approach and the 'value-driven' approach. (Note

---

*Mark Winter and Tony Szczepanek propose that p.m. link with Richard Normann's idea that the aim should be not so much to create value for the customer as to mobilise the customer to create their own value (Normann, R. [2001], *Reframing business: When the map changes the landscape*, New York: Wiley). Again, the emphasis is on the front end. Winter, M. and Szczepanek, T. (2007), "Projects and programmes as value creation processes: A new perspective and some practical implications" *International Journal of Project Management*, 26, 1, pp. 95–103.

[†]Lean Management, as we've seen, focuses on delivering added value too, but primarily through a reduction in waste–see for example Womack, J. P. and Jones, D.T. (2003) *Lean thinking*. Free Press: New York. I am not proposing a Lean agenda for project management but something less formulaic, more instinctive and generic.

items 3, 4 and especially 5 in the 'Adding Value' column.) The proactive, value-enhancing approach, on the other hand, fits well with the emphasis on value creation that came out of the *Rethinking Project Management Research* scoping exercise carried out around 2006–2008 (see Chapter 6, pp. 101–2)[5].

Based on what we've learnt already, the following areas offer opportunities for project management to implement this value-enhancing approach. (Note that much of the play focuses on the Front-End. This emphasis is partly because that's where the required activity generally is needed, and partly because most writing on project management tends to underplay the Front-End and concentrate on Execution.)

Governance and strategy

One of the first tasks in establishing the project or program is, as we've seen several times, to develop its strategy and ensure that this is as effectively aligned with the sponsor's aims and strategy as possible. Thus for Novartis, for example, the pharmaceutical company, the single most important piece of work of the project 'leader' at the time of 'candidate selection' – when the molecule begins its progression through, first, animals and, later, man to check for toxicity and to evaluate its therapeutic and commercial potential – is the preparation of the project strategy – in detail for the next few stages and years, broadly thereafter. The project leader, typically a clinician, in creating the project strategy, jointly with her or his team, is directly and formally shaping the development of the project definition.

The project management–governance relationship is central to effective strategy formulation, to good practice application, and to value enhancement[6]. Creating a good project management/governance relationship is not something that can be done by the sponsor alone – it needs effort from the project side as well. As a professional with specialist knowledge about the complex, challenging world of projects and programs, the p.m. is responsible for making this relationship work as effectively as possible*.

Seeing the project, or program, in its totality helps frame the sponsors' opportunities. A systems perspective is helpful in seeing what the elements are that, working as a whole, will influence the achievement of the overall goal: what the organisational, technical, social and other subsystems that interact are; how significant the interfaces are; what the control and regulatory mechanisms are; and what the best way to manage those interactions

---

*This is a focus slightly north and west (more senior and front-end oriented) than is typical of most project management scholars' views. For example, in a study of 34 p.m. 'experts', 17 Masters courses, and papers published in *PMJ* and *IJPM* between 1994 and 2006, Helgi Thor Ingason and Haukur Ingi Jónasson found an emphasis on strategic alignment, interpersonal relationship management and competence, resource management, and project planning and control: Ingason, H. T. and Jónasson, H.I. (2009), Contemporary knowledge and skill requirements in project management, *Project Management Journal*, 40, 2, pp. 59–70.

is. Who is best placed to do this analysis? It will rarely be the sponsor alone – it is unlikely that he or she will have either the knowledge or the time. It is most obviously the remit of project or program management.

Strategy at the portfolio level would most likely be developed by portfolio management, in conjunction with individual projects and programs, whether that is strategy for projects and programs falling within the existing portfolio, as in drug development, or with respect to the allocation of resources (people, equipment, money).

The IMEC program strongly recommended the real-options approach as a means of improving value. "The key insight of this approach is that uncertainty or volatility can increase the value of a project, as long as flexibility is preserved and resources are not irreversibly committed. As a result, the economic value of a project where it is still relatively unformed is often greater than the discounted present value of the expected future cash flows. Value is increased by creating options for subsequent sequential choices and exercising these options in a timely manner."* A corollary, which we have heard several times in this book, is to avoid making premature decisions. Avoid premature lock-in.

There will be times when the business decisions of the owner may not accord with preferred project management practice. Scope may not be fully worked out at the sanction decision gate and risks may remain high, but delaying for these reasons could lead to business problems – a fall in the owner company's share price, for example. The sponsor may decide that he needs to proceed in spite of his project team's recommendations. How then does the project team add value to the sponsor? This situation is analogous to that of a doctor advising his patient: the project manager needs to be professional: clear as to what good project and program management practice would recommend and advise accordingly. But he, or she, should try and understand the 'whole patient': understand for example the business drivers acting on the sponsor and develop and overlap the project's strategy with the sponsor's. Ultimately it is for the sponsor – the patient – to decide how much of this advice he will accept, within the compass of good governance principles.

Agile, as ever, is a special case. Insofar as Agile is essentially task oriented, it runs the risk of having weak governance at the project level. To counter this, Agile would seem to fit best under a strong overarching governance structure. (Some would even call such a structure program management, because it does literally pertain to/manage several [Agile] projects.)

Stakeholders

Stakeholders need mapping and actions will need taking to influence them in the project's favour. 'Project management' has a hugely important role in

---

*See Miller, R. and Lessard, D. R. (2000), *The strategic management of large engineering projects*, MIT Press: Cambridge, p. 110.

influencing and shaping stakeholder attitudes[7]. (*Vide* the US SST program – Chapter 3, p. 44 "the SST's management 'did not comprehend until it was too late that important societal changes could severely undermine their program' "; see too the conclusion of Part 2 on institution theory – Chapter 16, p. 224.)

Requirements and innovation

Poorly managed technical issues have historically been a major cause of project failure. Practice has improved considerably, however. Now the project management challenge is not so much to avoid major problems as to improve sponsor value[8]. While technical and commercial issues are often left to the technical and commercial experts in the project or parent organisation, the experienced project executive will want to review work done in shaping response to these issues, and will absolutely want to be involved in major decisions in these areas.

Technical definition should begin with requirements. This is one of the principle areas both of project risk and opportunity. The ability (or not) to successfully elicit requirements for a project, and for those requirements to then remain stable, is one of the major issues in reconstructing project management. If their stability cannot be assured then it may well make sense to go in the Agile direction and chop the scope into small pieces which, because they're short, means that the user can quickly test and use them. Assessing and strategising this takes time and skill; is the right group to do it Systems Engineering or Project Management, or a collaboration between the two (which is what, in a mature relationship, would most likely be the preferred way of working)?

Innovatory responses might be called for. Experiments designed. Again, is this best done by Engineering or Project Management, or a collaboration of the two? The project manager should have a strong impulse to ensure innovatory solutions are pursued and the instinct to manage the risk which they will inevitably raise – all this (off-line prototyping, Pre-Planned Product Improvement, etc.) being done within the constraints of the project life-cycle.

Typically, the project director and his or her team only really gets involved when the technical thinking needs integrating into the overall approach to the development of the project. In oil and gas Exploration and Production (upstream) projects, for example, subsurface geology and well-engineering are typically performed as discrete activities separate from the rest of the project. The project management team will be concentrating instead on developing the overall strategy and building the submission for the sanction approval stage-gate review.

On building projects such as large office developments, the project manager too often plays only a largely coordinating role, leaving the development of the project's technical definition to the architect. And the more 'signature' the architect, the more pronounced this is likely to be. There is a risk, and of course an opportunity, that the signature design that is then delivered may be excessively design-led and may not, in a broader sense, represent 'best value'. Addressing this takes personality! The trouble is, not only is there a conflict of roles and egos (the famous architects versus the nameless project managers), the very term 'value' is contested: whole life value; intangibles, such as name recognition; aesthetic response? It is genuinely difficult for all but the most assured project managers to prevail in mounting a challenge of this kind. But it's project management's responsibility to do so.

Commercial platform

Similarly, the project management team should have input into commercial as well as technical matters, beginning with the project strategy and the choice of contracting strategy: in particular, what functions to contract out (the flip side of what functions should be retained, possibly as core competencies), what resources are available, and how risk should be allocated. Should the work be let on a fixed price or cost reimbursable basis? What should the scope of services be – systems integration (as TRW pioneered – Chapter 3, p. 30), general contractor, project management? Is the form of contract appropriate? All these are matters which come under the purview of the 'Contracts and Procurement' function but, as they all can massively influence the way the project is managed, one would expect the project director (or equivalent) to sign them off.

Project and program leadership

To imagine that all this shaping, organising and challenging can be done somehow automatically is beyond credulity. Each action needs conceiving, explaining, and supervising. Articulating the project's vision – the 'future perfect'[9] – is not just a fundamental task but a core responsibility, particularly in the early stages when the idea of the project has yet to have acquired such currency. Charisma, drive, energy, negotiating capability – all give a project a focus, values and drive.

Ideally, therefore, *there should be someone acting as 'the single point of accountability'* for the project from the earliest stages of the project right through into taking some accountability at least for operational performance. (It may not be the same individual throughout: after all, Chief Executives come and go but the role continues.) Theoretically, in Lawrence and Lorsch terms, or as with the *Shusha*, this role provides the integration needed across all the interacting elements as soon as their interdependence becomes important. (In Thompson's language, 'reciprocal interdependence'.)[10] In practice, this may not happen, either because the organisation is not accustomed to doing so (which is not really excusable), or it lacks the necessary resources (which is understandable).

As we saw with Toyota, the composition of the project team should reflect all the relevant major subsystems and parties having a bearing on the design and development of the project. Forming and managing the team will not happen without executive action. Someone's got to orchestrate and manage the front-end activity(ies). Organising such *programaçao* is fundamentally squarely one of the core functions of project management.

Valuing time

Time is the most potent resource in projects. The way we engineer value into it is one of the most significant acts that the project management team can perform. It is what Goldratt was aiming for with Critical Chain project management, albeit primarily for small projects, but its importance as an operational idea that applies for all scales of project and program management cannot be over-emphasised. Understanding and influencing the rhythm of the project life-cycle to the sponsor's advantage is, I believe, a core competency of project management.

Identifying opportunities for fast-tracking (or simultaneous engineering or concurrent engineering – see Chapter 4, pp. 66–7) is a clear front-end activity. It should be undertaken for all projects and programs but particularly the larger ones: the impact can be huge, not just on schedule but on product quality and sales.

At the project level, particularly the major project level, questions on how relevant the elements of concurrent engineering are – integrated teams, parallel working, avoiding making decisions prematurely, design-for-manufacturability/buildability, product modelling – can be self-evidently important. At the program level, it may be more complex: there is scope here for mapping out how this would work for different kinds of program.

Control

Securing a funded and sanctioned budget is generally one of the principal objectives of the project's development phase and the resultant budget becomes, alongside the target completion dates, a major governing measure for the project's development. Planning is fundamental from Day 1 and never ceases until the project or program is complete. Calculating cash forecasts, agreeing budgets, allocating contingencies, assuring funding, and interacting with an evolving business plan: adjusting these to provide optimum terms for the project sponsor (or the supplier's contract manager) – all will require input from the project team.

Progress spend and forecast completion sums and dates need reporting regularly to the 'governance' board, as will the escalation of issues outstanding. Control systems and procedures, therefore, need establishing early in the project.

Risk needs modelling, appraising and acting upon. Only the key question is, risk from which perspectives? The cost and schedule efficiency ones may be less important in the project front end since they may well not yet have been fixed. (But not always – the project could have immutable deadlines or budgets established from the outset.) Risk will need appraising from a number of perspectives – operational performance, finance, social (community, sustainability, etc.). In particular it would seem sensible to assess them against the project's effectiveness measures and benefits.

All this has to be organised and the data analysed and actions agreed and acted upon from the earliest stages of the project. All are within the remit of project, or program, management for implementation and action. And the supply of budget and forecast completion data will often be the life-blood of portfolio management.

Benefits management and opex

Benefits management appears most obviously in program management, though it is by no means restricted to programs[11].

Benefits are assessed and play a big part in front-end decision-making in many types of projects as part of Cost–Benefit analysis: it is a major activity in transport projects, for example. But this is different from Benefits Management, where a 'benefits owner', who is generally external to the project, is identified, who has responsibility for defining the expected benefits, managing their 'harvesting' and feeding lessons from this project into future projects.

In oil and gas and mining, on the other hand, the output is simply a monetised product. There may indeed be quantifiable benefits other than these primary outputs (establish a strategic relationship with the host government or joint venture partner, for example) but these are likely to be very secondary and have nothing like the potency of the benefits and disbenefits we saw on the GCHQ New Accommodation Programme, for example (Chapter 5, pp. 93–4). In such a business, Benefits Management gains little obvious traction. There is one dimension, however, that corresponds directly with it: operations and maintenance. O&M performance is a real and tangible area of product and project benefit and the same principles of management apply – ownership; representation in strategy and design; quantification and feedback on performance. The enterprise might have rules and procedures for this (as the UK Treasury does for building, for example) but at an operational level, technical decisions will need making.

And this decision-making will need managing. We saw on Andrew (Chapter 5, pp. 79–80) how incentives can skew decisions – how capital cost may be reduced but at the expense of an unwanted reduction in operating performance. Who ultimately is responsible for achieving an appropriate balance? The project director? The sponsor? What intellectual formation (education) provides him or her with the perspective needed to shape this decision? The project team may not be the best place to make these technical decisions but, as always, their responsibility is to make sure that the correct integration takes place, when it's needed. They need to make sure the right input is provided. Having Operations represented on the project team helps. Doing VM/VE is another means.

In software projects, Agile is also a form of response to these questions: a series of very small projects with close developer–customer pairing, focussed on achieving some operational benefit.

Making decision relative to the long-term benefit of the project, or program, is absolutely the core of what professional project or program management should be about: building value through capital investment. Yet in reality very, very few project managers have either the educational experience or the training to do this. Nor the incentives: the organisation – the sponsor, the head of the project management function – rarely encourages or expects them to do it. As a result, it is not surprising that it often doesn't get done. Instead, the project management discipline too often delivers projects that are late, over budget or fail to meet their requirements. This failure is too often the direct result of not focussing on creating value. At least such is the case in the West, for the Japanese are already well down the track on this.

The Japanese Approach: Pursuing Innovation and Value

The global automobile industry has changed out of all recognition over the last 25 or more years. Cars are now reliable and comfortable, offering real value for money to an extent that they just weren't before. The force behind this has been Total Quality: making the brand irresistible to the customer; achieving very high reliability; building value; innovating. These qualities characterise the Japanese approach not just to manufacturing automobiles but to project management as well.

Japanese professional project management is organised under the banner of *Kaikaku* Project Management (KPM)*: explicitly, around the pursuit of innovation and value enhancement, as we saw in Chapter 4 in discussing the Japanese BoK. Much of KPM is reflected in the ethos argued for in this chapter, indeed in this book. For example, it stresses the importance of the sponsor and his strategy linkage with the project. It is shaped by a 'management of projects' perspective, not an execution-only one.

KPM approaches project and program management in the following sequence[12]:

1. Profile the project in terms of its mission. The mission may have implicit meanings as well as explicit ones.
2. Propose the best strategy option within the scope envelope that meets the program mission.
3. Design the 'structural', functional, and behavioural 'architecture' for the program within the limitations of the 'scheme' business plan, systems context, and Operations and Maintenance service requirements.
4. Design the human operating environment (*Ba*) to perform and deliver as required – active, knowledge-oriented, supportive, good communications.
5. Make decisions in sync with external market, internal enterprise, and embedded program and project drivers. Concurrent engineering is recommended.
6. Perform multi-dimensional value management.

Kaikaku means innovative reform.

The language might seem a little strange in places, but the approach is very close to what I would call 'better practice' or enlightened project management, as advocated in this chapter.

Not every project will need or want a project or program management approach that puts so much emphasis on innovation and value management. Sometimes just getting completion as quickly and cost-effectively as possible is all that is required. But philosophically, we should be prepared to adopt a style of management that is focussed on meeting or surpassing the customer's – the sponsor's – goals. It makes sense to do so.

Longer term, merely planning and monitoring isn't going to be enough. It won't help business and society address the challenges they face. Exploring this future world and the place of mop/p³m in it is the subject of the next and last chapter before we sum up.

References and Endnotes

1. APM (2006), *APM body of knowledge*, 5th edition, Association for Project Management: London, p. 14.
2. Morris, P. W. G., Patel, M. B. and Wearne, S. H. (2000), Research into revising the APM project management body of knowledge, *International Journal of Project Management*, 18, 3, pp. 155–164.
3. This book.
4. Male, S., Kelly, J., Gronqvist, M. and Graham, D. (2007), Managing value as a management style for projects, *International Journal of Project Management*, 25, 2, pp. 107–114.
5. Winter, M., Smith, C., Morris, P. W. G. and Cicmil, S. (2006), Directions for future research in project management: The main findings of a UK government-funded research network, *International Journal of Project Management*, 24, 8, pp. 638–649.
6. Helm, J. and Remington, K. (2005), Effective project sponsorship: An evaluation of the role of the executive sponsor in complex infrastructure projects by senior project managers, *Project Management Journal*, 36, 3, pp. 51–61; Miller, R. and Hobbs, B. (2005), Governance regimes for large complex projects, *Project Management Journal*, 36, 3, pp. 42–50; Reve, T. and Levitt, R. (1984), Organization and governance in construction, *Project Management*, 2, 1, pp. 17–25.
7. Aaltonen, K. and Sivonen, R. (2007), Response strategies to stakeholder pressure in global projects, *International Journal of Project Management*, 27, 2, pp. 131–141; Littau, P., Jujagiri, N. J. and Adlbrecht, G. (2010), 25 years of stakeholder theory in project management literature (1984–2009), *Project Management Journal*, 41, 4, pp. 17–29; Newcombe, R. (2003), From client to project stakeholders: A stakeholder mapping approach, *Construction Management and Economics*, 21, 8, pp. 841–848.
8. Lampel, J., Miller, R. and Floricel, S. (1996), Impact of owner involvement on innovation in large projects: Lessons from power plant construction, *International Business Review*, 5, 6, pp. 561–578.
9. Clegg, S., Pitsis, T., Marosszeky, M. and Rura-Polley, T. (2006), Making the future perfect: Constructing the Olympic dream, In: Hodgson, D. and Cicmil, S. (eds.) *Making projects critical*, Palgrave Macmillan: London.

[10] Lawrence, P. R. and Lorsch, J. W. (1967), The new management job: the integrator, *Harvard Business Review*, Nov.–Dec, pp. 142–151; Lawrence, P. R. and Lorsch, J. W. (1967), *Organisation and environment: Managing integration and differentiation*, Harvard University Press: Cambridge, MA; Wheelwright, S. C. and Clark, K. B. (1992), *Revolutionizing product development*, Harvard Business School Press: Cambridge, MA; Thompson, J. D. (1967/2003), *Organizations in action*, McGraw-Hill/New Brunswick: New York.

[11] Morgan, B. V. (1987), Benefits of project management at the front end, *International Journal of Project Management*, 5, 2, pp. 102–119.

[12] Ohara, S. and Asada, T. (2009), *Japanese project management*, World Scientific: Singapore, pp. 27–30.

'only connect' – the Age of Relevance[1]

As I walked away from a recent international academic project management conference, I realised that everything we'd been discussing was about means rather than ends. Indeed, not just not about ends; not even about application! The theories had been discussed largely only to explore different ways of interpreting what might be happening. We weren't even considering how the adoption of mop/p³m practices would lead to organisations' performance improving.

Connecting p.m. to Organisational Performance

But *is* organisational performance improving as a result of applying mop/p³m? Do we know? Have we even made the case that adopting project management does benefit organisational performance? It is far from being self-evident. It can be argued relatively easily that project management has made little impact, say on the development of blockbusters in the pharmaceutical industry. (Largely I believe because, as a discipline, it is being applied from too low a level and the development path is highly regulated and managed within the constraints of the portfolio*, though even more possibly because the scientists have failed to discover the requisite molecules.) The oil and gas sector has good projects and bad. (Factors affecting performance being location, owner capability/staffing , technical issues[2].)

---

*As a result, the scope of project management to do much other than planning (including strategy) and control is significantly diminished. It even is heavily limited influence over the choice of team members. The result is a discipline that is paltry and underrated compared with that of a clinician, which is the usual professional formation of the Project Leader. No wonder project management is not highly rated in drug development.

Reconstructing Project Management, First Edition. Peter W.G. Morris.
© 2013 John Wiley & Sons, Ltd. Published 2013 by John Wiley & Sons, Ltd.

Many ICT projects remain a horror-story (for example the computerisation of patient health records in the United Kingdom in the first decade of the 20th century[3]), again largely due to technical problems and requirements and people issues. And so the list goes on. But the message is clear: while there's been some improvement (in defence and in construction, for example[4]), project management as a discipline is neither yet reliable enough nor engaged enough in improving its clients' performance.

We have been driven into an inward-looking discussion of theories about, and techniques for, project and program management, partly because of the difficulty of relating mop/p³m to organisational performance. And the factors behind this difficulty are genuinely interesting, to the point that our analysis can become myopically self-absorbed. Not only are we dealing with non-repetitive undertakings, the uncertainties and risks specifically associated with creating – predicting, judging – the future can be huge. Project decisions are often particularly sensitive to the judgement and behaviours of individuals, to a much greater degree than is found in regular, non-project operations. And the breadth of issues that the project and program management community needs to address can be truly very wide.

It is this breadth of issues that leads to a corresponding breadth of theories and techniques. And then it is this very size and breadth that is intellectually both so attractive and potentially so absorbing but which can lead us to being more interested in the theories and techniques themselves than in the challenges and outcomes that we are really supposed to be addressing. The means rather than the ends.

The result is that while mankind faces some of the biggest, most serious and dangerous issues in its history, project and program management as a discipline are almost totally silent on addressing them. We should look, as a discipline, at this. What does the future hold?

The New Dystopia?

"Prediction is very difficult, especially if it's about the future"[5].

Over half of the world's population now live in cities, the majority served by inadequate infrastructure and poor housing. Indeed, the world's infrastructure needs a massive amount spending on it to repair or replace existing facilities let alone to provide new ones. A recent estimate is that over $40 trillion is required over the next 25 years[6] – only an indicative figure, obviously. The impact of global warming will exacerbate this need significantly: sea levels will rise substantially during this century, possibly by several metres, particularly if the Greenland glaciers melt, affecting many of the world's biggest cities. Storm damage will increase; our emergency [project] response capability will need strengthening. The environment will, for many, become of overwhelming importance.

CO_2 emissions need dramatically reducing: the UK's 2008 Climate Change Act, for example, requires greenhouse gas emissions to be 80% below 1990 levels by 2050. The biggest emitter is the existing built stock –

how should this problem be addressed[7]? The equivalent of a new Manhattan program, according to a UK Government Chief Scientist, but I doubt it. Changing behaviours has to be part of the solution but really we need a bigger, more comprehensive approach. (Perhaps 'Transition theory' – see pp. 274–5)

Energy represents an enormous series of challenges – from new carbon capture and storage technology to massive new oil and gas exploration and production in remote and difficult locations. Brazil alone is likely to spend a trillion dollars – half of its GDP – on developing its deep-water oil and gas fields between 2011 and 2021. ("A moon shot under the ocean", as *The Economist* put it[8]: but Apollo was only a few per cent of America's GDP back in the 1960s.) Meanwhile, energy shortages are already beginning to affect our quality of life.

The world's population, currently around 6 billion, will probably peak at about 9 billion by 2050. Demographic changes will increasingly impact society, from the way we need to design and use buildings (more flexible and adaptive to accommodate evolving user needs), through finding employment and decent lives for our citizens, to the funding of old age and dealing with pressures from immigration. Crucially, the proportion and absolute numbers of old people will triple, from 670 million in 2005 to 2 billion by 2050. This will put enormous burdens on younger people, e.g. higher taxation, shared accommodation, increased health care, carer provision, etc. (Yet our projects are already suffering skills shortages.)

Feeding 9 billion people will be a major challenge: we shall need to produce about 70% more food than we do today, in conditions where land and water are scarce; where because of climate change "people, regions and countries dependent on agriculture look vulnerable"[9]; where fertilisers and transport are either in short supply and/or are expensive. No wonder achieving food security is a major worry for many governments. Worse affected will be the developing world.

More than 3 billion people live in 'water-stressed' or 'water-scarce' countries, 85% in Sub-Saharan Africa. Meanwhile, over-fishing is eradicating the primary protein source of 1 in 5 people while simultaneously making many of them poorer. Pollution and climate change are adding to the wasting of the seas. Mass fish extinction is an acute possibility.

We are using up our raw materials at a rate which is simply unsustainable. (And I do not believe the answer lies in space colonisation.) Alternatives have to be found.

Violence, be that revolution (the Arab Spring of 2011, writ larger), terrorism, or even warfare (clandestine, proxy or open) will probably increase, be more potent and more intrusive. Damage and disruption are as likely to be to our energy, food and retail, transport, and financial systems as much (initially) as on our physical selves.

New diseases are likely as well, as more aggressive strains of existing ones such as tuberculosis, malaria, and cholera. Medical research meanwhile roars away, helping us to live longer, thereby costing society more, to the point that many of us can hardly afford our medication or old-age care. Much efficacious treatment is unavailable to the poor who can't afford to pay for it.

There will be continuing major developments in our understanding of the brain. We will be able to use new knowledge of genetics to synthesise new biotechnological systems.

Expect big developments in robotics, including at the nano level, and more and more data and information: a ubiquity of intelligence. There will be an increasing convergence of technologies – NBIC: nano–bio–info–cogno). Electronics, computing and communication technologies should continue to develop rapidly, powering enormous increases in product performance, improving productivity and profoundly changing social behaviours. More and more data will become available, bringing new business and general opportunities, and threats. As we rely increasingly on massive interconnected information networks, so we will become more vulnerable to systemic collapse. By the middle of the century (2045, to be exact[10]), 'technological singularity' is forecast to have arrived: the point where our machines could have more intelligence than their human masters – an idea that, while mocked, nevertheless points to a likely, if disturbing, future for modern mankind which, after all, has only been around for about 40,000 years, all of which happened to be during an unusually long spell of warm weather for our planet.

Carlota Peres has suggested that since the Industrial Revolution there has been a series of 'technological ages', each lasting about 50 years: canals and textiles; steam, railways; steel, electricity and heavy engineering; oil and mass production; and ICT/digital[11]. I have grave doubts as to the validity and usefulness of such schematisation but it is curiously engaging, not least if one were to hypothesise that we are, in reality, now entering a new age of eco-management. How we manage society's relations with the ecology of the planet is unquestionably *the* dominant issue now facing mankind. Not that other things will somehow go on the back boiler. Maybe 'robotic man', post-singularity, post eco-accommodation, will be the age which will follow in 50 years' time?

Anyway, ecologically if not socially or technologically, the outlook for say 2050 – just one professional generation away – is challenging, with plenty of opportunity for our quality of life to deteriorate. But will it? For some, undoubtedly, it will worsen, but for others, it may be better. *The Economist* is certainly upbeat: "There is every chance that the world in 2050 will be richer, healthier, more connected, more sustainable, more productive, more innovative, better educated, with less inequality between rich and poor and between men and women, and with more opportunity for billions of people"[12]. And if you look back 50 years, well, life was far from wonderful. We had the Cold War and the arms race (bringing with it of course the development of systems project management). We were emerging from an era (1914–1945) in which over 100 million people had been killed as a result of state actions. Standards of living were far off from today's. Our health was worse. The computer had just begun to be used. The jet airliner was just arriving. It was a different world.

We don't know what the future really will be – a nuclear explosion in London? – but we can have some pretty good ideas, and we should be prepared. More, we should be building paths and driving forward in ways which positively help create a better, fairer way of life; a more equitable planet.

The Role of MoP/P³M

"When you come to a fork in the road, take it"[13].

How might project, program or portfolio management help society to address these and similar challenges, and how will these changes affect the practice of mop/p³m? It will be no surprise to see that most of the direction is set via policy decisions taken prior to portfolio, program or project management kicking-in, for these are essentially implementation disciplines. Nevertheless, as we argued in Chapter 20, mop/p³m does have a value-enhancing role inputting implementation perspectives to policy creation and strategy formulation, and in this way it can positively contribute to addressing society's challenges. And as regards the affects on practice, mop/p³m will undoubtedly be influenced by the developments coming at us in the coming decades. Many if not most will be in applications in front-end areas, calling on the responsibilities and capabilities typical of the PMO, leveraging the promise perhaps of institutional theory, structuration theory and transition theory, utilising the insights of leadership, technology management and human behaviours. The emphasis should be on impact and relevance. So, what have we?

Portfolio management

Portfolio Management is, as we've seen, about managing the allocation of resources in terms of the opportunities available, the risks posed, and the potential returns offered. Its application will differ depending on how decision-making is organised in the parent entity: specifically, albeit only partially, on whether the portfolio is in the private sector or the public.

But it is actually more than this. It's said of Steve Jobs that one of his decisive skills as CEO of Apple was his judgement on which of the many, many options to develop next Apple should *not* do. His judgement in managing the overall Apple portfolio, in selecting the features to incorporate, was critical to the success of the company[14]. The management of the portfolio of drug candidates in the pharmaceutical company's development pipeline is similarly an active, 'deliberate' activity. In these private sector cases, managers are able to use portfolio management as a decision-making tool, one that has direct impact on the portfolio itself. Portfolio management can be *the* critical strategic, project-governance, management function.

Unfortunately, many of the public sector environmental and infrastructure issues that we now face are not so susceptible to portfolio management unless and except where the economy is centrally planned. Portfolio management of, for example, putative built environment projects currently is more reactive, more a record of the state of development being progressed by third parties than an aid to policy-making or project planning and execution[15]. (This could well change in the future with the application of spatial and other computer modelling techniques – though their application is more to do with political will than technology.) Projects emerge, often unasked, invariably uncoordinated. In 2011, the UK published its National Infrastructure Plan: the Plan begins with a list of major projects. There is no process covering how such proposed projects should be brought into a potential existence. (An earlier attempt by the Government's Chief Scientist in 2010 to take a comprehensive look at the UK's land usage between now and 2050 was hushed-up by the government as being too hot politically[16].) State-directed/ planned economies, however, offer much greater scope for planned (portfolio) management[17]. The opportunity for portfolio management to assist us prioritise our responses to society's challenges depends on the extent to which there is operational linkage with the implementing agencies – or more specifically, a governance and management structure for achieving this.

Program management

Once projects begin to have money spent on them, then all the skills and knowledge that we've encountered in this book become relevant, although with great care. Many of the projects, whether in front-end development or in downstream execution, will need program management coordination. Part of this surely will be to identify the types of projects that could usefully be initiated and which should be managed programmatically. Reducing CO_2 emissions from the existing building stock, for example. But how should this be done? What is the theoretical basis showing us how to deploy such program management intervention? The problem for program management is that it has generally lacked a distinctive theoretical, methodological base. In practice, as we've seen, it can mean several quite different things: multi-project management, platform management in product development, strategic change management.

Sir Edwin Chadwick deployed a multi-path approach to improving public health in the mid-19th century (Chapter 2, p. 18), doing everything from engineering projects to writing social history and drafting legislation and regulations. Frank Geels' transition theory now offers a much more sophisticated analysis of the range of factors that we can now see should be included in designing any large-scale system change, which is what we are faced with in multiple instances in the decades ahead. Geels' work on the upgrading of the Dutch sewerage system and on the introduction of reinforced concrete offers not just a direct update on Chadwick's approach but an exciting, robust intellectual framework that program management might usefully adopt in addressing issues arising from and within 'the new dystopia'[18]. Geels is particularly interested in change in socio-technical systems, with change being coordinated via actors and institutional rules (cognitive, normative or regulatory), as opposed to the more common sectoral perspective: airline booking, ticketing and checking-in as opposed to the airline industry as a whole, for example. The technical dimension is important: it provides "a certain hardness" which is often the lever of large system change.

Geels begins with innovation occurring in technological niches, elements of which are gradually linked together from and across the different institutional levels – micro (niche), meso (regime) and macro (landscape) – in new socio-technic configurations (Figure 21.1). Changes in landscape, in

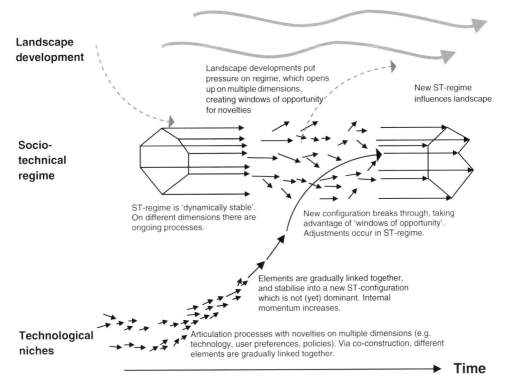

Figure 21.1   Geels' transition theory model.
Source: © Geels, F. (2004), From sectoral systems of innovation to socio-technical systems: Insights about dynamics and change from sociology and institutional theory, *Research Policy*, 33, (6–7) September 2004, pp. 897–920, Fig. 9.

turn, feed back, putting 'pressure on regimes' and 'open[ing]-up on multiple dimensions'[19]. Within such a framework, program management has a change management role to play: goal-driven – highly teleological – emphasising the work required in the front-end design stages – both process and product – on things like governance, strategy, platform design, technology management, innovation, stakeholder management, improving value, identifying and managing risk, resourcing, budgeting, regulation, planning and programming, and utilising learning. (Quite a list: let's not underestimate the challenge.)

Geels' transition theory is potentially particularly apposite for addressing sustainability issues[20]: a melding with program management could be enormously powerful in mobilising mop/p^{3}m towards the needs of the 'eco age'.

Project management

The challenge in the management of projects and the contribution it offers is not in the development of technical solutions *per se* but in the assembling and driving forward of those emerging solutions. The prize will be being able to address context and improve value. Project management practice will vary therefore – Agile will be quite different from mega-projects, for example. Nevertheless we can suggest some characteristics of projects relevant to 'tomorrow's world'.

The tools, the context and the concepts by which we manage projects over the next 30–50 years will be richer and more powerful. But how effectively we use them will be down to us, and our future colleagues. It will probably have the following flavour.

Control will use integrated, intelligent models more fluently – configuration management and BIM, n-CAD, etc.: risk will be modelled far more readily, from more perspectives. Organisational forms will be more fluid, agile and adaptive – more, and better, network-based. This should help improve the leveraging of supply chain capabilities. Strategy and governance will be the linchpin. The provision of funding and the organisational consequences that flow (better control, greater performance responsibility, etc.) could be major forces affecting development. Context will be understood and shaped, in a richer 'organizational ecology', multiple, integrated way; political will will be crucial. Increasingly, the Level 3 perspective will be recognised as important: institutional theory will become operationally useful, scenario planning will become more common. Evidence of innovation will be a central driver. Designers will emphasise resilience and adaptability more.

Management is not easy; it is particularly hard when you are being driven by urgency and fluid and uncertain requirements; where you have to be very careful over resource usage, particularly people: "Projects, ultimately, are managed by people. Not systems, not contracts – people"[21]. It is people (and their robotic assistants) who will implement these changes. In this more fluid, networked, agile world, management may well be becoming less clunkily formal, more instinctive. And as we develop thus, so the task of the Level 3 institutional mop/p³m support will be to enable and ensure this instinctiveness. Arguably one of the best ways of doing this is through training and education. So organisational learning and development – the two being linked in ways they have not really been to date – will become more important. But again, more plurally, not in so singularly a 'one-club' way (in a golfing sense) as it has tended to be to date: like self-organising learning and knowledge communities; greater use of coaching and mentoring; learning-led assurance reviews; supply-chain wide, learning-informed, value management workshops.

It is in interpreting and guiding such developments that the project management (mop) academic community has a distinctive contribution, providing sound knowledge and scholarly thinking. It is in performing in a role, working to an extent partly instinctively but supported from a solid educational base, that the practitioner really makes a difference. Maybe the future is ours after all.

References and Endnotes

[1] The phrase is the novelist E.M. Forster's from *Howards End*.
[2] A 2012 unpublished survey of the sector identified governance, technology, contracts and procurement and organisation and people as issues – particularly in the independents and National Oil Companies.

3 Cross, M. (2011), Original vision of NHS IT programme "will never be realised", report says, *British Medical Journal*, 342, 4, p. 3125; *The Guardian* (Thursday 22 September 2011), NHS told to abandon delayed IT project: £12.7bn computer scheme to create patient record system is to be scrapped after years of delays, p. 4; National Audit Office (2004), *Major IT Procurement: The impact of the Office of Government Commerce's initiatives on departments and suppliers in the delivery of major IT-enabled projects*, Report to the Comptroller and Auditor General HC 877: Session 2003–2004, London.

4 National Audit Office (2004), *Ministry of Defence: Major projects report 2004*, Report to the Comptroller and Auditor General HC 1159-l: Session 2003–2004, London; National Audit Office (2005), *Improving Public Services through better construction*, Report by the Comptroller and Auditor General HC 364-1 Session 2004–2005, London; National Economic Development Office (1991), *Guidelines for the management of major construction projects*, NEDO: London.

5 Attributed to Nils Bohr, Yogi Berra and Sam Goldwyn.

6 KPMG (2011), Global infrastructure group, London.

7 Williams, J. (2012), *Zero carbon homes: A road map*, Earthscan/Routledge: London.

8 *The Economist* (November 5, 2011), pp. 87–89.

9 Franklin, D. and Andrews, J. (2012), *Megachange: The world in 2050*, The Economist Publications: London, p. 103.

10 Kurzweil, R. (2005), *The singularity is near*, Penguin: Harmondsworth.

11 Peres, C. (2002), *Technological revolutions and financial capital*, Edward Elgar Publishing: London.

12 Franklin, D. and Andrews, J. (2012), *op. cit.*: 12: p. xiv.

13 Yogi Berra again.

14 Isaacson, W. (2011), *Steve Jobs*, Little, Brown Book Group: New York.

15 HM Treasury (2011), *National infrastructure plan 2011*, TSO: Norwich.

16 Government Office for Science (2010), *Land use futures: Making the most of land in the 21st century*, The Government Office for Science: London.

17 Bremmer, I. (2010), *The end of the free market: Who wins the war between states and corporations?* Viking: New York.

18 Geels, F. W. (2006), The hygienic transition from cesspools to sewer systems (1840–1930): The dynamics of regime transformation, *Research Policy*, 35, 7, pp. 1069–1082; Geels, F. W. and Deuten, J. J. (2006), Local and global dynamics in technological development: A socio-cognitive perspective on knowledge flows and lessons from reinforced concrete, *Science and Public Policy*, 33, 4, pp. 265–275.

19 Geels, F. (2004), From sectoral systems of innovation to socio-technical systems: Insights about dynamics and change from sociology and institutional theory, *Research Policy*, 33, pp. 897–920.

20 Kemp, R., Loorbach, D. and Rotmans, J. (2007), Transition management as a model for managing processes of co-evolution towards sustainable development, *The International Journal of Sustainable Development and World Ecology*, 14, 1, pp. 78–91; Loorbach, D. (2010), Transition management for sustainable development: A prescriptive, complexity-based governance model, *Governance: An International Journal of Policy, Administration, and Institutions*, 23, 1, pp. 161–183; Meadowcroft, J. (2009), What about the politics? Sustainable development, transition management, and long term energy transitions, *Policy Science*, 42, pp. 323–340.

21 Morris, P. W. G. (1994), *The management of projects*, Thomas Telford: London, p. 303.

Part 4

Summa

Reconstructing Project Management, First Edition. Peter W.G. Morris.
© 2013 John Wiley & Sons, Ltd. Published 2013 by John Wiley & Sons, Ltd.

Summary and Conclusions

In summary, what, then, do we have? First, the story and the scope. Second some conclusions.

Man has been managing projects since he first walked on this Earth. Project Management, as a formal discipline for doing this, emerged in the first half of the 20th century, becoming articulated as a coherent, useful discipline in the US missile programs around 1952–1955, later boosted by the US Department of Defense in the early 1960s and demonstrated with immense panache on the Apollo program. Professional project management societies part-formalised the discipline in various bodies of knowledge beginning in the mid-1980s; unfortunately, in PMI's case at least, not positioning its BoK to represent the knowledge needed to manage projects but as that which is supposedly unique to project management, in the process missing out several essential elements vital to the role of the discipline, particularly in the front-end.

Up until this point, theory could hardly have been said to have under-pinned the discipline – what there was (systems thinking, OR and organisa-tional integration) may have helped a few but was not essential to its practice (with possibly the exception of project planning). Around the early 1990s, however, organisation theorists became interested in projects as temporary organisations and the project-based approach became increasingly popular. This perspective has continued to thrive ever since. Meanwhile, a review of the data on project overruns in the late 1980s suggested a broader paradigm for the discipline: 'the management of projects', one where the project organisation is the unit of analysis, where context, the front-end, technology, people and the commercial basis of the project's development and delivery are included, as well as the traditional control topics.

Reconstructing Project Management, First Edition. Peter W.G. Morris.
© 2013 John Wiley & Sons, Ltd. Published 2013 by John Wiley & Sons, Ltd.

During the 1990s, project management began to be seen not just from individual projects' perspectives but as an enterprise issue – of organisational capability. With the arrival of personal computing, planning was no longer the major preoccupation it had been. Maturity models became popular as p.m. competency and capability assessment tools (although none were really fit-for-purpose.) People were recognised as key, theoretically never to be compromised (though in practice they inevitably were): project and program management training and education, and competence and career development, became more important to organisations, and individuals. Program management began to be seen as an important, qualitatively distinct, but important, sibling discipline to project management. New theories and practices emerged, like Critical Chain and Agile, but with limited practical impact. Critical management theorists criticised the discipline on a number of grounds (that arguably collectively made little or no difference to practice).

Meanwhile, despite some academic misgivings, standards and guidelines continued to be published, contextualisation being left in the hands of practitioners, which is reasonable, but with little guidance on how to do this, which is not. Curiously, the ethos of at least the more important of such guidelines – *PMBOK® Guide* and *ISO 21500* – seems to be 'plan and then put on cruise control'. There are few guides which emphasise influencing context or adding value, the Japanese being the major exception.

Ours is an age facing major challenges – the number of people on the planet exceeds its carrying capacity (but by what margin we're not sure). This is leading to all kinds of difficulties – global warming, resource shortages, security threats – although there is progress too: medicine, ICT, higher standards of living. As a result, the ways we practise mop/p^{3}m are changing: we see pressures for improved communications, greater agility, an enabled, more instinctive application of good practice; growing recognition of the importance of good governance and the value of institutional support, and of the role of project management in influencing – even, to the extent practicable, in managing – its environment.

But meanwhile our organisations still find it difficult to learn how to manage projects better, despite all the knowledge available. Coaching might help, but who is to coach? Academia too rarely involves itself in practice – practice too often being the servant of theory, not theory practice. Project managers don't really think about sponsors' needs hard enough. Improving our impact demands new skills, knowledge and behaviours. But the tools are at hand: more collaborative, action research, better sponsorship, better coaching, job-swaps, even more ubiquitous electronic intelligence, thoughtful professional leadership. Shouldn't these be the hallmarks of today's tomorrow: of the Age of Relevance?

So much for the rhetoric. What about the specifics of managing projects?

The little girl, who had stayed with us since the Introduction, raised her eyes from the can of orange juice and ceased concentrating. "The challenge", she said, sounding tired and rather emotional, "is that doing project management and theorising about organisations are two quite different things.

The first sees itself as real and in the business of producing results, while the second, with its meta-questioning, is often incapable of providing definitive answers. What," she continued, "have we learnt from all this reading and writing and talking about the discipline of managing projects, about 'reconstructing project management'?" And with that and a little flourish she upped and left, but not before leaving me her notes, which I've edited slightly, as follows. I leave discussion of their robustness and utility to you.

Conclusions for the Discipline

- Project management is neither a well-defined term nor a well-defined discipline, some seeing it as essentially limited to execution (post-requirements definition). Its Bodies of Knowledge are often inconsistent (though of the three major ones – PMI's, APM's, and the Japanese – PMI's is the one conceptually most out of line with the other two.)
- Similarly, program management suffers from several variants of its core definition (of a collection of inter-related projects having shared objectives and possibly shared resources) – for example shared technology or commercial platforms or as vehicles for achieving strategic change. All would benefit from a more rigorous theoretical base. Geel's transition theory offers promise for program management as large scale change management. It is far from clear what the difference would be between projects connected as a program against sub-projects connected as part of a large project.
- The 'management of projects' is a more useful and comprehensive conception of the discipline. 'mop' covers management's role in overseeing the shaping of project and program goals and strategies and the development of the project definition in its front-end.
- Management of the project therefore needs to be competent both at managing front-end development as well as down-stream execution.
- The project as an organisational entity should be the focus, the unit, of analysis – it is, afterall, the vehicle for effecting change and for realising investment goals.
- The front-end is the most important part of the project offering the greatest opportunities to add, or to destroy, value.
- The product development process (often termed the project life-cycle) is the one feature of project organisations that differentiates them from non-project organisations.
- There *is* a discipline of managing projects, knowledge of which can be taught and published as guidelines, though such guidance will need contextualising both in its articulation and its application. (All management knowledge is contextual.) Our knowledge of managing projects is, epistemologically, social science. Its application, recognising the range of topics involved, is science, technology, and art.
- Many different elements from many different types of knowledge areas are needed in managing projects. Hence there is not one single theory of projects and their management. Rather, the discipline is pluralistic with several different epistemological bases – positivist, interpretive, critical

realist, etc. It is also highly teleological. (And although there are plenty of theories of project management, we should expect no laws – Griseri.)

- Projects are 'invented not found'. Project management does not always operate in a realist ontology. In the front-end we are 'inventing the future': more a 'becoming', nominalist ontology. As we move into execution there is a shift to a more 'being', realist position.

- The methodology upon which our knowledge of managing projects is derived should be appropriate to the type of knowledge. Terms should be clearly defined and appropriate. The analysis and synthesis leading to any findings should be transparent and cogent.

- Project management's core function is integration. Its focus is holism (integrating to achieve success at the overall system level). Its work is typically interdisciplinary. The project manager is 'the single point of integrative accountability' in achieving the outcome desired by the sponsor; and, to the extent possible, other stakeholders.

- Project management should be as instinctive as possible (Dreyfus and Dreyfus) but if not applied over an educated base then project instinct can prove fallible and decisions can be seriously wrong.

- Of the many stakeholders influencing the front-end and down-stream execution, the sponsor is the most critical. It is his, or her, money (or budget) that is being drawn on.

- The sponsor represents the governance regime that defines the way the project is to be managed.

- Projects' clients may formally be their sponsors but morally the client is also Operations or 'the user'.

- The sponsor potentially can have the most influence on the conduct of the project (as, for example, in the way Stage-Gate reviews are run) yet probably has had the least amount of formal training in the discipline of managing projects/ project management of anyone senior on the project.

- The two 'essential' project stage-gates are where approval is given, one, to initiate development work, and two, at Sanction, to proceed with execution. (A third gate lies between these two and confirms the scope that is to be developed for sanction approval.)

- The project development strategy should be available, even if only in outline, at the first of these two stage-gates. This strategy should marry into the sponsor's business strategy (and hence project business case) and should be kept up-to-date. An Execution Strategy should be ready for review and sign-off at Sanction.

- A 'real options' approach should be considered in the early stages of project shaping as a means of preserving up-side flexibility and avoiding premature lock-in.

- Overlapping stages ('concurrency') is likely to be inefficient and dangerous to the health of the project (unless carefully phased by work package). Concurrent Engineering, on the other hand, should be beneficial to the project.

- The project strategy – in fact the whole project – should be subjected to periodic value management review. The aim should not necessarily be to reduce cost but to improve 'bang for the buck' – value. 'Peer Assists' perform a similar function. The benefits expected of the project or program and their 'owners' should be identified and lessons on their ultimate achievement fed back.

- Effectiveness measures (reflecting what the sponsor really wants to get out of the project) should be elaborated in addition to the cost, schedule and scope goals. These latter are efficiency measures and are, at best, means, not ends in themselves. (Generally they are not seen as subservient to effectiveness goals but they almost certainly should be.) Elaborating performance measures can be a major, but very important, task.
- Sustainability has arisen as a fundamental goal in recent years. It is not always obvious how best it should be operationalised, however (e.g. in the extractive industries).
- The time period to which, and over which, goals, and their evaluation, relate should be specified. (Commissioning, 3-year, 5, 10, Whole life, etc.)
- Projects typically seek outcomes as well as outputs. (Outcomes are not, as some have claimed, the preserve of programs.) In general, project management does not do a good enough job at relating its efforts to business performance outcomes.
- While knowledge and process/structure has to 'fit' the organisation's environment – its milieu – project/ program management should, within reasonable boundaries, seek to influence and shape this environment to its benefit. The academic community in particular tends to focus on means rather than impact (ends); on sub-discipline issues and theories rather than overall project or program management. This should change.
- People are central to the successful development and implementation of a project: leadership ('inventing visions of the future', and getting people to work for that future), at all levels (distributed); teamwork; influencing; communications; decision-making. Institution theory is an as-yet under-utilised basis for combining both the effects of structure/ process (capabilities) and actors (competence and intent). While we should never compromise on people, we often are forced to do so.
- Competencies (knowledge, skills, and behaviours) are role-specific. They deepen with experience. (Experience is not a competence.) Capabilities are the organisational infrastructure: the systems, processes, structure, technology, 'rules' that people operate under, e.g. contract conditions.
- Forecasting the enterprise's future staffing competencies' and capabilities' needs requires an assessment not just of the characteristics of upcoming projects but of the environment they will be operating in.
- Projects typically involve the in/on-boarding of new people and organisations. Procurement is often a key function therefore. Project management, and 'Contracts & Procurement', should seek to achieve this on as aligned a basis as possible (to the sponsor's goals and strategies).
- Estimating and Contracts and Procurement are part of the discipline of the management of projects and should not be seen as organisationally (too) separate.

- Estimating is crucial: it sets many of the project's targets. Beware cognitive distortion (Flyvbjerg).
- Eliciting requirements, facilitating innovation, and managing technology – engineering and design – also all bear hugely on the success of the project and should be positively helped by effective management. Project management has a role in establishing the project's requirements.
- Risk is a fundamental characteristic of projects. Risk is the possibility of a negative event occurring. Distinguish this from uncertainty caused by 'unknowns'. Risks should be modelled not just for the usual efficiency targets (scope, schedule, budget) but also for the project's effectiveness measures; as well as for other key factors such as credit-worthiness. Look at the up-sides associated with risks.
- Change needs very careful management. Post-Sanction it should be avoided since it will be highly disruptive. Pre-Sanction it can, properly guided, stimulate an improved implementation proposal.
- Information tools seem likely to bring some significant changes in the way projects are managed: data collection on operational usage, BIMs in construction, additive manufacturing, visualisation.
- Network-based organisations are likely to become more common.
- Trust is important but never believe it is forever or unconditional. (Iago's lesson.)
- Agile does not conform to traditional concepts of project management. It may work on certain types of project but essentially it is a form of task management.
- Critical Chain is appropriate primarily for small projects.
- Beware the proffered simplicity of maturity measures: are the right topics being assessed, are the measures too simplistic? Do organisations' p.m. abilities sometimes decline? (In reality they quite often do.)
- Project and Program Directors and Managers will often have to adapt their behaviours to fit the context they are working in. They should not let the execution-driven behaviour typical of most project management be a bar to their management of the front-end where creative, innovative input is needed and with it different skills and styles.
- Learning from projects is difficult. Lessons-learned registers can too easily be boring. Knowledge needs an owner. Motivation is important to stimulating learning. Tacit knowledge is the richest kind but can be hard to access. (People forget and may not be available as much as desired; problems with structure and comprehensiveness.) It generally helps to use one of the many review or communications points in the project to facilitate this access.

All this sounded frightfully serious so I added a personal addendum:

- Be positive and determined. But enjoy the humour in life. "That day is wasted on which we have not laughed" (the sun dial at Madresfield, an old English country house!).

Appendices

Appendix 1: Critical Success Factor Studies

CSF studies identify the factors that cause projects and programs to succeed or fail. In other words, they show what factors need to be managed in order for project management to be successful in performing its delivery function.

What the following data show is that these factors consistently arise from the areas of strategy/governance, technology/requirements, commercial, organisation, control and people, hence demonstrating the broad range of subjects, topics, functions and disciplines that managers of projects need to address. (A further observation is the high incidence of Governance and the relatively low incidence of Control – which is particularly interesting given that Control was, as we saw in Part 1, the primary driver behind the creation of the discipline, and of course still is the overriding ethos of what one might term 'basic project management'.

Reconstructing Project Management, First Edition. Peter W.G. Morris.
© 2013 John Wiley & Sons, Ltd. Published 2013 by John Wiley & Sons, Ltd.

Study		Study findings	Topic area
Marschak, Glennan and Summers (1967)[1]	A dozen US aero-engine, fighter, bomber and missile radar projects	Problems of technical and environmental uncertainty. Two opposite extremes of development strategies compared: the flexible (broad specifications until uncertainty is reduced) and the inflexible (assuming clear future requirements and hence having detailed specifications).	Strategic, environment, technology
Wilson (1969)[2]	UK power stations (36 in total, 70 units)	The study identified delays in commissioning were caused by adverse site conditions, manufacturing difficulties, design faults, and labour problems.	Environment, technology, people
Sapolsky (1972)[3]	Polaris	Outstanding leadership and (suspect value of) PERT.	Control, people
Wilson (1973)[4]	Concorde	A detailed account of what happened on the project.	
Murphy et al. (1974)[5]	650 aerospace, construction and other projects	Problems: (1) coordination, (2) relevant and agreed success criteria, (3) structure and control, (4) project difficulty, (5) funding pressure, (6) project uniqueness, and (7) team build-up.	Strategic, organisational, people
Mansfield and Wagner (1975)[6]	R&D projects in 16 firms	Success rate varied considerably. Success is greater if economic evaluation is formally carried out early in the project and if marketing is integrated with R&D.	Strategic
Edmonds (1975)[7]	Rolls-Royce RB211 aero-engine	Inadequate project appraisal, fascination with technology, unhealthy regard given to government backing, unrealistic contract conditions and pricing, underestimation of technical challenges—all conspired to create a disaster which should have been foreseen.	Strategic/governance, technology/requirements
Gerstenfeld (1976)[8]	11 successful projects, 11 unsuccessful ones and 10 innovations, all in West Germany	Unsuccessful projects difficult to stop: need for more formal reviews. Successful projects more likely to be demand-pull in origin.	Control
Mason et al. (1977)[9]	US nuclear power industry	Major areas of difficulty: design/ construction interfacing, construction methods, coordination and communication, manpower and productivity.	Environment, strategic/governance, technology/requirements, commercial, organisational, control, people
National Economic Development Office (1976)[10]	12 petrochem plants, six power stations	Construction projects took longer in the UK than abroad and absorbed more man-hours because of better industrial relations.	People

Study		Study findings	Topic area
Seamans and Ordway (1977)[11]	Apollo and the US Energy Research Development Administration	Management of state-of-the-art technology, funding, selling, support, manpower, planning, communication, visibility, decentralisation, risk, control and flexibility.	Technology, funding, organisational, control, people
Hall (1980)[12]	London's third airport, M25, Concorde, BART, Sydney Opera House	More politically sensitivity regarding costs and benefits is required.	Front-end planning
Hopkins (1980)[13]	148 companies' R&D projects	Fifty percent (50%) of companies achieved success on 67% of new products. Failure caused by poor market research, timing and technical problems.	Marketing, technology/ requirements, commercial
Rycroft and Szyliowicz (1980)[14]	Aswan High Dam	The predominant factor was politics. Decision taking was primarily 'satisficing' (muddling through) rather than 'optimising'.	Governance
Szyliowicz (1980)[15]	Eregli iron and steel works (Turkey)	While the project was ostensibly analysed and managed very rationally, the underlying assumptions were often weak or false. Political considerations were dominant.	Governance
Canaday (1980)[16]	US nuclear power industry	Huge cost increase due to a suspect cost reduction incentive system (regulation), underestimation of technical and organisational difficulties, inflation, increased safety requirements, weak contracting policies.	Environment, strategic/ governance, technology, commercial, organisational
Monopolies and Mergers Commission (1981)[17]	UK (nuclear) power industry	Advocates national site agreement, firm-price contracting, design contracts, broad-based manufacturing competition, with performance guarantees, unified overall control.	Governance, technology, commercial, organisational, people
General Accounting Office (1981, 1983, 1985)[18]	Two coal liquefaction programmes, submarine construction, six weapon systems	Dangers of premature commitments to contracting (coal liquefaction), poor QA and shipyard productivity (submarines), inadequate attention to production matters and programme uncertainty (weapon systems). Concurrency is discussed in detail.	Governance, technology, commercial, organisational, control
Cooper (1982)[19]	R&D projects in 103 firms	Success rate on average=59%. Use of marketing resources more significant than amount of R&D.	Marketing
Feldman and Milch (1982)[20]	Dallas-Fort Worth, London, Milan, Montreal, New York, Paris, Toronto and Vancouver airports	Problems of technocratic business-oriented decision making (on airports) in democratic, urban societies about a complex, uncertain business (air traffic) are explored. A bias towards phased development with greater recognition of due process is displayed.	Environment, strategic, governance, technology

(Continued)

Study		Study findings	Topic area
Paul (1982, 1983)[21]	12 Third World development projects	Political commitment, capital, technology and leadership are key 'enabling conditions', strategic management embraces management strategy, structure, process and environment in a congruent way, government plays a key role.	Environment, strategic, governance, technology, people
Myers and Devey (1984)[22]	55 US and Canadian process plants	Extent of project definition (including regulatory requirements) focussed on project responsibility drawn from all relevant corporation divisions, single project manager, concurrency dangerous, cost-plus contracts result in slippage.	Environment, governance, technology, commercial, organisational, control, people
Balachandra and Raelin (1984)[23]	51 high-tech R&D projects	Success/failure factors: (1) top management support, (2) rate of new product development, (3) probability of technical success, (4) technological route, (5) project manager as project champion, (6) technical/ marketing relations, (7) end use, (8) project team, (10) life- cycle of product, (11) internal competition, and (12) cost schedules.	Strategic, marketing, governance, technology, organisational, control, people
Bignell and Fortune (1984)[24]	Humber Bridge, RB211	General descriptions of why projects fail: problems of communication, control and the individual.	Organisational, control, people
Horwitch (1984)[25]	The US synfuels program	Broader aspects of technology management, hostile political environment, absence of champions, changing corporate strategies not taken sufficiently into account. Need for broader top management 'project' training	Environment, strategic/ governance, control, people
Feldman (1985)[26]	Charles de Gaulle, third London airport and Concorde	Frequency of political change created uncertainty; political and industrial pressure kept Concorde going, despite objective evidence against doing so.	Governance, commercial
Pugh (1985)[27]	71 UK aerospace projects	Cost overruns are a function of poor project definition, concurrency and risk	Technology, organisational, control
Might and Fisher (1985)[28]	103 system development projects	Success is a function of (1) organisation design, (2) authority of project manager, and (3) clarity of success variables.	Requirements, organisational, control, people
Baker, Green and Bean (1986)[29]	211 R&D projects	110 projects successful technically or commercially. Success factors: (1) experienced general management involvement, and (2) extent goals well defined.	Requirements, people
Whipp and Clark (1986)[30]	Study of the Rover 3500 automobile	A longitudinal case study showing how different company cultures clashed, market definition was confused, design and production were poorly integrated, industrial relations were poorly handled and why major innovative leaps forward are so risky.	Strategic, technology, commercial, organisational, people

Study		Study findings	Topic area
National Audit Office (1986)[31]	Case studies of 12 large, expensive UK defence projects	Technical uncertainty led to underestimation, the significant likelihood of cost growth with concurrency, the danger of poor project definition and weak design control, difficulties of estimating the work associated with software, over-optimism of contractors, problems of cost control, and the cost growth follows from interrupted funding.	Technology, organisational, control
Baker *et al.* (1983)[32]	Survey of 650 aerospace and construction managers	The following factors are important: (1) clear goals, (2), goal commitment of project team, (3) on-site project manager, (4) adequate funding to completion, (5) adequate project team capability, (6) accurate initial cost estimates, (7) minimum start-up difficulties, (8) adequate planning and control techniques, (9) task (vs. social) orientation, and (10) absence of bureaucracy.	Requirements, organisational, control, people
Morris and Hough (1987)[33]	Eight case studies of large projects from different sectors	The following influence success: (1) project objectives, (2) technical uncertainty and innovation, (3) socio-political factors, (4) duration urgency, (6) financial contract and legal issues, and (7) implementation (organisation, supply chain, people).	Environment, strategic, technology, commercial, organisational, people
Pinto and Slevin (1987)[34]	Survey of 418 PMI members related to 400 projects	The study identified 10 factors, including: (1) clarity of goal and general direction, (2) top management support, (3) clarity of project schedule/plan, (4) client consultation, (5) personnel issues, including recruitment, selection, and training of the team, (6) adequate technology to support the project, (7) client acceptance of finished product, (8) monitoring and feedback, (9) adequate channels of communication, and (10) adequate trouble-shooting expertise.	Strategic/ governance, technology/ requirements, commercial, organisational, control, people
Jaselskis and Ashley (1988)[35]	75 construction projects	27 factors grouped into four headings affect outstanding project performance: (1) project manager's capabilities, (2) experience and authority, (3) the stability of project team, and (4) project planning and control effort.	Control, people
Pinto and Slevin (1989)[36]	Survey of 151 R&D projects	Critical success factors: (1) top management support, (2) client consultation, (3) personnel recruitment, (4) technical tasks, (5) client acceptance, (6) monitoring and feedback, (7) communication, (8) trouble-shooting, (9) characteristics of the project team leader, (10) power and politics, (11) environment events, and (12) urgency.	Technical, organisational, control, people

(Continued)

Study		Study findings	Topic area
Drezner and Smith, (1990)[37]	107 US weapons systems programs from 1950 to 1990	Although there are large variations in the duration of programs in each decade, the time to design and develop programs has lengthened. There is no single policy option that would reduce the length of the acquisition cycle.	
Drezner et al. (1993)[38]	197 US weapons systems programs	Cost growth running consistently at about 20%. No single causal factor can be identified.	
The Standish Group (1994)[39]	Survey of 365 IT executive managers	Success factors: (1) user involvement, (2) executive management support, (3) clear statement of requirements, (4) proper planning, (5) realistic expectations, (6) smaller project milestones, (7) competent staff, (8) ownership, (9) clear vision and objectives, and (10) hard-working and focussed staff. Factors that may result in projects being challenged: (1) lack of user input, (2) incomplete requirements and specifications, (3) changing requirements and specifications, (4) lack of executive support, (5) technology incompetence, (6) lack of resources, (7) unrealistic expectations, (8) unclear objectives, (9) unrealistic time frames, and (10) new technology. Factors causing projects to be impaired and ultimately cancelled: (1) incomplete requirements, (2) lack of user involvement, (3) lack of resources, (4) unrealistic expectations, (5) lack of executive support, (6) changing requirements and specifications, (7) lack of planning, (8) didn't deed it any longer, (9) lack of IT management, and (10) technology illiteracy	Governance, technology/ requirements, commercial, organisational, control, people
Cooper and Kleinschmidt (1995)[40]	New Product Development (NPD) survey representing 135 companies	Critical Success Factors: (1) a high-quality new product process, (2) a clear, well-communicated new product strategy for the company, (3) adequate resources for new products, (4) senior management commitment to new products, (5) an entrepreneurial climate for product innovation, (6) senior management accountability, (7) strategic focus and synergy (i.e. new products close to the firm's existing markets and leveraging existing technologies), (8) high-quality development teams, and (9) cross-functional teams.	Strategic/ governance, technology/ requirements, organisational, control, people

Study		Study findings	Topic area
Tishler *et al.* (1996)[41]	Defence projects in Israel: statistical analysis of 110 defence projects executed in Israel over the last 20 years	Critical for project success in a military environment: (1) the perceived urgency of the project's output; (2) the customer follow-up team members' professional qualifications, sense of responsibility for project success and the stability of key personnel; (3) proven technological feasibility at the start of a project; (4) attention to design considerations (produceability, quality, reliability and design to cost) in the early phases of development; and (5) the professional qualifications and team spirit of the development team.	Strategic, governance, technology, organisational, control, people
GAO (1996)[42]	US Department of Energy program management	Between 1980 and 1996, 31 of DoE's 80 major system acquisitions were terminated prior to completion, after expenditures of over $10 billion. Of the 15 projects completed, three had yet to be used for their intended purpose. Funding and contractor management are central to improved performance	Governance, commercial, control
Whittaker (1999)[43]	Information Technology (IT) projects: survey of 176 respondents	The study found that the three most common reasons for IT project failure are (1) poor project planning, (2) a weak business case, and (3) a lack of top management involvement and support.	Governance, control, people
Sumner (1999)[44]	Enterprise-wide Information Management Systems Projects	Success factors: (1) Support of senior management, (2) redesign of business processes to "fit" what the software will support, (3) investment in user training, (4) avoidance of customisation, and (5) use of "business analysts" with both business knowledge and technology knowledge.	Environment, strategic/ governance, technology/ requirements, commercial, organisational, control, people
Holland and Light (1999)[45]	Enterprise Resource Planning (ERP) projects: 8 case studies	Strategic factors include: (1) legacy systems, (2) business vision, (3) ERP strategy, (4) top management support, (5) project schedule and plans. Tactical issues include: (1) client consultation, (2) personnel, (3) Business Process Change (BPC) and software configuration, (4) client acceptance, (5) monitoring and feedback, (6) communication, and (7) trouble shooting.	Governance, strategy, technology, commercial, organisational, control, people

(Continued)

Study		Study findings	Topic area
Yeo (2000)[46]	Survey of 92 Singapore-based IS respondents.	Three spheres of influence: (i) process, (ii) context and (iii) content. It also identified the main failure factors belonging to each sphere of influence as follows: (i) Process driven issues: (1) Underestimate of timeline, (2) Weak definitions of requirements and scope, (3) Inadequate project risk analysis, (4) Incorrect assumptions regarding risk analysis, (5) Ambiguous business needs and unclear vision;, (ii) Context-driven issues: (1) Lack user involvement and inputs from the onset, (2) Top-down management style, (3) Poor internal communication, (4) Absence of an influential champion and change agent, (5) Reactive and not pro-active in dealing with problems; (iii) Content-driven issues: (1) Consultant/vendor underestimated the project scope and complexity, (2) Incomplete specifications when project started, (3) Inappropriate choice of software, (4) Changes in design specifications late the project, (5) Involve high degree of customisation in application.	Environment, strategic/governance, technology/requirements/scope estimation, commercial, organisational, control, people
Miller and Lessard (2000)[47]	60 large engineering projects	The key issues underlined in this study are the competence of sponsor in terms of the ability to deal with exogenous turbulence (political, economic, social) and endogenous (partnership and contractual issues)	Environment, strategic, governance, technology, commercial, organisational, control, people
Somers and Nelson (2001)[48]	86 senior IS executives	The following factors are critical for ERP implementation: (1) Top management support, (2) project team competence, (3) interdepartmental cooperation, (4) clear goals and objectives, (5) project management, (6) interdepartmental communication, (7) management of expectations, (8) project champion, (9) vendor support, (10) careful package selection, (11) data analysis and conversion, (12) dedicated resources, (13) use of steering committee, (14) user training on software, (15) education on new business processes, (16) Business Process Reengineering, (17) minimal customisation, (18) architecture choices, (19) change management, (20) partnership with vendor, (21) use of vendors' tools, and (22) use of consultants.	Organisational, control, people

Study		Study findings	Topic area
Cooke-Davies (2002)[49]	136 (mainly) European projects executed 1994–2000 by 23 organisations.	12 factors critical to project success: (1) adequacy of company-wide risk management education, (2) maturity of risk allocation processes, (3) maintaining a visible risk register, (4) maintaining an up-to-date risk management plan, (5) documenting organisational responsibilities on the project, (6) shortening project duration, (7) allowing changes to scope only through a mature scope change control process, (8) maintaining the integrity of the performance measurement baseline, (9) maintaining an effective benefits delivery and management process, (10) matching projects to corporate strategy and business objectives, (11) establishing metrics for direct feedback on project performance and success and, (12) continuous improvement through "learning from experience".	Organisational, control, people
Flyvbjerg et al. (2003)[50]	22 road and rail projects in Sweden, 10 rail transit projects in the US, 13 metro projects in UK, and 253 other projects	Cost overrun best explained by "strategic misrepresentation", cost underestimation and overrun pay-off. Underestimation of cost and overestimation of demand is due to lack of accountability and risk negligence in promoters' decision making (under-estimation looks good in cost-benefit studies).	Governance, commercial, control,
Altshuler and Luberoff (2003)[51]	General analysis of US urban projects	Focuses on support role of US government and business. Good on political theory, empty on mop: all pastry, no filling!	Political environment, governance
Grün (2004)[52]	Multiple case studies (Olympic Games, university hospitals, and a huge wind energy converter)	Focus on the sponsor–the 'owner'–and his ability to formulate and deal with change of goals, manage basic configuration, influence the socio-political environment, and provide an appropriate managerial capacity and management structure.	Strategic, governance, socio-political environment, technology, organisation, control
National Audit Office (2004)[53]	Information Technology (IT) projects	Causes of failure in IT-enabled projects: Lack of (1) clear link between the project and the organisation's key strategic priorities including agreed measures of success, (2) clear senior management and leadership, (3) effective engagement with stakeholders, (4) skills and proven approach to project management and risk management,	

(Continued)

Study		Study findings	Topic area
		(5) understanding of and contact with the supply industry at senior levels in the organisation. Also, (6) evaluation ofproposals driven by initial price rather than long-term value for money (especially securing delivery of business benefits), (7) too little attention to breaking development and implementation into manageable steps, and (8) inadequate resources and skills to deliver the total delivery portfolio.	
National Audit Office (2005a)[54]	Construction projects: 10 case studies	(1) Scheduling, (2) developing and supporting capable clients, (3) basing design and decision making on "whole life value", (4) appropriate procurement and contracting strategies, (5) working collaboratively through fully integrated teams, and (6) evaluating performance and embedding project learning.	Strategic, governance, commercial, organisational, control, people/ learning
National Audit Office (2005b)[55]	UK defence projects: survey of 140 respondents, 19 case studies and 30 interviews	Four main levels must function as a coherent whole: (1) establishing and sustaining the right cultural environment, (2) creating clear structures and boundaries, (3) measuring progress and making decisions, and (4) reporting to enable strategic decisions.	Environment, organisational, control
Sun and Wing (2005)[56]	NPD projects in the Hong Kong toy industry: survey of 51 respondents in eight toy companies	A group of success factors identified for each phase of NPD. In *idea generation and conceptual design*: (1) clearly defined target market, (2) innovativeness of the product to the market, (3) leadership of project leader, (4) support by R&D skilled people, (5) ideas generation by brain-storming, (6) cross-functional co-operation, and (7) flexibility and responsiveness to change. In *definition and specification*: (1) implementation of quality standards, (2) clear project goal, and (3) consider issues at early stage. During *prototype and development*: internal communication within the project team. During *commercialisation* the most important factors are (1) delivery of the NP to customers on time, (2) the right time to launch and (3) competitive product cost.	Marketing, requirements, innovation, organisational, control, people
GAO (2006)[57]	53 DoD major weapon acquisition programs	Lack of alignment between corporate and project/program technology strategies.	Strategic, governance, technology

Study		Study findings	Topic area
Kappelman *et al.* (2006)[58]	IT projects – 19 experts and a survey of 55 IT project managers	Early Warning Signs of IT project failure mainly centres around people and process risks. *People-related risks* identified are: (1) lack of top management support, (2) weak project manager, (3) no stakeholder involvement and/or participation, (4) weak commitment of project team, (5) team members lack requisite knowledge and/or skills, (6) subject matter experts are overscheduled. *Process-related risks were*: (1) lack of documented requirements and/ or success criteria, (2) no change control process (change management), (3) ineffective schedule planning and/or management, (4) communication breakdown among stakeholders, (5) resources assigned to a higher priority project, and (6) no business case for the project.	Governance, technology, commercial, organisational, control, people
Cooper and Kleinschmidt (2007)[59]	NPD projects: survey of 161 German, Danish and North American business units	Four key drivers: (1) a high-quality new product process, (2) the new product strategy for the business unit, (3) resource availability, and (4) R&D spending levels.	Strategic, governance, organisational
Meier (2008)[60]	US government defence and intelligence agency large-scale acquisition programs.	(1) Overzealous advocacy, (2) immature technology, (3) lack of corporate technology roadmaps, (4) requirements instability, (5) ineffective acquisition strategy, (6) unrealistic program baselines, (7) inadequate systems engineering, and (8) workforce issues.	Technology/ requirements, commercial/ procurement, organisational, control, people
Toor and Ogunlana (2009)[61]	Large-scale construction projects in Thailand.	(1) Project planning and control, (2) project personnel, and (3) involvement of client are critical for project success.	Governance, control, people
Yu and Kwon (2010)[62]	Urban regeneration projects in Korea. Survey of 122 experts.	(1) Minimisation of conflict between stakeholders, (2) balanced adjustment between public and private interests, (3) good communication and information sharing, and (4) cooperativeness of stakeholders on a project.	Environment, strategic/ governance, stakeholders, organisation (communications), people
Tabish and Neeraj Jha (2010)[63]	Public sector construction projects in India. Survey of 105 professionals.	(1) Awareness of and compliance with rules and regulations, (2) pre-project planning and clarity in scope, (3) effective partnering among project participants and (4) external monitoring and control.	Governance, commercial (supply chain), control

(Continued)

Study		Study findings	Topic area
Lind (2011)[64]	Survey of 116 IT projects at firms in the US	(1) Size of the project, (2) clarity of goals and mission, (3) availability of required technology and (4) client acceptance of the project.	Strategic/ governance, technology/
Busi et al. (2011)[65]	Instrumentation and Control projects in South African petrochemical industry. Survey of 110 respondents	(1) Appropriately skilled people and trained personnel, (2) understanding of the technology, (3) proper documentation of decisions and (4) change management.	Strategic/ technology, control, people
Songer and Molenaar (1997)[66]	World Bank projects. Survey related to 178 projects.	(1) Monitoring, (2) coordination, (3) design, (4) training and (5) institutional environment.	Strategic/ governance, technology, organisational, control, people
Others		Knight (1976)[67]; Cooper (1999)[68]; Clarke (1999)[69]; GAO (1999)[70]; Kog et al. (1999)[71]; Jang and Lee (1998)[72]; Crawford (2000)[73]; Chan et al. (2001)[74]; Fincham (2002)[75]; Dvir et al. (2003)[76]; Nguyen et al. (2004)[77]; Williams (2004)[78]; Bing et al. (2005)[79]; Finney and Corbett (2007)[80]; Chow and Cao (2008)[81]; Jacobson and Choi (2008)[82]; Koutsikouri et al. (2008)[83]; Lam et al. (2008)[84], Rai et al. (2009)[85]; Kulatunga et al. (2009)[86]; Chan et al. (2010)[87]; Zhao et al. (2010)[88]; Chan et al. (2010)[89]; Dulaimi et al. (2010)[90]; Nasir and Sahibuddin (2011)[91]	

References

[1] Marschak, T., Glennan, T. K. and Summers, R. (1967), *Strategies for R&D: Studies in the microeconomics of development*, Springer-Verlag: New York.

[2] Wilson, A. (1969), *Committee of inquiry into delays in commissioning CEGB power stations*, TSO: London.

[3] Sapolsky, H. (1972), *The Polaris system development: Bureaucratic and programmatic success in government*, Harvard University Press: Cambridge, MA.

[4] Wilson, A. (1973), *The Concorde fiasco*, Penguin Books: Harmondsworth.

[5] Murphy, D. C., Baker, B. N. and Fisher, D. (1974), *Determinants of project success*, National Technical Information Services: Springfield, Virginia 22151, USA, Accession No. N-74-30392, 15 September, 1974.

[6] Mansfield, E. and Wagner, S. (1975), Organisational and strategic factors associated with probabilities of success and industrial R&D, *Journal of Business*, 48, p. 2.

[7] Edmonds, M. (1975), Rolls-Royce, In: Hague, D. C., Mackenzie, W. J. M. and Barker, A. (eds.) *Policy and private interests: The institutions of compromise*, Macmillan: London.

[8] Gerstenfeld, A. (1976), A study of successful projects, unsuccessful projects and projects in progress in West Germany, *IEEE Transactions on Engineering Management*, EM-23, 3, pp. 116–123.

[9] Mason, G. E., Larew, P. E., Bocherding, J. D., Okes, S. R. and Rad, P. F. (1977), *Delays in nuclear power plant construction*, United States Research and Development Administration, E (11-1)-4121, Washington, DC.

[10] National Economic Development Office (1976), *Report on engineering construction performance*, TSO: London.

[11] Seamans, R. and Ordway, F. I. (1977), The Apollo tradition: An objective lesson for the management of large-scale technological endeavours, *Interdisciplinary Science Reviews*, 2, pp. 270–304.

[12] Hall, P. (1980), *Great planning disasters*, Weidenfeld & Nicolson: London.

[13] Hopkins, D. S. (1980), *New products: Winners and losers*, Conference Board: New York.

[14] Rycroft, R. W. and Szyliowicz, J. S. (1980), *Decision making in a technological environment: The case of the Aswan High Dam*, University of Denver: Denver, CO.

[15] Szyliowicz, J. S. (1980), *Planning, managing and implementing technological development projects: The case of the Eregli iron and steel works*, University of Denver: Denver, CO.

[16] Canaday, H. T. (1980), *Construction costs overruns in electric utilities: Some trends and implications*, Occasional paper no. 3, National Regulatory Research Institute, Ohio State University, November.

[17] Monopolies and Mergers Commission (1981), *Central Electricity Generating Board: A report on the operation by the Board of its system for the generation of supply of electricity in bulk*, TSO: London.

[18] General Accounting Office (1981), *Controlling federal costs for coal liquefaction program hinges on management and contracting improvements*, PSAD-81-19, Washington, DC; General Accounting Office (1983), *status of major acquisitions as of September 30 1982*, GAO/NS IAD-83-32, Washington, DC; General Accounting Office (1985), *why some weapon systems encounter production problems while others do not: Six case studies*, GAO/NSIAD-85-34, Washington, DC.

[19] Cooper, R. G. (1982), New product development success in industrial firms, *Industrial Marketing Management*, 11, pp. 215–223.

[20] Feldman, E. J. and Milch, J. (1982), *Technocracy versus democracy*, Auburn House: Boston.

[21] Paul, S. (1982), *Managing development programs: The lessons of success*, Westview Press: Boulder, CO; Paul, S. (1983), *Strategic management of development programmes*, International Labour Office: Geneva, Switzerland.

[22] Myers, C. W. and Devey, M. R. (1984), *How management can affect project outcomes: An exploration of the PPS database*, Rand Corporation, N-216-SFC, Santa Monica, California.

[23] Balachandra, R. and Raelin, J. A. (1984), When to kill that R&D project, *Research Management*, 4, July–August, pp. 30–33.

[24] Bignell, V. and Fortune, J. (1984), *Understanding systems failures*, Open University Press/Manchester University Press: Manchester.

[25] Horwitch, M. (1984), The convergence factor for successful large-scale programs: The American Synfuels experience as a case in point, In: Cleland, D. I. (ed.) *Matrix management systems handbook*, Van Nostrand Reinhold: New York.

[26] Feldman, E. J. (1985), *Concorde and dissent: Explaining high technology failures in Great Britain and France*, Cambridge University Press: New York.

[27] Pugh, P. G. (1985), *Who can tell what might happen? Risk and contingency allowances*, Paper presented at the Royal Aeronautic Society Management Studies Group, Spring Convention.

[28] Might, R. J. and Fischer, W. A. (1985), Role of structural factors in determining project management success, *IEEE Transactions on Engineering Management*, 32, 2, pp. 71–77.

[29] Baker, N. R., Green, S. G. and Bean, A. S. (1986), Why R&D projects succeed or fail, *Research Management*, 29, November–December, pp. 29–34.

[30] Whipp, R. and Clark, P. (1986), *Innovation and the auto industry*, Francis Pinter: London.

[31] National Audit Office (1986), *Ministry of Defence: Control and management of the development of major equipment*, Report by the Controller and Auditor general, TSO: London.

[32] Baker, B. N., Murphy, D. C. and Fisher, D. (1983), Factors affecting project success, In: Cleland, D. I. and King, W. R. (eds.) *Project management handbook*, Van Nostrand Reinhold: New York.

[33] Morris, P. W. G. and Hough, G. H. (1987), *The anatomy of major projects*, John Wiley and Sons: New York.

[34] Pinto, J. K. and Slevin, D. P. (1987), Critical factors in successful project implementation, *IEEE Transactions on Engineering Management*, 34, 1, pp. 22–27.

[35] Jaselskis, E. and Ashley, D. B. (1988), Achieving construction project success through predictive discrete choice models, *Proceedings of the 9th World Congress Project Management*, Association of Project Managers, September, 4–9, 1988, Glasgow, Scotland, pp. 71–85.

[36] Pinto, J. K. and Slevin, D. P. (1989), Critical success factors in R&D projects, *Research Technology Management*, 32, January–February, pp. 31–35.

[37] Drezner, J. A. and Smith, G. K. (1990), *An Analysis of Weapons Acquisition Schedules*, Rand Corporation, RM-3927-ACQ Santa Monica, CA.

[38] Drezner, J. A., Larraine, J. M., Hess, R. W., Hough, P. I. and Norton, D. M. (1993), *An Analysis of Weapons Acquisition Cost Growth*, Rand Corporation, MR-291-AF Santa Monica, CA.

[39] The Standish Group (1994), *The CHAOS Report*, http://www.standishgroup.com.

[40] Cooper, R. G. and Kleinschmidt, E. J. (1995), Benchmarking the firm's critical success factors in new product development, *Journal of Product Innovation Management*, 12, pp. 374–391.

[41] Tishler, A., Dvir, D., Shenhar, A. and Lipovetsky, S. (1996), Identifying critical success factors in defense development projects: A multivariate analysis, *Technological Forecasting and Social Change*, 51, 2, pp. 151–171.

[42] GAO (1996), *Department of Energy: Opportunities to improve management of major system acquisitions*, Report to the Chairman, Committee on Governmental Affairs, U.S. Senate. GAO/RCED-97-17. Government Printing Office: Washington, DC.

[43] Whittaker, B. (1999), What went wrong? Unsuccessful information technology projects, *Information Management & Computer Security*, 7, 1, pp. 23–29.

[44] Sumner, M. (1999), Critical success factors in enterprise wide information management systems projects, *Proceedings of the Americas Conference on Information Systems*, Milwaukee, WI, pp. 232–234.

[45] Holland, C. P. and Light, B. (1999), A critical success factors model for ERP implementation, *IEEE Software*, 16, 3, pp. 30–36.

[46] Yeo, K. T. (2000), Critical Failure factors in information system project, *International Journal of Project Management*, 20, pp. 241–246.

[47] Miller, R. and Lessard, D. R. (2000), *The strategic management of large engineering projects*, MIT Press: Cambridge.

[48] Somers, T. and Nelson, K. (2001), The impact of critical success factors across the stages of enterprise resource planning implementations, *Proceedings of the 34th Hawaii International Conference on System Sciences*, Hawaii.

49 Cooke-Davies, T. (2002), The "real" success factors on projects, *International Journal of Project Management*, 20, 3, pp. 185–190.

50 Flyvbjerg, B., Bruzelius, N. and Rothengatter, W. (2003), *Megaprojects and risk: An anatomy of ambition*, Cambridge University Press: Cambridge.

51 Altshuler, A. and Luberoff, D. (2003), *Mega-Projects: The changing politics of urban public investment*, Brookings Institution Press: Washington, DC.

52 Grün, O. (2004), *Taming giant projects*, Springer: Berlin, Germany.

53 National Audit Office (2004), *Major IT procurement: The impact of the Office of Government Commerce's initiatives on departments and suppliers in the delivery of major IT-enabled projects*, Report to the Comptroller and Auditor General HC 877: Session 2003–2004, TSO: London.

54 National Audit Office (2005a), *Improving Public Services through better construction*, Report by the Comptroller and Auditor General HC 364-I Session 2004–2005, TSO: London.

55 National Audit Office (2005b), *Driving the successful delivery of major defence projects: Effective project control is a key factor in successful projects*, Report to the Comptroller and Auditor General, HC 30 Session: 2005–2006, TSO: London.

56 Sun, H. and Wing, W. C. (2005), Critical success factors for new product development in the Hong Kong toy industry, *Technovation*, 25, 3, pp. 293–303.

57 GAO (2006), *Assessment of selected major acquisition systems*, GAO-06-626, General Accountability Office: Washington, DC.

58 Kappelman, L., McKeeman, R. and Zhang, L. (2006), Early warning signs of it project failure: The dominant dozen, *Information Systems Management*, 23, 4, pp. 31–36.

59 Cooper, R. G. and Kleinschmidt, E. J. (2007), Winning businesses in product development: The critical success factors, *Research-Technology Management*, 50, 3, pp. 52–66.

60 Meier, S. R. (2008), Best project management and systems engineering practices in pre-acquisition practices in the federal intelligence and defense agencies, *Project Management Journal*, 39, 1, pp. 59–71.

61 Toor, S. and Ogunlana, S. (2009), Construction professionals' perception of critical success factors for large-scale construction projects, *Construction Innovation: Information, Process, Management*, 9, 2, pp. 149–167.

62 Yu, J. and Kwon, H. (2010), Critical success factors for urban regeneration projects in Korea, *International Journal of Project Management*, 29, 7, pp. 889–899.

63 Tabish, S. and Neeraj Jha, K. (2011), Identification and evaluation of success factors for public construction projects, *Construction Management and Economics*, 29, 8, pp. 809–823.

64 Lind, M. R. (2011), Information technology project performance: The impact of critical success factors, *International Journal of Information Technology Project Management*, 2, 4, pp. 14–25.

65 Busi, F., Barry, M.-L. and Chan, A. (2011), Critical success factors for instrumentation and control engineering projects in the South African petrochemical industry, *Technology Management in the Energy Smart World (PICMET)*, 2011 Proceedings of PICMET '11, Portland, OR, pp. 1–8.

66 Songer, A. D. and Molenaar, K. R. (1997), Project characteristics for successful public-sector design-build, *Journal of Construction Engineering and Management*, 123, 1, pp. 34–40.

67 Knight, G. (1976), *Concorde: The inside story*, Weidenfeld & Nicolson: London.

68 Cooper, R. G. (1999), From experience: The invisible success factors in product innovation, *Journal of Product Innovation and Management*, 16, 2, pp. 115–133.

[69] Clarke, A. (1995), The key success factors in project management, *Proceedings of a Teaching Company Seminar*, London, December.

[70] General Accounting Office (1999), *Best practices: Better management of technology can improve weapon system outcomes*, GAO/NSIAD-99-162, General Accounting Office: Washington, DC.

[71] Kog, Y. C., Chua, D. K. H., Loh, P. K. and Jaselskis, E. J. (1999), Key determinants for construction schedule performance, *International Journal of Project Management*, 17, 6, pp. 351–359.

[72] Jang, Y. and Lee, J. (1998), Factors influencing the success of managing consulting projects, *International Journal of Project Management*, 16, 2, pp. 67–72.

[73] Crawford, L. (2000), Profiling the Competent Project Manager, *Project Management Research at the Turn of the Millennium: Proceedings of PMI Research Conference*, 21–24 June, 2000, Paris, France, pp. 3–15. Sylva, NC: Project Management Institute.

[74] Chan, A., Ho, D. and Tam, C. (2001), Design and build project success factors: Multivariate analysis, *Journal of Construction Engineering and Management*, 127, 2, pp. 93–100.

[75] Fincham, R. (2002), Narratives of success and failure in systems development, *British Journal of Management*, 13, pp. 1–14.

[76] Dvir, D., Raz, T. and Shenhar, A. (2003), An empirical analysis of the relationship between project planning and project success, *International Journal of Project Management*, 21, 2, pp. 89–95.

[77] Nguyen, L. D., Ogunlana, S. O. and Lan, D. T. X. (2004), A study on project success factors in large construction projects in Vietnam, Engineering, *Construction and Architectural Management*, 11, 6, pp. 404–413.

[78] Williams, T. (2004), Assessing and building on the underlying theory of project management in the light of badly overrun projects, *PMI Research Conference*, 2004, London. Newton Square: PA: Project Management Institute.

[79] Bing, L., Akintoye, A., Edwards, P. and Hardcastle, C. (2005), Critical success factors for PPP/PFI construction projects, *Construction Management and Economics*, 23, 5, pp. 459–471.

[80] Finney, S. and Corbett, M. (2007), ERP implementation: A compilation and analysis of critical success factors, *Business Process Management Journal*, 13, 3, pp. 329–347.

[81] Chow, T. and Cao, D. (2008), A survey study of critical success factors in agile software projects, *Journal of Systems and Software*, 81, 6, pp. 961–971.

[82] Jacobson, C. and Choi, S. (2008), Success factors: Public works and public-private partnerships, *International Journal of Public Sector Management*, 21, 6, pp. 637–657.

[83] Koutsikouri, D., Austin, S. and Dainty, A. (2008), Critical success factors in collaborative multi-disciplinary design projects, *Journal of Engineering, Design and Technology*, 6, 3, pp. 198–226.

[84] Lam, E., Chan, A. and Chan, D. (2008), Determinants of successful design-build projects, *Journal of Construction Engineering and Management*, 134, 5, pp. 333–341.

[85] Rai, A., Maruping, L. M. and Venkatesh, V. (2009), Offshore information systems project success: The role of social embeddedness and cultural characteristics, *MIS Quarterly*, 33, 3, pp. 617–641.

[86] Kulatunga, U., Amaratunga, D. and Haigh, R. (2009), Critical success factors of construction research and development, *Construction Management and Economics*, 27, 9, pp. 891–900.

[87] Chan, A. P. C., Lam, P., Chan, D., Cheung, E. and Ke, Y. (2010), Critical success factors for PPPs in infrastructure developments: Chinese perspective, *Journal of Construction Engineering and Management*, 136, 5, pp. 484–495.

[88] Zhao, Z., Zuo, J., Zillante, G. and Wang, X. (2010), Critical success factors for BOT electric power projects in China: Thermal power versus wind power, *Renewable Energy*, 35, 6, pp. 1283–1291.

[89] Chan, D., Chan, A., Lam, P. and Wong, J. (2010), Identifying the critical success factors for target cost contracts in the construction industry, *Journal of Facilities Management*, 8, 3, pp. 179–201.

[90] Dulaimi, M. F., Alhashemi, M., Ling, F. and Kumaraswamy, M. (2010), The execution of public–private partnership projects in the UAE, *Construction Management and Economics*, 284, pp. 393–402.

[91] Nasir, M. and Sahibuddin, S. (2011), Critical success factors for software projects: A comparative study, *Scientific Research and Essays*, 6, 10, pp. 2174–2186.

Appendix 2: 'Characteristics of Successful Megaprojects or Systems Acquisitions'

The following were published by the National Research Council (1999) in its document, *Improving Project Management in the Department of Energy* (National Academy of Sciences: Washington D.C.) as a stand-alone guide to the principles of managing successful megaprojects or systems acquisitions. The list is interesting for several reasons. Here are three.

1. Consider the breadth of topics covered: it is much greater than PMI's *PMBOK® Guide*.
2. Note that (albeit its date is 1999), it is very execution biased – there is nothing about the challenges of working at the institutional level to build capability and competence across the enterprise.
3. Beware methodology: at first sight – under the NRC banner (header and footer) – we are led to think that this list is the result of a research program. In fact, we learn (p. 106) that it is only "based on the collective experience of more than a dozen highly knowledgeable professionals in large-scale projects".

And here's the list. Compare it with the list in Table 18.1.

Conditions Essential to Success

- Project sponsors know what they need and can afford; benefits are clearly defined and understood by all participants.
- The project has a champion in the owner's organisation, whose position and influence enable him or her to affect behaviour and performance in the owner's organisation.
- The sponsor/owner/user is clearly focused on the successful completion of the project throughout the life of the project.

Reconstructing Project Management, First Edition. Peter W.G. Morris.
© 2013 John Wiley & Sons, Ltd. Published 2013 by John Wiley & Sons, Ltd.

- Open communications, mutual trust, and close coordination are maintained between owner/users and project management.
- Project managers are experienced professionals dedicated to the success of the project. Each demonstrates leadership, and possesses the requisite technical, managerial, and communications skills brought into the project early.
- Contracts and contract incentives are clear, unambiguous and appropriate to the performance objectives.
- Risks are borne by the parties most able to manage, control, or reduce them. Therefore, owners bear the risks related to site conditions, external factors, and overall scope of the work; contractors bear risks for their own efficiency and performance in fulfilling the terms of the contract; owners and contractors work together to minimize minimise total project risks rather than shifting them from one to the other.
- Accountability for project success or failure is understood to be the responsibility of named key individuals.

Conditions Important to Success

- Contracts are awarded on the basis of value, not just cost. Value includes demonstrated capability, experience, leadership, initiative, accepted projected schedule, and other factors directly related to the successful performance of the work.
- Each party engaged in the project knows who that party's customer is, what that customer is buying in both quantity and quality, and when that customer expects delivery.
- The project organisation and mission are clearly defined and understood by everyone. The roles and responsibilities for each key person are published and the chain of command is clearly defined.
- The depth, stability, and time commitments by key personnel are appropriate for the project to ensure low turnover in management and key technical positions.
- Key project personnel from all participating entities are trained in public affairs, public information, effective communications, and information management.
- A partnering arrangement is used, in which owners, users, contractors, stakeholders, regulators, and public representatives are brought together at the outset to come to consensus on the tasks that must be accomplished and the roles and responsibilities of each.
- The public and stakeholders understand and accept the purpose of the project, the types of technologies to be employed, the processes used to award contracts, and the past relationships of the contractors with the local labour force, suppliers, and vendors.
- Acceptance, concurrence, and buy-in are obtained from all stakeholders based on their being well-informed and involved in the decision-making process leading up to the start of the project. Stakeholder acceptance is high throughout the project maintained by proper control of the work, good communications, diplomacy, and consideration.

- The project has a single information technology standard and agreed-upon protocols that have been published and are understood and observed by all.
- Contract types and terms are appropriate to the risks and to the allocation of risks between the parties.
- Adversarial relations are avoided through good contracts, good communications, and teamwork, from the earliest stages of the project.
- The project is relatively immune to external factors that could affect the scope, mission, quality, cost, or duration of the project.

Index

Actors are identified by having their first names listed, commentators only their initials. Only real nouns have initial capitalisation.